I Didn't Do It for You

I Didn't Do It for You

HOW THE WORLD BETRAYED
A SMALL AFRICAN NATION

MICHELA WRONG

HarperCollins*Publishers*

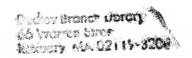

HarperCollins books may be purchased for educational, business, or sales promotional use. For information, please write: Special Markets Department, HarperCollins Publishers, 10 East 53rd Street, New York, NY 10022.

Picture Credits: p.1 Ferdinando Martini © British Library; p.2 Haile Selassie © Ullstein Bilderdienst/AKG Images, London; John Spencer © Michela Wrong; p.3 Pankhurst family © Museum of London/Heritage Images, Sylvia and son © Topham, Sylvia with medal © Popperfoto; p.4 Army crackdown © Research and Documentation Centre, Asmara, GIs © Don; p.5 Crippled Fighter © Jenny Matthews; p.6 Lessons under the trees © Jenny Matthews, Asaias Afwerki © Jenny Matthews; p.7 Aftermath of Battle © Jenny Matthews, Liberation © Ministry of Information, Asmara; p.8 Badme Village © Michela Wrong, Restored steam engine © Nick Lera

Originally published in 2005 in Great Britain by Fourth Estate.

FIRST EDITION

Printed on acid-free paper

Library of Congress Cataloging-in-Publication Data
Wrong, Michela
    I didn't do it for you : how the world betrayed a small African nation / Michela Wrong.
        p. cm.
    Originally published: Great Britain : Fourth Estate, 2005.
    Includes bibliographical references and index.
    ISBN 0-06-078092-4
    1. Eritrea—History—Revolution, 1962–1993. 2. Eritrea—History—1993–
3. Eritrea—Foreign relations. I. Title
DT397.W76 2005
963.507—dc22   2004060774

05 06 07 08 09 RRD 10 9 8 7 6 5 4 3 2 1

To Elena and Silvia Harty, as promised

Contents

Foreword

It was well past midnight, and in Cairo airport's transit lounge it was clear most passengers had already entered the trance-like state of passivity that accompanies long-distance travel. Outside, in the fluorescent glare of the hallway, a trio of stranded Senegalese women traders, majestic in their colourful *boubous*, were shouting, with operatic volume, at the Egyptian airport staff behind the counter, who were responding with an icy silence that said more about Arab attitudes towards black Africa than direct insults ever could. But here in transit, eyes had glazed over, the energy had leached from the air. A group of Nigerian youths, whose clothes gave off the nose-tickling aroma of dried fish, lay slumped in the plastic orange scoop seats, spines turned to jelly. They were being messed around by EgyptAir staff, who couldn't be bothered to check them in to the airport hotel their tickets entitled them to. They seemed past caring, anger had long since given way to exhaustion.

A few seats away, a middle-aged Pakistani businessman was fighting the prevailing mood of stupefied indifference. Visiting cards at the ready, he was in defiantly chatty mode, and was taking the fact that the airline had mislaid his luggage in his stride. ('EgyptAir no good,' he confided. 'Hmm, yes, I know.') He worked for a company that manufactured soap powder, he said, and constantly travelled the African continent and the

Middle East, sizing up possible markets for his multinational.

'And you, what do you do?'

'I'm a journalist. I'm writing a book about Eritrea. That's where I'm going now.'

His brow furrowed, he must have misheard. 'You are writing a book about Algeria?'

'No, not Algeria. Eritrea.'

'Nigeria?' He was floundering now.

'No.'

A wild guess. 'Al-Jazeera?'

'No, no. Eritrea.' Enunciating the word with the exaggerated lip movements of a teacher addressing a class with special needs, I searched for some explanatory shorthand. 'You know. Small country on the Red Sea. Used to be part of Ethiopia. It's only two hours' flight from here. I'm waiting for my connection.'

There was a brief silence. This seasoned traveller looked both flummoxed and embarrassed. 'I'm sorry. But I've simply never heard of the place.'

It was a conversation that was to keep recurring throughout the four years I spent writing this book. Mention that I was researching Eritrea and the reaction would be a sympathetic nod and ruminative silence as the other person tried to work out whether I was talking about a no-frills airline, a Victorian woman novelist or, perhaps, some obscure strain of equine disease. The same ignorance, I discovered, extended to the written record. I read books on Art Deco and Modernist architecture that made no mention of the city of Asmara, containing some of the world's most perfect examples of the styles. I waded through weighty accounts of the American intelligence services that either never referred to the fact that Eritrea had been the site of one of Washington's key listening posts, or dismissed it in a few paragraphs. I dusted off biographies on the suffragette Sylvia Pankhurst that treated her involvement in the Horn of

Africa as little more than an eccentric coda to her life. Whether in conversation or in print, Eritrea rang few bells.

At first, it amused me, then, slowly, it began to irritate. My annoyance grew in parallel with my knowledge, for the deeper I delved, the clearer it became that Eritrea had never been the obscure backwater suggested by the polite, blank expressions. Its narrative was entwined with those of colonial empires and superpowers, its destiny had engaged presidents in the White House and leaders in the Kremlin. It had obsessed emperors who believed themselves descended from Solomon and preoccupied dictators who took the Fascist salute.

I became increasingly defensive as I staked out claims to relevance, spelling out links and liaisons obliterated by the passage of time. 'You know, we ran the place for ten years,' I would say, talking to a fellow Briton. Or, if it was an American: 'Actually, the US had one of its most important spy bases in Eritrea during the Cold War.' As for the Italians, I heard myself speaking with the fixed-eye testiness of someone lost in a private obsession: 'It was your oldest colony, your "first-born". Didn't they teach you that at school?'

History is written – or, more accurately, written out – by the conquerors. If Eritrea has been lost in the milky haze of amnesia, it surely cannot be unconnected to the fact that so many former masters and intervening powers – from Italy to Britain, the US to the Soviet Union, Israel and the United Nations, not forgetting, of course, Ethiopia, the most formidable occupier of them all – behaved so very badly there. Better to forget than dwell on episodes which reveal the victors at their most racist and small-minded, cold-bloodedly manipulative or simply brutal beyond belief. To act so ruthlessly, yet emerge with so little to show for all the grim opportunism; well, which nation really wants to remember that?

The problem, as the news headlines remind us every day,

is that while the victims of colonial and Cold War blunders do not pen the story that ends up becoming the world's collective memory, they also don't share the conquerors' lazy capacity for forgetfulness. Any regular Western visitor to the developing world will be familiar with that awkward moment when a local resident raises, with a passion and level of forensic detail that reveals this is still an open wound, some injustice perpetrated long ago by the colonial master. Baffled, the traveller registers that the forgotten massacre or broken treaty, which he has only just discovered, is the keystone on which an entire community's identity has been built. 'Gosh, why are they still harping on about that?' he thinks. 'Why can't they just move on? We have.' It is a version of the 'Why do they hate us so much?' question a shocked America asked in the wake of September 11. Eritrea's story provides part of the answer to that query. It is very easy to be generous with your forgiving and forgetting, when you are the one in need of forgiveness. A sense of wounded righteousness keeps the memory sharp. Societies that know they have suffered a great wrong have a disconcerting habit of nursing their grievances, keeping them keen through the decades.

It's hard to think of another African country that was interfered with by foreign powers quite so thoroughly, and so disastrously, as Eritrea. 'My country has a lot of history,' an Eritrean academic once told me. 'In fact,' he added, with lugubrious humour, 'that's all it has.' And like Palestine and Rwanda, East Timor or Northern Ireland, Eritrea's fate has illustrated the truth that, when it comes to destabilizing regions and disturbing the sleep of nations, size doesn't matter. When the UN laid the groundwork for a guerrilla war by ushering in a flawed constitution in the 1950s, Eritrea's population was slightly more than 1 million. When Ethiopia's mighty army lost to what its leadership had dismissed as 'a handful of bandits' in 1991, Eritrea's residents numbered less than the population of Greater Manchester today. A province no larger than a medium-sized American state, most of it uninhabited, raised a clamour so loud that in the end it could no longer be ignored. Eritrea's history resembles one of those expensive television advertisements in which the flicking of a tiny cog leads to the toppling of a spanner, which sends windscreen wipers gyrating, a tyre rolling – a long chain of cause and effect which eventually climaxes in the revving of a sleek family car. Grudges which are not addressed acquire a momentum all their own, shuddering across continents and centuries until they erupt in a thunderous roar. 'Small' conflicts, left to fester long enough, have an uncanny way of bringing down empires.

Written as Eritrea faces one of the toughest challenges in its history, this is a book about betrayal, repeated across the generations, and how the expectation of betrayal can both create an extraordinary inner strength and distort a national psyche, sending a community down strange and lonely paths.

Despite the glazed-eye reactions I encountered, Eritrea's history is not an abstruse irrelevance. The peculiar mulishness of its citizens did not spring spontaneously into existence. Eritrea's seeming idiosyncrasies have entirely logical roots, roots that – to the Western reader – reach surprisingly close to home. They cannot be unpicked from the vainglorious dreams of a 19th-century Italian man of letters, an English suffragette's clash with snooty British officialdom and the pantomime antics of several generations of GIs. In a new era of superpower intervention abroad, in which the War on Terror has replaced the fight against Communism as justification for Western military adventurism, the old cliché about the dangers of forgetting history's mistakes holds truer than ever. Part and parcel of our own story, Eritrea should serve as a lasting cautionary tale, encouraging us to pause and ponder before rushing in. We forget the roles we play in such far-off outposts at our peril.

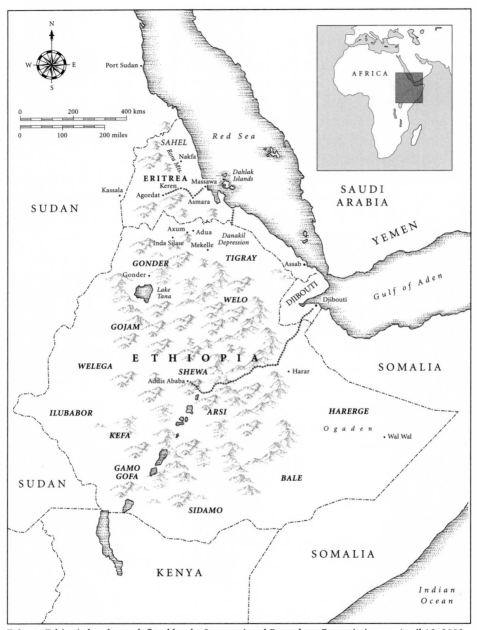

Eritrea–Ethiopia border as defined by the International Boundary Commission on April 13, 2002

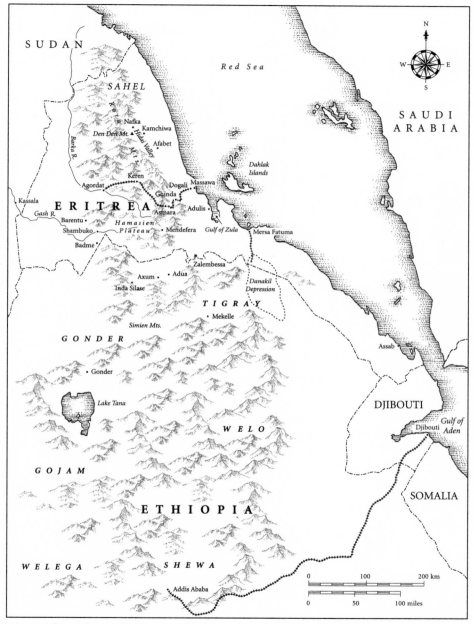

Eritrea–Ethiopia border as defined by the International Boundary Commission on April 13, 2002

CHAPTER 1

The City Above the Clouds

'That thin air had a dream-like texture, matching the porcelain-blue of the sky, with every breath and every glance he took in a deep anaesthetising tranquillity.'

Lost Horizons, James Hilton

Whenever I land in Asmara, a novel read in adolescence comes to mind. It tells the story of a small plane whose pilot turns hijacker. Crash-landing in a remote part of the Himalayas, he dies of his injuries before he can explain his bizarre actions to the dazed passengers. They emerge from the wreckage to be greeted by a wizened old monk, who leads them to a citadel hidden above the peaks, a secret city whose existence has never been recorded on any map. They are welcomed to Shangri-La, where, breathing the chill air that wafts from the glaciers and surveying the world from a tremendous height, they begin reassessing their lives with the same calm detachment and cosmic clarity as the monks. But as time goes by, they learn they must make a terrible choice. They can stay in Shangri-La and live forever, for their hosts have discovered something approaching the secret of eternal youth. Or they can plunge back into the hurly-burly of the life they knew and eventually die as ordinary mortals, grubbing around down on the plains.

Flying in from Cairo, where even during an early-morning stopover the air blasted radiator-hot through the open aircraft door, one always had the sense of landing in a capital located where, by rights, it had no place to be.

Even in the satellite photos Eritrea, a knobbly elongated triangle lying atop Ethiopia, its giant neighbour to the south, seems an inhospitable destination, a landscape still too raw for human habitation. The route the planes follow takes you over mile upon relentless mile of dun-coloured desolation, made beautiful only by the turquoise fringe where sand meets sea; a beauty that you know would evaporate if ever you ventured down to sea level to brave the suffocating heat. A spray of islands, the Dahlak, show only the faintest dusting of green. The rolling coastal sands, which show up from outer space as a strip of pearly-pink, run from the port of Massawa north-west to the border of Sudan. To the south-east, where a long, thin finger of land points towards Djibouti, the rock turns a forbidding black. Volcanic lava flows have created a landscape grimmer than the surface of the moon. This is the infamous Danakil Depression, said to be the hottest place on earth, where summer temperatures touch heights feared by even the whippet-thin Afar tribesmen. Behind this flat coastal strip, the land billows up to form a magnificent escarpment, the ripples of hills deepening into jagged waves of dark rock, a giant crumple of mountain creased by empty ravines and bone-dry river beds. It is only in the triangle's western corner, where Eritrean territory juts and bulges into northern Ethiopia, that rivers – the Gash and the Barka – flow all year round. Here in the western lowlands, the gradient finally levels off, wrinkles smooth away and the arid sands cede to the deep green that spells rain, the shade of trees, the blessing of crops.

But it is not the bleakness, but the altitude that makes Asmara's location as improbable as that of Shangri-La. The

Italians who colonized Eritrea at the tail end of the 19th century fled the stifling heat of the Red Sea by heading into the ether, up towards the *kebessa*, or central highlands. Coming in to land on the wide Hamasien plateau, there is none of the familiar routine of diving through a carpet of white fug to emerge in another, greyer reality. Defying the laws of gravity, planes bound for Asmara certainly go up, but to passengers aboard they barely seem to bother coming down. You hardly have time to register the neat concentric rings – like worm's trails – left by farmers' terracing, the rust-coloured plain, the long white scratches of roads, before the wheels hit the ground. At 7,600 ft – a mile and a half high – the capital lies at the same heady altitude as many of Europe's lower ski stations. The crisp mountain air is so thin, landing pilots must slam on their brakes and then keep them, shrieking, in place, to prevent their aircraft overshooting the tarmac. Take-offs seem to go on forever, as the accelerating plane lumbers down the tarmac in search of air resistance, finally achieving just enough friction for the rules of aerodynamics to kick in before it careers into the long grass.

Go to the edge of the escarpment, on the outskirts of town, and you will find yourself on the lip of an abyss. You are at eye level with eagles that launch themselves like suicides into the void, leaping into a blue haze into which mountain peaks, far-off valleys and distant sea all blur. At this altitude, only the most boisterous clouds succeed in rising high enough to drift over the city. Pinned down by gravity, they form instead a sulky cumulus eiderdown that barely shoulders the horizon. So for much of the year, the sky above Asmara is clear blue – a delicate cornflower merging into a deep indigo that holds out the promise of outer space. In Western cities at night, the orange glow of street lamps washes out the stars. Asmara, where hotels issue rooms with 20-watt bulbs to keep overheads down, gives out so little light that the reclining crescent moon seems almost within reach at the bottom of the street, and the constellations glisten in your upturned face. When you fly out of the capital in the dark, it feels as though you have quit the earth to soar, like some half-bird, half-man creature from Greek mythology, straight into the Milky Way itself.

With so little between you and the sun, the light possesses an alpine sharpness, so bright it almost hurts, chapping lips and creasing faces. While you shiver in the underlying chill, your skin tans at breakneck speed. Locals handle the contrast in temperatures with a practised twitch of their white cotton shawls, while the outsider finds himself neurotically dressing and undressing as he moves from icy shade to scorching sun and back again. At midday, when the light beats so strongly it numbs the senses, colours are washed away, and the world becomes an over-exposed landscape of black and white. On the street, people become dark silhouettes: here an old woman, wrapped in a *shemmah*, hides below an umbrella, there a student walks in a familiar Eritrean attitude, exercise book brandished before him to shield his eyes. The light is so white it seems to etch the thinnest of black borders around everyday objects and carve their mark upon the retina, like a nuclear flash outlining a body against a wall.

Up on the plateau, lungs must labour a little to pump enough oxygen into the bloodstream. New arrivals tend to find that alcohol rushes disconcertingly quickly to the head. Down a few of the local Melotti beers here and you can stagger to bed as drunk as if you'd worked your way through a couple of bottles of wine. When combined with the constant awareness of the vertiginous fall on the outskirts of town, the rarefied air, some say, lends itself to a certain giddiness in temperament. 'We're at 2,500 metres and it's my belief that those 2,500 metres have a lot to answer for,' a Swiss expatriate told me, struggling to explain the national hot-headedness, a tendency to quick reactions and extreme measures. 'The lack of oxygen, it has an effect on people's brains. Maybe people here go a little bit crazy.'

Asmara itself is a giant monument to colonial folly. When it came to architecture, the Italians lost their heads. Outside the capital, the slopes are dotted with modest stone shacks and grass-roofed rondavels. But in Asmara, pride of Benito Mussolini's short-lived second Roman empire, the architects of the 1930s unleashed the full, incongruous force of their Modernistic creativity.[1] Their cinemas were Art Deco palaces built for the worship of the glamorous gods of Hollywood, their streamlined apartment blocks paid tribute to the twin cult of speed and technological progress. Fascism's architects designed petrol stations that looked like aircraft in mid-flight and office blocks that resembled space rockets surging into orbit. With port-hole windows and jutting prows, their factories conjured up visions of vast ocean liners cresting the waves, their shopping arcades curved around hillocks like locomotives shrieking round a turn.

These days, the vigorous designs have lost their clean-cut certainty. The pastel-coloured buildings, painted in the soft apricot, pink and pistachio tones of melting Neapolitan ice cream, are shabby, plaster peeling in great scabs from their exteriors. Red-eyed pigeons coo above broken water pipes and the rusted *persiane* shutters hang akilter in their grooves. Nonetheless, draped in billowing blankets of bougainvillea, scattered with red-blossomed flamboyants, doused in the purple petals of the jacaranda, Asmara is undoubtedly the most beautiful capital on the continent.

Its beauty has a sombre tinge, for it has been premised on tragedy. No enlightened conservationist ever set out to preserve Asmara from the over-excited developers who spoiled downtown Nairobi or turned Lagos into a tangled mess of motorways and bridges in the 1970s. Conflict kept Asmara locked in time, creating in the process an accidental architectural treasure. While entrepreneurs with more money than sense ripped the hearts out of other colonial African cities, the economic stagnation that came with Eritrea's long war of secession against Ethiopia proved more effective than any neighbourhood campaign ever could at preserving Asmara's pure lines.

Such stultification has bestowed a Toy Town dinkiness upon the capital, the city that time forgot. On my first visit, I felt as though I had walked into a world in which my Italian grandfather would have felt completely at home, an Italy I had only ever glimpsed in family photo albums, because it has ceased to exist in Europe. Perhaps the nostalgia of that borrowed memory went some way to explaining the sudden happiness that gripped me whenever I returned, as tangible as the aroma of *berbere* spices permeating the streets. The Fiat 500 bubble car, known affectionately as the Topolino, might have disappeared from Rome's streets, but it still bowled valiantly – if rather slowly – along Asmara's avenues. Asmarinos drove museum pieces not because they were admirers of classic cars but because, for decades, no new cars were imported. Every Asmara café served the same stubby brown bottles of unlabelled beer. Since expensive foreign lagers rarely reached these parts, why bother identifying the only brand in town? In the little barber shops old men wearing the same pinched Borsalino hats and woollen waistcoats that once hung in my grandfather's closet exposed their jugulars to cut-throat razors, while their friends perched gossiping behind them. The term 'blue-collar' has become such an intellectual abstraction in the West, it gave me a jolt to see that workmen in Eritrea actually wore blue overalls. As for the white-collar business suits displayed in tailors' dusty windows, they were as quaintly old-fashioned as the hand-painted shop signs, with their approximate, impressionistic English: 'Fruit and Vagatables', 'Pinut Butter', 'Lubricunt', 'Draiving School', 'Computer Crush Course'.

Those who travel around Africa will be familiar with the mental game of 'Spot the Colonial Inheritance'. Is that Angolan secretary's failure to process your paperwork the result of Mediterranean inertia, fostered by the Portuguese, or a symptom of the bureaucratic obfuscation cultivated by a Marxist government? Is the bombast of a West African leader a legacy of a French love of words, or a modern version of the traditional African village palaver? Which colonial master left the deeper psychological mark: Britain, France, Portugal or Belgium? There are places where the colonial past seems to have left only the most cosmetic of traces on a resilient local culture, and places where the wounds inflicted seem beyond repair. In the river city of Kisangani, where I saw destitute Congolese camping in the mouldering villa built for the ruthless explorer Henry Stanley, rooms intended for pianos and chandeliers holding scores of families who washed out of buckets, I had a sense of a host body rejecting a badly-applied graft. White man's culture had been imposed with such bullying force, its buildings had never seemed to uncomprehending locals more than meaningless hulks, as surreal and totemic as the motorbike helmet Che Guevara once saw being proudly sported by a tribal chieftain in the equatorial forest. In Eritrea, the opposite seems the case: the graft has taken – so well, indeed, that the new skin has acquired a lustre all its own. 'So you're half Italian, are you?' Eritreans would say when I mentioned my parentage. 'Then half of you belongs here.' At weekends, the plains around Asmara are dotted with groups of cyclists in indecently tight shorts who whiz past grazing goats: the Italians left behind one of their favourite sports. The twittering swallows dive-bombing the steps of the Catholic church of Our Lady of the Rosary, whose bells compete for attention with the muezzin's call and prayers from the Orthodox cathedral, would not look out of place swooping over a honey-coloured Tuscan piazza. When school-girls tumble out of school they wear *grembiulini*, the coloured aprons once ubiquitous in Italian playgrounds. At the marble-countered bars, where bottles of Eritrean versions of Campari, Fernet Branca, Martini and Pernod form a stained-glass display, hissing Gaggia machines pour out cappuccinos and espressos so strong they are little more than a brown dab at the bottom of a doll's cup. '*Come sta?*' one coffee-drinker asks another, '*Andiamo, andiamo,*' call the ticket touts at the bus station, '*Va bene, dopo,*' shrugs the unsuccessful beggar ('All right, later') and little children scream *''Tilian, tilian'* ('Italian') – followed by a hopeful '*bishcotti*' ('bishcuits') – at the sight of an un-familiar face, whether Japanese, Indian or American.

Whether one is watching the evening *passeggiata* along Asmara's Liberation Avenue, when hundreds of dark-haired youths stroll arm-in-arm past gaggles of marriageable girls, eyes meeting flirtatiously across the gender divide; or observing the Sunday ritual in which bourgeois Eritrean families, bearing little cakes and little girls – each fantastically ribboned and ruched – pay each other formal visits, it's impossible to view these as alien colonial rituals. Maybe it was the similarity between the Eritrean mountains and the rugged landscape of the *mezzogiorno*, or maybe the fact that so many southern Italians, Arab blood coursing through their veins, are actually as dark as Eritreans. But the colony never felt quite as un-remittingly foreign to the Italians as Nigeria did to the British, Mali to the French or Namibia to the Germans. Something here gelled, and the number of light-skinned *meticci* (half-castes) left behind by the Italians is abiding evidence of that affinity.

Which is not to suggest that this liaison is a source of simple congratulation. Quite the opposite. Eritreans flare up like matches when they talk about the abuses perpetrated during the Fascist years, when they were expected to step into the gutter rather than sully a pavement on which a white man walked. 'If you did the slightest thing wrong, an Italian would give you a good kicking,' one of the white-haired Borsalino-wearers recalled, his eyes alight with remembered fury. But this is the most ambivalent of hostilities. Eritreans remember the racism of the Italians. But they know that what makes their country different from Ethiopia, their one-time master to the south, what made it impossible for Eritrea to accept her allotted role as just another Ethiopian province, is rooted in that colonial occupation which changed everything, forever. The Italian years are, simultaneously and confusingly, both an object of complacent pride and deep, righteous anger. 'Italy left us with the best industrial infrastructure in the world. Our workers were so well-educated and advanced, they ran everything down in Ethiopia,' Eritreans will boast, only to complain, in the next breath, that Fascism's educational policies kept them ignorant and backward, stripped of dignity. 'Fourth grade, fourth grade. Our fathers were only allowed four years of education!' So central is the Italian experience to both Eritrea and Ethiopia's sense of identity, to how each nation measures itself against the other, that during the war of independence the mere act of eating pasta, Eritrean President Isaias Afwerki once revealed, became a cause of friction between his rebel fighters and their guerrilla allies in northern Ethiopia, a dietary choice laden with politically-incendiary perceptions of superiority and inferiority.[2]

But the history that obsesses Eritrea is rather more recent. Once, on a visit to Cuba, I was fascinated to see, displayed at the national museum with a reverence usually reserved for religious icons, Che Guevara's asthma inhaler and a pizza truck that had been raked with bullets during a clash between

Castro's men and government troops. Before my eyes, mundane objects were becoming sanctified, events from the still-recent past spun into the stuff of timeless legend. I had never visited a country that seemed so in thrall to its own foundation story. But then, that was before I went to Eritrea.

Arriving in 1996 to write a country survey for the *Financial Times*, I became intrigued by the extent to which Eritrea's war of independence had been woven into the fabric of thought and language. The underdog had won in Eritrea, confounding the smug predictions of political analysts in both the capitalist West and communist East, and the vocabulary itself provided a clue as to why outsiders had got it so wrong. A lot of concepts here came with huge, if invisible, capital letters. There was the Armed Struggle, as the 30-year guerrilla campaign launched in the early 1960s against Ethiopian rule was universally known. There was the Front or the Movement, both ways of referring to the Eritrean People's Liberation Front (EPLF), the rebel group that eventually emerged as main challenger. There was the Field, or the Sahel – the sun-blasted region bordering Sudan where the EPLF turned soft civilians into hard warriors. There were the Fighters or *tegadelti*, the men and women who fought for the Movement, and the Martyrs, Fighters who did not live long enough to witness victory. There was the Strategic Withdrawal, not to be confused with retreat (Eritreans *never* retreat) – that testing moment in 1977 when the EPLF, facing a crushing onslaught by a Soviet-backed Ethiopian army, pulled back into the mountains. Above all, there was the Liberation and its conjugations ('I was Liberated', 'We Liberated Asmara', 'This hotel was Liberated'), the glorious day in 1991 when Ethiopian troops rolled out and Eritrea finally became master of its fate. The street names being introduced by the new government: Liberation Avenue, Heroes Street, Revolution Avenue, Knowledge Street were part of the same phenomenon.

The language itself left precious little room for a critical distance between speaker and subject, no gap where scepticism could crystallize.

The bright murals painted on Asmara's main thoroughfares were the equivalent of the Bayeux tapestry, commemorating a time of heroes that still spread its glow. They showed young men and women sporting no-fuss Afros, thigh-length shorts and cheap black sandals, the pauper's military kit. They crouched in the mountains, shooting at silvery MiG jets, or danced in celebration around camp fires. The murals' original models strolled below, older now, weighed down by the more pedestrian, if equally tricky challenges posed by building a new nation-state. Meeting in the street, two male friends would clasp hands, then lean towards each other until right shoulder banged into right shoulder, body bounced rhythmically off body. When vigorous young men did it, they looked like jousting stags, when old comrades did it, they closed their eyes in pleasure, burrowing their heads into the crook of each other's necks. Peculiar to Eritrea, the shoulder-knocking greeting originated in the rural areas but became a Fighter trademark, and it usually indicated shared experiences rarely spoken about, never to be forgotten. The women Fighters – for women accounted for more than a third of the Movement – were also easily spotted. Instead of white shawls, they wore cardigans. Their hair was tied in practical ponytails, rather than intricately braided in the traditional highlands style. They looked tough, weathered, quietly formidable.

'Eritrea's a great place, if you have a penchant for tragedy,' a British doctor on loan to one of the government ministries quipped. The titles of the standard works on Eritrea, displayed in the windows of every bookshop, told you everything about a national familiarity with suffering, a proud community's capacity for teeth-gritting: *Never Kneel Down*, *Against All Odds*,

Even the Stones are Burning, A Painful Season and a Stubborn Hope. Reminders of loss were everywhere. Over the age of about 40, most Westerners become familiar with the sensation of carrying around with them a bevy of friendly ghosts, the spirits of dead relatives and lost comrades who whisper in their ears and crack the occasional joke. In Eritrea, the wraiths crowded around in their multitudes, threatening to engulf the living. During the Armed Struggle, which claimed the unenviable title of Africa's longest war, Eritrea probably lost between 150,000 and 200,000 to conflict and famine. Some 60,000 Fighters died before the regime in Addis Ababa, toppled by a domestic rebel movement in league with the EPLF, agreed to surrender its treasured coastline. Given Eritrea's tiny population, this amounted to 1 in 50. Visiting Eritrean homes, one came to anticipate the sideboard on which a blue-fringed 'Martyr's Certificate', issued in recognition of a family that paid the ultimate sacrifice, held pride of place; the framed degree papers and graduation photographs testifying to skills a serious-looking son or daughter would now never put to the test. The Struggle had affected every family, it could not be escaped. Perhaps this explained why the Martyrs' Cemeteries scattered around the country were usually, behind the defiant paintings of Kalashnikov-toting warriors, neglected and overgrown. Who needed to tend graves, when the memory of the dead was so very present?

This was a nation of citizens with bits missing. Often, at the end of a conversation, I would rise to my feet only to register, as the man I had been talking to escorted me to the door, that he walked with the lunging awkwardness of someone with a wooden leg. The hand I was shaking, I'd realize, was short of a finger or two, the eye that had failed to follow my movements, or was watering painfully, was probably made of glass. The capital was full of young men and women on crutches, one empty trouser leg flapping in the breeze. If they were lucky, they sat at the controls of motorized wheelchairs, provided by a government mindful of the debt it owed its *tegadelti*. Of an evening in Asmara, you could sometimes spot a lone amputee whizzing down Martyrs' Avenue at breakneck speed, determinedly propelling his wheelchair towards Asmara's nightspots with two flailing walking sticks; an African skier without snow.

It was difficult not to be moved. It was difficult not to be admiring. My reaction was far from unique. When it came to falling for Africa's 53rd and newest state, hundreds of well-intentioned Westerners had already beaten me to it.

There is a breed of expatriate that seems particular to the Horn of Africa. Foreigners who, quite early in their travels, discovered Ethiopia or Eritrea and fell in love, with all the swooning, uncritical absolutism of youth. Perhaps they had ventured elsewhere in Africa and didn't like what they found: the inferiority complexes left by an oppressive colonial past, menacing hints of potential anarchy, the everyday sleaze of failing states. Then they came to the Horn and were swept away by the uniqueness of the region's history, the sophistication of their Ethiopian and Eritrean friends. They marvelled at the dedication of puritanical leaderships trying to do something more creative than fill Swiss bank accounts, and became True Believers. 'Ah yes, so-and-so. He has always been a Friend of Ethiopia,' you would often hear officials in Asmara and Addis say. 'Have you read so-and-so's book? She's a true Friend of Eritrea.' The rebels-turned-ministers had grasped a vital truth. True Believers are worth a hundred spokesmen to guerrilla organizations and the cash-strapped governments they go on to form. Sharing the religious convert's belligerent frustration with those who have not seen the light, quicker than the locals to detect a slight, they are tireless in defending the cause. During their time in the bush, both the EPLF and Ethiopia's

Tigrayan People's Liberation Front (TPLF) had acquired a coterie of them: hard-working Swedish aid workers, idealistic human rights activists, self-funded journalists and left-wing European parliamentarians. They had remained loyal during the hard times and now revelled in the sight of their old friends, once regarded as tiresome nuisances by Western governments, holding executive power on both sides of the border.

By the time I left Asmara, I was well on the way to joining their ranks. Looking back, I know I would have been less susceptible to Eritrea's tragic charms had I spent less time reporting on the horrors of central Africa. Having gorged on gloomy headlines, I was hungry for what seemed increasingly impossible: an African good news story. I was used to guerrilla groups who raped, pillaged, even – occasionally – ate their victims, whose gunmen were despised by the communities they claimed to represent. In Eritrea you could hear the hushed awe in civilians' voices when they talked about the demobilized Fighters who had won them independence and were now trying to build a society freed from the stifling constraints of tribe, religion and gender. As a white woman, I was used to being shooed to the front of queues, paid the exaggerated respect that spoke of generations of colonial browbeating. It gave me a perverse thrill to hear an Eritrean student confess that he and his fellow citizens suffered from a superiority complex towards outsiders. In other African nations, I was accustomed to being refused interviews by government ministers terrified by the possibility that they might show some spark of individual intelligence that could later be judged to have undermined the omniscient Big Man. Here ministers not only spoke to me, they strayed with confidence outside their official briefs and showed a disconcerting habit of wanting to discuss Samuel Pepys and Charles Darwin. I was used to writing about supplicant African governments moaning over conditions placed on aid by the World Bank and International Monetary Fund, dependent on Western approval for every policy change. These men told me, in tones that brooked no dissent, that having won independence on its own, Eritrea would decide its development programme for itself. The advice of strangers was neither wanted nor needed: self-reliance was the watchword.

In the Field, the EPLF had eschewed ranks, and the personality cults that were de rigueur elsewhere in Africa were regarded with fastidious disapproval. What a relief, after seeing portraits of Moi and Mobutu above every shop counter, to hear an Eritrean, driving past a window displaying a rare photograph of Eritrean President Isaias Afwerki, 'tsk' disapprovingly and say: 'I really don't like that.' Rather than building a palace, Isaias still lived in a modest Asmara home donated by the government. He wore simple safari suits, not Parisian couture. Visiting journalists were granted interviews within a day of arrival (in my years of visiting I had four); here was none of the scripted inaccessibility of the leader hiding behind his fawning courtiers. As for the blaring motorcades favoured by his contemporaries, shoppers on Liberation Avenue would sometimes register with a start that the man they had just passed, walking quietly along on his own, was their head of state. Isaias was in the habit of rising from the table at the end of official receptions and – to the horror of scrambling bodyguards – asking guest presidents to join him on one of his unscheduled strolls around Asmara. While foreign investors raved about the absence of official corruption, the stiff-backed integrity of those in government, Western capitals hailed Isaias and his freshly-instated friend across the border, Ethiopian Prime Minister Meles Zenawi, as forming the core of a new group of principled leaders spearheading a much-needed African Renaissance. The two men had worked together as rebel leaders – they were rumoured to be distantly related – and future cooperation seemed assured. With this visionary duo at the helm, what could go wrong? The Horn seemed destined for an unprecedented era of stability and prosperity.

The country was awash with Soviet and American weaponry, yet crime was almost unknown. The most dangerous thing that could happen to you in Asmara after dark was to stumble on a piece of broken paving. Ironically, a capital that had witnessed so much violence was blessed with an extraordinary tranquillity, it breathed peace in time with the cicada's rhythmic rasp. Asmara was certainly the only African city in which not only was I regularly offered lifts by strangers, but I accepted them without hesitation. I joined diners who gestured me over to their tables in restaurants and cleared a seat for customers who decided, off their own bat, that they fancied sharing a coffee. As for begging, it was regarded as below Eritrean dignity. I saw a persistent beggar boy being given a reproving cuff round the ear from an ex-Fighter mortified by the impression he was making on a visitor. One's expectations were always being turned on their head. 'Have you got any local money?' a handsome Eritrean student who had shared my flight asked as we were about to leave the airport terminal. Before I had time to mutter a refusal, he had extracted a banknote from his wallet: 'Here, take this for the taxi. You can pay me back later.' It was a typically Eritrean moment: in one of the world's poorest nations, I had just become the scrounger.

Journalists are mocked for using their taxi drivers as political barometers. But the conversation between airport terminal and city centre can prove more insightful than any diplomatic briefing. I was accustomed to the standard African taxi man's dirge. It started with a whinge about economic hardship, moved to a caustic assessment of both the president and opposition's shortcomings, and climaxed in a prediction – usually horribly prescient – of just how awful things were about to get.

16

In Eritrea, the first taxi driver I met turned out to be one of Eritrea's longest-serving ex-Fighters. Ministers booked for interview strode past me in reception to knock shoulders with him and pat him on the back. He not only thought the president was a hero, he knew exactly what needed to be done to rebuild a war-shattered country. But then, so did every Eritrean I met. In truth, conducting a range of interviews began to feel like an exercise in futility. Whether minister, business-man, waiter or farmer, everyone seemed to think along identi-cal lines. But this didn't sound like regurgitated propaganda. The need for self-reliance, the miracles that could be worked through discipline and hard work, the importance of learning from Africa's mistakes: such beliefs had been hammered out during committee meetings and village debates, for the EPLF was passionately committed to grassroots discussion. I had the uncanny feeling that I was speaking to the many mouths of one single, Hydra-headed creature: the Eritrean soul.

By God, they were impressive, though it has to be said that one rarely experienced a fit of uncontrollable giggles. The self-deprecating, surreal hilarity I had come to appreciate in central Africa as the saving grace of lives lived in grotesque disorder was absent here: Eritreans did dour intensity better than they did humour. Their wiry physiques – the result of not years, but generations of going without – spoke of iron control. Their personalities were as starkly defined as the climate itself, stripped of fuzzy edges. If you made the mistake of flippantly challenging one of their black-and-white certainties, you could feel the shutters coming down, as they withdrew into prickly, how-could-you-expect-to-understand-us censoriousness.

A refrain kept running through my head, a catchphrase from a British sitcom of the 1970s. 'I didn't get where I am today . . .' a beetle-browed magnate would intone at the start of every sweeping pronouncement. Eritrea, it seemed to me, had its

17

I DIDN'T DO IT FOR YOU

own, unarticulated version of the uncompromising mantra. 'I didn't spend 10/20/30 years at the Front to be patronized by a foreigner/kept waiting by a bureaucrat/messed around by a traffic cop,' it ran. Extraordinary suffering brought with it, I guessed, a sense of extraordinary entitlement that easily tipped over into chippiness. 'Why are Eritreans so bad at saying "thank you"?' I once asked an ex-Fighter friend. I was feeling slightly irritated at receiving the classic Eritrean reaction to a gift chosen with some care: an expressionless grunt, followed by the quick concealment of the unopened present, never to be mentioned again. 'I bet it's because they feel it's below their dignity.' My friend launched into a long explanation as to how, in rural communities, a peasant was expected automatically to share anything he received with the village. This democratic practice had been maintained at the Front, he said, so gifts had little meaning. In any case, showing emotion – whether happiness or grief – was regarded as a sign of weakness, simply not done. Even saying 'please' seemed demeaning, a form of begging. The explanation continued, various theories were explored, until finally my friend paused and added, almost as an afterthought, 'Anyway, there's a feeling that we fought for 30 years and no one helped us, so why should we thank anyone? We don't owe thanks to anyone.'

Even that small admission felt like a major insight, because Eritreans, famous for their reserve, do not like to talk about themselves. Whether they spoke in Italian – the Western language of the older generation – or English, taught to the young, it was always a struggle persuading an Eritrean to drop the collective 'We' and experiment with a self-indulgent, egotistical 'I'. The flow of words would slow to a dribble and dry up. For the *tegadelti*, in particular, it went against every lesson of community effort and shared sacrifice learnt at the Front. A curious monument taking shape on one of Asmara's main roundabouts captured those values. Celebrating its victory, any other new government would have ordered a statue: of its leader, a tableau of freedom fighters depicted in glorious action, or a symbolic flaming torch. The Eritreans chose instead an outsize black metal sandal, a giant version of the plastic *shidda* worn by hundreds of thousands of Eritreans who could afford neither leather nor polish. Ridiculously cheap, washable, long-lasting, the Kongo sandal – as it was known – was the poor man's boot, perfect symbol for an egalitarian movement. It must be the world's only public monument to an item of footwear.

My survey done, I took the image of Eritrea away with me, a memory to be treasured and coddled, summoned when bleakness loomed. I was not alone in finding that with Eritrea as an example, Africa seemed a little less despairing, a touch more hopeful. If Eritrea, with its devastating history, could pull it off, surely other nations might too?

Then True Believerdom took a tumble. In May 1998, to general astonishment, Eritrea and Ethiopia went back to war, after a minor dispute over a dusty border village escalated into mass mobilization on both sides. The much-trumpeted friendship between Isaias and Meles had counted for little: the two leaders were no longer talking. Ethiopia accused Isaias of being a megalomaniac, Eritrea regarded the new war as proof that Ethiopia had never digested the loss of its coast and was bent on reconquest. Defying an Ethiopian flight ban, I flew to Asmara with a group of journalists, our chartered Kenyan plane taking a looping route via Djibouti and over the waters of the Red Sea to lessen the chances of being shot down. At the end of a buttock-clenching trip, we landed to find Eritrean helicopters crouched on the tarmac of an airport that had just been bombed by Ethiopian jets. Foreign embassies were scrambling to evacuate their nationals, the BBC's World Service was telling British citizens to leave while they still could.

The mood in town was bewildering: every Asmarino I met was convinced they would win this new war, albeit at the highest of prices, every foreign journalist believed they must lose. The Eritreans' unshakeable certainty was exasperating, a positive handicap during a crisis that might require for its solution the murky skills of diplomacy, an ability to conceive of shades of grey. As ever, the community stood grimly united. 'Eritrea is not made of people who cry,' said an old businessman who had just waved goodbye to a son going off to fight. 'We did not want this, but once it comes we will do whatever our country requires.' The Eritrean capacity for speaking with one voice was beginning to sound a little creepy to my ears, as depressing as the belligerent warmongering blasting from television screens in Addis Ababa. In its chiming uniformity, it had a touch of *The Stepford Wives*.

Two years later, after at least 80,000 soldiers had died, the doubters were proved correct. With Ethiopian forces occupying Eritrea's most fertile lands to the west and a third of Eritrea's population living under UNHCR plastic sheeting, a peace deal was signed and a UN force moved in to separate the two sides. The war had been a disaster for Eritrea. But True Believers, already seriously questioning their assumptions, were about to be dealt a final, killer blow. In September 2001, President Isaias arrested colleagues who had dared challenge his handling of the war – including the ex-Fighters who had been closest to him during the Struggle – and shut down Eritrea's independent media, a step even the likes of Mugabe, Mobutu and Moi had never dared, or bothered, to take. So much for Africa's Renaissance. Many of the ministers whose independent musings had so impressed me were now in jail, denied access to lawyers. Plans to introduce a multiparty constitution and stage elections were put on indefinite hold, bolshie students sent for military training in the desert where no one could hear their views. Aloof and surrounded by sycophants, Isaias clearly had no intention of stepping down. As it gradually became clear that this was no temporary policy change, Eritrean ambassadors stationed abroad began applying for political asylum, members of the Eritrean diaspora postponed long-planned returns. As for the economy, who was going to invest now that the country's skilled workers were all in uniform, the president had fallen out with Western governments, and relations with Ethiopia, Eritrea's main market, were decidedly dodgy? No one cuffed the beggars on Liberation Avenue any more, because the beggars were not chirpy urchins but the old, left destitute by their children's departure for the front.

Far from learning from the continent's mistakes, Eritrea had turned into the stalest, most predictable of African clichés. What was striking was how far the waves of despair and outrage at this presidential crackdown travelled. For the journalists, diplomats, academics and aid workers who followed Africa, this felt like a personal betrayal, because it had destroyed the last of their hopes for the continent. Had this happened in Zambia or Ivory Coast, we would have shaken our heads and shrugged. Because it had taken place in Eritrea, special, perverse, inspiring Eritrea, we raged. 'How could they, oh, how *could* they?' I remember an Israeli cameraman friend moaning over lunch in London's Soho. This from a man who could not have spent more than a fortnight in Eritrea in his life.

Somewhere along the line, it wasn't yet clear where, the True Believers must have missed the point. They had failed to register important clues, drawn naive conclusions, misinterpreted key events. The qualities we had all so admired obviously came with a sinister reverse side. Had we mistaken arrogant pig-headedness for moral certainty, dangerous bloody-mindedness for focused determination? I had become intrigued by the Eritrean character, I realized, without digging very far into the circumstances in which it had been forged. 'They carry their history around with them like an albatross,' a British aid worker who had spent years with the EPLF had once warned me, but at the time I had not grasped her meaning. What was it in the country's past, I wondered, that had given rise to such stubborn intensity, so invigorating in some circumstances, so destructive in others? What had made the Eritreans what they were today, with all their extraordinary strengths and fatal weaknesses?

Even the most determined optimist has his moment of reckoning. An instant when he is forced to admit the society he sanctified is far darker, more convoluted, yes, on occasions downright *nasty* – than he was ready to admit. Increasingly, I found my mind wandering back to an incident I had once witnessed on Knowledge Street, round the corner from the sandal monument. Walking past a moving bus, I had noticed that the passengers were in uproar. At the heart of the storm of gesticulation sat a wizened old grandmother. The bus drove by and I heard it brake suddenly behind me, the doors open, the sound of an object hitting the pavement, the doors close, and then the bus disappeared into the night. Turning, I was astonished to see that the old woman, whom I guessed to be in her seventies, had been hurled horizontally out of the door – probably by the other passengers. Certainly, no one had interceded on her behalf. Maybe she had been very rude to the conductor, maybe she was a well-known fare dodger. Tempers, I knew, frayed fast in Eritrea. But I was astonished to witness an incident of this kind in Africa, where respect for old age runs so deep. That collective ejection was the kind of unsettling event that made you wonder if you had ever understood anything at all.

CHAPTER 2

The Last Italian

'When the white snake has bitten you, you will search in vain for a remedy.'

A 19th-century rebel leader warns Eritrean chiefs against the Italians

The old man lunged for his wooden cane and began flailing about around our feet. A moment earlier, the yard had seemed at peace, its occupants lulled to near coma in the heat, which lay upon us with the weight of a winter blanket. Now a deafening cacophony of clucks, squawks and screeches was coming from under the trestle bed on which Filippo Cicoria perched. From where I sat, I could see a blur of scuffling wings, stabbing beaks and orange claws. Two of his pet ducks were battling for supremacy. This was a cartoon fight, individual heads and wings suddenly jutting from the whirlwind at improbable angles. I kicked feebly in the ducks' direction. 'No, no,' grunted Cicoria, jabbing rhythmically with his cane. 'You have [jab] to hit them [jab, jab] on the head [jab].' The squawks were rising in hysteria, but his broken leg, pinned and swollen, was making it difficult to manoeuvre into a position where he could deliver a knock-out blow. 'That's enough, you stupid bastards ... THAT'S ENOUGH.' There were two loud shrieks as the cane finally hit home and the duo fled for safety, leaving a small deposit of feathers behind.

Feathers, I now saw, lay everywhere. A breeze from the sea, a narrow strip of turquoise behind him, lifted a thin layer of white down deposited by the pullets cheeping softly in the hutches above his bed. A dozen muscovy ducks dozed in the shade, their gnarled red beaks tucked under wings, while at the gates grazed a gaggle of geese. The air was rich with the acid stink of chicken droppings. The man, it was clear, liked his fowls. But not half as much as he liked old appliances. Cicoria's scrapyard, perched on the last in the chain of islands that forms the Massawa peninsula, held what had to be the biggest collection of obsolete fridges and broken-down air-conditioning units in the whole of Africa. Testimony to man's losing contest with an unbearable climate, the boxes were stacked in their scores, white panels turning brown in the warm salt air. They lay alongside piled sheets of corrugated iron, abandoned car parts, ripped-up water fountains, discarded barbecues and ageing fuel drums. Chains and crankshafts, girders and gas cylinders, tubes and twists of wire, all came in the same rich shade of ochre. The entire junkyard was a tribute to the miraculous powers of oxidization. Once, Cicoria had been Mr Fix-It, the only man in Massawa who knew how to repair a hospital ice-maker, tinker with a yacht's broken engine or get a hotel's air conditioning running. Now, hobbled by a fall and slowed by emphysema, he was just Mr Keep-It, struggling for breath inside a man-made mountain of rust.

I had telephoned from Asmara, keen to meet a man who I had been told personified a closing chapter of colonial history. 'He's the last one in Massawa,' an elderly Italian friend in the capital had said. 'When all the other Italians left, he stayed, through all the wars. He can't come up to Asmara now, the air's too thin for him.' When Cicoria lifted the receiver, I heard a farmyard chorus of honks and clucks, so loud I could barely make out his words. He had sounded ratty, but not openly hostile. 'Is there anything you'd like me to take him, since you haven't seen him for a while?' I asked my friend. 'Errr . . . No.' 'Well, I'll just pass on your best wishes, shall I?' I suggested. 'Yes, hmmm, that would be nice.' The reticence was puzzling.

The Italians have a word for those who fall in love with Africa's desert wastes, putting down roots which reach so deep, they can never be wrenched up again. We say 'gone to seed', or 'gone native'. The Italians call them the *insabbiati* – those who are buried in the sand – 'people', as Cicoria pronounced with lip-smacking relish, 'completely immersed in the mire'. At 77, Cicoria was happy to count himself amongst their ranks and indeed, when I'd arrived for my appointment with Massawa's last Italian, my gaze had initially flitted to him and skated on, looking vainly for a white face. Cicoria was as dark as a local, evidence of a lifetime spent working in the sun and the squirt of Eritrean blood that ran in his veins, inheritance of an Eritrean grandparent. A skinny wreck of a man, wearing a T-shirt that drooped to reveal his nipples, he sat hunched on the bed he had ordered to be carried out of his house and deposited in the centre of his metalwork collection. 'In there, I felt like a beast in a cage, out here, at least I can swear at my animals.' They say men's ears keep growing when everything else has stopped, and in Cicoria's case it seemed to be true. The onslaught of the years had turned his face into a gargoyle of ears, nose and missing teeth. Shrunken by time, this once-active man had gathered on the table before him what he clearly regarded as the bare necessities of human existence: two telephones, a roll of toilet paper and a slingshot.

He was as ravaged and pitted as the port itself. Massawa is a town with two faces. At the setting of the sun, when everyone heaves a sigh of relief, it becomes a place of hidden recesses and mysterious beauty, the lights playing softly over warm coral masonry. Tiny grocery shops, their walls neatly stacked with shiny metallic packets of tea and milk powder, soap and oil, glow from the darkness like coloured jewels. As the cafés under the Arabic arcades spring into life, naval officers in starched white uniforms sit and savour the cool evening air, watching trucks from the harbour chugging their way along the causeways, taking grain back to the mainland. Crouched in alleyways, young women sell hot tea and hardboiled eggs, the incense on their charcoal braziers blending with the pungent smell of ripe guava, the nutty aroma of roasting coffee and an occasional hot blast from an open sewer. But in the squinting glare of daytime, when only cawing crows and ibis venture out into the blinding sun, Massawa is just an ugly Red Sea town, scarred by too many sieges and earthquakes.

The town's geographical layout – two large islands linked to the mainland by slim causeways built by the 19th-century Swiss adventurer Werner Munzinger – always meant it was an easy town to hold, a difficult place to conquer. In the Second World War, a defiant Italian colonial administration had to be bombed into submission by the British and the port was then crippled by German commanders who scuttled their ships in a final gesture of spite. When the EPLF guerrilla movement first tried to capture Massawa from the Ethiopians in the 1970s, its Fighters were mown down on the exposed salt flats. Thirteen years later, the rebels succeeded, but the town took a terrible hammering in the process. Pigeons roost in the shattered blue dome of the Imperial Palace, shrapnel has taken hungry bites out of mosques and archways, walls are pitted with acne scars. Near the port, a plinth that once carried a statue of the mounted Haile Selassie, pointing triumphantly to the sea he worked so hard to claim on Ethiopia's behalf, stands decapitated. The Marxist Derg regime that ousted him tried to destroy the statue, the EPLF made a point of finishing the job. Occasionally, you'll come across a building in the traditional Arab style, its intricately-carved wooden balcony slipping gradually earthwards. But some of Africa's most grotesque modern buildings – modern pyramids of glass and cement – leave you wistful for what must have been, before the bombs and artillery did their work on the old coral palazzo. The hand-written sign propped next to the till of a mini-market round the corner from Cicoria's workshop captures what, in light of Massawa's history, seems an understandable sense of foreboding. 'Our trip – long. Our hope – far. Our trouble – many' it reads.

Cicoria had lived through it all, surviving each military onslaught miraculously unscathed. 'Once, they were shooting and one person dropped dead to the left of me, one was killed to the right and I was left standing in the middle. I've always had the devil's own luck.' He'd come to Massawa in the 1940s, a 15-year-old runaway escaping an unhappy Asmara home. 'My mother had died and I never got on with my dad. I hated my father terribly. He was an ignorant peasant.' His grandfather had been one of the area's first settlers, a constructor dispatched by Rome to build roads and dams in an ultimately fruitless attempt to win the trust of Abyssinian Emperor Menelik II. 'My family has a chapel in Asmara cemetery. You should visit it.' Cicoria must have inherited from his grandfather some technical skill that drew him to the shipyards, where Italian prisoners-of-war and Russian, Maltese and British operators – 'the ones who'd gone crazy in the war' – were repairing damaged Allied battleships. After the machinists clocked off, the boy would sneak in and mimic their movements at the lathes. 'I learnt how to make pressure gauges, spherical pistons and starter machines. No one ever taught me anything, I just watched and learnt. I can make anything, just so long as it's black and greasy,' he boasted.

This was the talent that had allowed him to play the inglorious role of Vicar of Bray, adapting smoothly to each of Eritrea's successive administrations. When Massawa's other Italians were evacuated, Cicoria's skills meant he was too valuable to lose. Under the British, he worked on the warships, under the Ethiopians he was summoned to repair damaged artillery and broken domestic appliances. 'All the Derg officers used to bring me their fridges to repair.' When the Eritrean liberation movement started up, he claimed, he turned fifth columnist and joined an undercover unit, using his privileged access to sabotage the Ethiopian military machine. 'I'm one of theirs. I'm *Shabia*, a guerrilla.' But his eyes darted shiftily away when I pressed for details.

One quality his survival had certainly not relied upon was personal charm. As his Eritrean wife, a statuesque woman of luminous beauty, prepared lunch, I began to grasp what lay behind the hesitation in my Italian friend's voice. Cicoria, it turned out, was good at hate. During a career in which I had interviewed many a ruthless politician and sleazy businessman, I had rarely met anyone, I realized, harder to warm to. His malevolence was democratically even-handed – he loathed just about everyone he came into contact with, the sole exception being the British officials who had recognized his skills all those decades ago. The American officers he had worked for had been 'crass idiots', the Ethiopians hateful occupiers. He despised his contemporaries in Asmara – my friend, it emerged, was a particular object of scorn – for not bothering to learn Tigrinya ('a bunch of illiterates'). Modern-day Eritreans were useless, cack-handed when it came to anything technical. His life had been a series of fallings-out with workmates and relatives, most of whom were no longer on speaking terms. Perhaps they'd been alienated by Cicoria's weakness for drink, or his habit of taking a new wife whenever he tired of an existing mate. 'It's not legal, but if you knew my life history, you'd understand.' Leafing through a smudged photo collection he pointed to a first wife ('as black as coal – can't stand the sight of me'), a daughter ('that bitch'), a brother ('a real shit') and a son ('nothing in his head'). The 16-year-old son running errands around the yard scored little better. 'Look at him. Strong as an ox,' he shook his head pityingly. 'But he's got no brain, no brain at all.' Even the muscovy ducks were viewed with jaundiced eyes. 'My fondness for them only goes so far. Then I eat them.' Only the latest of the many wives, whose face lit up with extraordinary tenderness when it rested upon him, won grudging praise. 'She's a good woman. Incredibly strong,' he said, watching admiringly as she manoeuvred a fridge out of the house. 'But she's too old for me now. What I really need is a nice 19-year-old.' Most depressing of all, Cicoria really did not seem to like himself – 'I've always been a rascal, a pig when it comes to women, and I drink too much' – while clearly finding it impossible to rein in a fury that kept the world at bay.

His view of Eritrea's future was bleak. 'This war is never ending. Believe me, these imbeciles will be fighting each other till the end of time.' Ill-health had deprived him of his one pleasure – his joy at hearing the stalled and obsolete revving back into life – and gravity pinned him at sea level. With the loss of his beloved lathes, which lay exasperatingly out of reach, something had died. 'I used to have high hopes,' he muttered, 'but this fall has been the last blow. Now I can't see things improving.' He had been to Italy for hospital treatment the year before and the trip, his first to the ancestral motherland, had been a revelation. He was now planning a permanent move there, he said, once he found a buyer for the scrapyard. I nodded, but found it impossible to imagine. The *insabbiati* do not travel well. Transposed, too late in life, to Europe's retirement homes, they fade away, pale and diminished, smitten by the syndrome Italians call '*mal d'Africa*'. Far better to sit sweltering in this Red Sea cauldron, king of all he surveyed, compliant family at his beck and call.

Before saying goodbye, I put the question that had been niggling me. 'What's the slingshot for?' His eyes lit up: 'Any moment now, a crow will land on that telephone line. I'm a very good shot, but the bastards are canny. If you watch, as soon as my hand moves towards the slingshot, he'll be off.' We waited. On cue, a crow landed on the line. 'Now watch.' Cicoria's hand travelled smoothly across the table to the sling-shot. The crow cocked its head. With impressive speed, he lifted the weapon and fired. But the bird had already taken off, flapping its glossy black wings across the translucent waters of the Red Sea. Cicoria shook his head. 'Bastard.'

A crabby geriatric, surrounded by the detritus of 20th century civilization, hating the world. With Cicoria, I felt, I had tasted the sour dregs of an overweeningly ambitious dream. The Italians who established their Eritrean capital in Massawa in 1890, the officials in Rome who fondly believed Africa's original inhabitants were destined to wither away, ceding their land to a stronger, white-skinned race, could never have imagined that their bracing colonial adventure would splutter to this bad-tempered, seedy end.

They had come to the Horn with grandiose plans, buoyed by the bumptious belief – shared by all Europe's expanding powers at that time – that Africa was an unclaimed continent, theirs not only for the taking but for the carving up and sharing out amongst friends. It was an assumption that held true nowhere in Africa, but least of all when applied to what was then known as Abyssinia, the ancient Ethiopian empire that lay hidden in the Horn's hinterland, beyond a wall of mountain.

By the mid-19th century, Abyssinia had experienced 100 years of anarchy, its countryside devastated by roaming armies, its weak emperors challenged by power-hungry provincial warlords, or *rases*. Its shifting boundaries bore little relation to those of Ethiopia today. The empire had lost most of its coastline to the Turkish Ottomans in the 16th century, had been pushed from the south by Oromo migrations and was facing infiltrations from the west by the Egyptian army and the Mahdi's Dervish followers in Sudan. But Emperor Yohannes IV, operating out of the northern province of Tigray, looked to a glorious ancestral past for inspiration. Steeped in legends of the vast Axumite kingdom which had stretched in ancient times from modern-day east Sudan to western Somaliland, he dreamt of rebuilding a great trading nation which would roll down from the highlands and spill into the sea, a Christian empire in a region of Islam. Blessed with a sense of historical and religious predestination, he was unimpressed by clumsy European attempts to muscle in on the region. 'How could I ever agree to sign away the lands over which my royal ancestors governed?' he once protested in a letter to the Italians. 'Christ gave them to me.'

Italy first placed its uncertain mark on the Red Sea coastline in 1869, when Giuseppe Sapeto, a priest acting on behalf of the shipping company Rubattino, itself serving as proxy for a cautious government, bought the port of Assab from a local sultan. The trigger for the purchase was that year's opening of the Suez Canal, which was set to transform the Red Sea into a vital access route linking Europe with the markets of the Far East. Bent on capitalizing on anticipated trade, Britain had already claimed Aden, the French had established a foothold in what is today Djibouti, while Egypt had bought Massawa from the Turks. As the European nation geographically closest to the Red Sea, as the birthplace of the great Roman and Venetian empires, Italy felt it could not stand idly by as its rivals scrambled to establish landing stations and trading posts along the waterway.

But commercial competition was never Italy's sole motivation for planting its flag in what would one day be Eritrea. The 19th century had seen a bubbling up of scientific curiosity in Africa, with geographical societies sending a succession of expeditions to explore the highlands and establish contact with Abyssinia's isolated monarchs. Many never returned, cut to pieces by hostile tribesmen. But those who did brought back wondrous tales of exotic wildlife and bizarre customs. Their reports fired the imaginations of Italy's writers, parliamentarians and journalists, who talked up Rome's 'civilizing mission', its duty to bring enlightenment and Catholicism to a region blighted by the slave trade and firmly in the clutches of the Orthodox Church. 'Africa draws us invincibly towards it,' declared one of the Italian Geographical Society's patrons. 'It lies just under our noses, yet up until now we remain exiled from it.'[1]

Beneath the idle intellectual curiosity lay some sobering economic realities. Italy had only succeeded in uniting under one national flag in 1870, having thrown off Bourbon and Austrian rule. A very young European nation was struggling to meet the aspirations created during the tumultuous Risorgimento. Italy had one of the highest birth rates in Europe. Emigration figures reveal how tricky Rome found feeding all these voracious new mouths. Between 1887 and 1891, to take one five-year example, 717,000 Italians left to start new lives abroad, most of them heading for Australia and the Americas. The number was to triple in the early 1900s. Italy, a growing number of politicians came to believe, needed a foreign colony to soak up its land-hungry. At worst, a territory in the Horn of Africa could serve as a penal colony, taking the pressure off Italy's prisons. At best, it would provide Italian farmers with an alternative to the fertile, well-watered territories they sought across the Atlantic.

No one who has visited Eritrea and northern Ethiopia today, no one who has experienced the punishing heat of the coastal plains and seen the dry river beds, would strike his hand to his forehead and exclaim: 'Just the place for our poor and huddled masses!' But then, Italy's African misadventure was always based on an extraordinary amount of wishful thinking. The priest who bought Assab claimed the volcanic site, one of the bleakest spots on God's earth, bore a striking resemblance to the north Italian harbour of La Spezia or Rio de Janeiro. Colonial campaigners conjured up visions of caravans trundling through Red Sea ports and new markets piled high with Italian manufactured goods, although explorers had already registered that the peasants of Abyssinia were virtually too poor to trade. ('The Abyssinians go barefoot and it will be hard to persuade them to use shoes ... A thousand metres of the richest fabrics would be more than enough to meet the Abyssinians' annual needs,' worried one.)[2] Italian politicians who toured the highlands would rave about 'truly empty' lands lying ready for the taking,[3] although more discerning colleagues noted that every plot, however seemingly neglected, had its nominal owner. Geographical precision was sacrificed in favour of the rhetorical flourish: 'The keys to the Mediterranean', one foreign minister famously, bafflingly, assured parliament, were to be found 'in the Red Sea'.[4] Ignorance sets the imagination free. When it came to their own internal affairs, Italy's lawmakers were too well-versed in the gritty detail of domestic politics, too answerable to their constituencies, to

indulge in flights of fancy. When it came to Africa, however –
continent of doe-eyed beauties, noble warrior kings and
peculiar creatures – even the pragmatists let their imaginations
run free.

Assab proved something of a false start. After an initial
flurry of excitement, it lay undeveloped and unused, as Italian
politicians vacillated over the merits of a colonial project. Then,
in 1885, British officials gave Italy's foreign policy a kick,
inviting the Italians to take Egyptian-controlled Massawa. The
debt-ridden regime in Cairo was on the verge of collapse
and the British, new masters in Egypt, were anxious not to
see a power vacuum develop which could be filled by the
French, their great rivals in the scramble for Africa. They
helpfully explained to Italian naval commanders exactly where
the Egyptian cannon were positioned, allowing the port to be
captured without loss of life.

Massawa's capture left Italy in control of a stretch of the
coast. But with their men succumbing to heatstroke, typhoid
and malaria, the Italians knew the boundaries of their fledgling
colony would have to be extended into the cool, mosquito-free
highlands if it was ever to amount to anything. They began
edging their troops up the escarpment, claiming lowland towns
whose chiefs had little love for Ras Alula, Emperor Yohannes'
loyal warlord and ruthless frontier governor. It was at a spot
called Dogali, 30 km inland, that Alula decided to draw a
line in the sand in 1887, his warriors virtually wiping out an
advancing column of 500 Italian troops. But, distracted by a
major Dervish attack, Yohannes was in no position to press
home his advantage. When the Abyssinian emperor was killed
in battle and his crown claimed by his rival to the south-east,
the King of Shewa, the Italians seized the opportunity to scale
the Hamasien plateau, marching to Asmara and into the highlands of Tigray.

The colony baptized 'Eritrea' after *Erythraeum Mare* – Latin for 'Red Sea' – was beginning to take shape, and in the capital Massawa, Italian administrative offices sprang up alongside the classical Turkish and Egyptian buildings. Backed by King Umberto, always one of Italy's most enthusiastic colonialists, the government initially entrusted the territory to Antonio Baldissera, a general with a reputation for ruthlessness. Registering that Italy could not afford to keep a standing army in Eritrea, Baldissera turned Massawa into a military recruitment centre for what he referred to as 'the inferior races'. Stripped of farming land by their new rulers, Eritrean youths had little option but to sign up as *ascaris*, ready to fight Rome's colonial wars at a fraction of the price of an Italian soldier.

Rome's *primogenito*, its colonial first-born, was hardly the earthly paradise parliament – deliberately kept in the dark by both King and cabinet – had been led to expect. This was a military regime built on bullying and fear. Playing a clumsy game of divide and rule, in which he tried to turn local chieftains against the new Emperor of Abyssinia, Menelik II, while professing eternal friendship, Baldissera filled Massawa's jail with suspected traitors and would-be defectors. When his officers met resistance, they resorted to enthusiastic use of the *curbash*, a whip made of hippopotamus hide that flayed backs raw. But the Italian public would have remained blithely unaware of the true state of affairs, had it not been for a scandal that exploded in the press in March 1891.

Ironically enough, the controversy was triggered by the government of the day. It had grown uneasy at what it was hearing from Massawa, where a formerly trusted Moslem merchant and a tribal chieftain had been sentenced to death for treason. Smelling a rat, Rome ordered an inquiry into the activities of Eteocle Cagnassi, Eritrea's secretary for colonial affairs and Dario Livraghi, head of the colony's native police force, who promptly fled. From exile in Switzerland, Livraghi penned a detailed confession, which he sent to a Milan newspaper. Just why the police chief should choose to thus expose himself remains unclear. But the editors of *Il Secolo* were so alarmed by Livraghi's account, they ordered their journalist on the ground to carry out his own investigation before they dared print a word. His findings caused a sensation.

Rich Eritrean notables, including respected holy men, were regularly disappearing at night, never to be seen again. Their fate was an open secret in Massawa, reported journalist Napoleone Corazzini. Arrested by Livraghi's policemen, they were being shot, clubbed and stoned to death and immediately buried in shallow graves on the outskirts of town. Others had been tortured to death in prison, arrested not for genuine security reasons but because corrupt Italian officials were greedily intent on confiscating their assets. Lists of intended victims had been found in Cagnassi's office and Livraghi had personally carried out many of these extrajudicial killings. Corazzini, something of a tabloid hack, painted a grotesque scene: a Moslem cleric begging for mercy before a freshly-dug grave; Livraghi, cackling like a maniac, firing repeatedly into the old man; the police chief smoking calmly as the pit was filled and finally trotting his horse cheerfully over the mound to ensure the earth was packed nice and tight.

Having published Corazzini's account, the newspaper felt it was safe to run Livraghi's story, which presented an even grimmer picture. On top of what the journalist had described as 'routine assassinations', the Italians were using terror to keep locally-recruited Eritrean warriors loyal to the new colonial regime. Officially, suspected waverers were led to the border with Abyssinia and 'extradited'. In fact, Livraghi revealed, they ended up in mass graves, slaughtered on the orders of Massawa's military command. At least 800 'rebels' had been killed in this way, sending a blood-curdling lesson to anyone thinking of following their example.

For decades, a barely-interested Italian public had lazily taken it for granted that Italy was doing good in Africa, its enlightened administrators lifting a heathen people out of the primeval slime. The Massawa scandal exposed colonialism at its most bestial. With every day that passed, new revelations about life in Eritrea – including a shocking account of how Italian officers had jokingly drawn lots for the five attractive widows of a murdered victim, then carted them off by mule – were being published in the press. Ordinary Italians were beginning to wonder why so many soldiers' lives had been lost setting up a colony in which atrocities were apparently commonplace. The newspapers demanded an investigation, reluctant to believe their own articles. Aware that its fledgling African policy faced a test more dangerous than any military confrontation, Rome announced the establishment of a royal inquiry. And this was where Ferdinando Martini, ruthless humanist, pragmatic sophisticate, the iconoclast who ended up saving the establishment, entered the picture.

The son of a comic playwright, Martini came of aristocratic stock. He was born in Florence, a city whose inhabitants regarded themselves, in many ways, as guardians of Italian culture. As a liberal member of parliament for the Tuscan constituency of Pescia and Lucca, he was to be returned to parliament a total of 13 times. By the time the old magic finally failed and he lost his seat, held without interruption for 45 years, he was 77 years old and inclined to regard retirement as a blessing. But any 19th-century gentleman worthy of the title prided himself on being a polymath and, for Martini, a political career always went hand-in-hand with literature. Following in his father's footsteps, he was to produce a steady stream of light comedies, erudite speeches and witty articles, taking time out from the political manoeuvrings and backroom bargaining associated with Montecitorio, the parliament in Rome, to run and edit several literary newspapers.

When it comes to history, those who write with ease enjoy an unfair advantage over ordinary mortals. They may be slyly self-promoting or subtly manipulative without us fully realizing it. Time has placed forever out of reach the ultimate litmus test, in which we hold their version of events up against our own memories and spot the inconsistencies. Because their words are what the records retain, because the gaps in their accounts left by the embarrassing and discreditable cannot always be filled, we see them largely as they intended to be seen.

No politician ever mastered his own legacy more effectively than Martini, thanks to the huge body of work he left behind. This was someone who felt compelled to put pen to paper every day, even if it was only to record a mocking paragraph in his diary or dash off an affectionate note to his daughter. The screeds of elegant copperplate draw the portrait of a man both irreverent and perceptive, capable of acknowledging his own failings while deriving huge amusement from those of others. They chime with the posed portrait photographs which show the author, eyebrows raised, high-domed head tipped quizzically to one side, challenging the camera. 'He is balding, and this bothers him,' reads the entry in a light-hearted bio-graphical dictionary of the day.[5] It describes an acid-tongued perfectionist, who liked to boast that his intellectual independ-ence had won him the enmity of every political party. 'He is blessed with an incisive mind and a lively turn of phrase. But if he judges others harshly, he is no less exacting with himself.' Martini comes across as a charismatic maverick, hard to dislike, a fact that makes his unexpected role as apologist for white supremacy all the more insidious.

He had started out as one of the fiercest critics of Italy's African adventure, arguing that a European nation which had itself only just thrown off the yoke of foreign rule was, in trying to subjugate a foreign people, guilty of the worst kind of hypocrisy. Why invest in Massawa's infrastructure, when Italy's poor south itself stood in crying need of development? After the Dogali massacre, Martini stuck his neck out by refusing to hail the slaughtered men as heroes and demanding the immediate recall of Italian troops, on the grounds that remaining in Eritrea was 'neither the policy of a daring nation nor a wise people'. So, by asking Martini and several other well-known anti-colonial campaigners to be part of the seven-man team assigned to investigate the goings-on in Massawa, Rome was signalling its honourable intentions to a suspicious public. With Martini as vice-chairman of the royal inquiry, how could there possibly be a cover-up?

Setting off from Naples, the team spent eight weeks touring the colony. Travelling by mule, they interviewed Eritrean chiefs and Italian officials, took notes on climatic conditions and analysed local trade. For the inquiry's remit went far beyond investigating the alleged human rights abuses. The Massawa scandal had highlighted the need for an authoritative appraisal of Eritrea's economic and strategic potential. It was time Rome decided exactly what it wanted of its Red Sea colony.

It was a potential turning point in Eritrean history. Given Martini's reputation for forthrightness and the doubts he had voiced about the colony's *raison d'être*, his left-wing colleagues in parliament and Italian voters had no reason to expect anything other than a stringently impartial account. The level of trust placed on his shoulders makes what transpired that much harder to forgive.

For when the team published its conclusions in November,

editors' mouths dropped open. In their first, 9,000-word report, the inquiry members meekly accept the excuses made by the military commanders they had questioned in Rome and Massawa. Damning journalistic accounts are brushed to one side, as are Livraghi's confessions, the product, the report hints, of an unhinged mind. With the exception of less than a dozen executions ordered during a crisis by Baldissera, who had helpfully explained that 'it was necessary to strike terror into those barbarians to make them submit', the team finds no evidence of night-time assassinations. It sympathizes with the general for the pressures he came under, finding that the colony's existence 'really was under threat'. As for the 'supposed massacres' of entire Eritrean military units, these 'did not take place'. There might have been a couple of incidents in which rebels being escorted to the border – a mere 16, rather than 800 – had been shot. But, adopting an approach favoured in many a rape trial, the team prefers to blame the victims, whose failure to cooperate with their captors brought their fate upon themselves. Another convenient scapegoat was the Eritrean police force, which apparently had a problem grasping the concept of military discipline.

The very wording of the inquiry's extraordinary conclusion, with its wealth of unconscious racism, tells us everything we need to know about the team's philosophical point of departure. 'If, in some isolated case, an abuse was committed, it can only be attributed to the savage temperament of the indigenous policemen necessarily entrusted with carrying out orders, and to the victims themselves,' it reads. 'Neither the [military] command nor any colonial officials can be held responsible.' In the light of these findings, it was hardly surprising that a Massawa court absolved both Cagnassi and Livraghi, while sentencing two Eritrean police chiefs to long prison sentences. Newspapers which had called for an Italian withdrawal from Eritrea were left flailing, the parliamentary debate on the matter – despite some sarcastic speeches by anti-colonial deputies – sputtered to an anti-climax, without a vote. The system had protected its own and, as several Italian officials revealed in memoirs published long after events, the mass killings and frenzied executions of suspected troublemakers swiftly resumed in Eritrea.[6]

The second report the team drafted represents, at least as far as the former anti-colonials on the team were concerned, a further betrayal of principle. Rejecting the sceptical accounts of previous visitors, Martini and his colleagues hail Eritrea as a 'fertile and virgin land . . . stretching out its arms to Italian farmers'. The colony, they say, is ideally placed to serve as an eventual outlet for Italy's émigrés. To that end, Rome should concentrate on consolidating Eritrea's borders, improving relations with local chiefs, replacing the military command with a civilian administration and attracting the peasant landowners who will form the backbone of a vibrant Italian community. Not an inch of acquired territory should be surrendered.

By simultaneously burying a scandal that threatened to rock the government and bestowing its blessing on Italy's African daydreams, the inquiry had effectively granted a faltering colonial project a new lease of life. On this, the first of Martini's two key encounters with Eritrea, the supposed freethinker had played a central role in a shameless whitewash which not only ensured Massawa's atrocities quietly faded from view, but guaranteed the colony survived to be fought over another day.

Why did Martini do it? Why did he risk his reputation by putting his name to what a historian of the day described as 'an incredible, medieval document, which should have been confiscated as an apologia for the crime . . . A sickening defence of assassination'?[7]

Any journalist is familiar with the sensation of being

'nobbled' by the target of an investigation. Starting out on a story in a state of hostile cynicism, his views falter as one interviewee after another put their cases with impassioned sincerity. The trust placed in the journalist is so unwavering, the hospitality so warm and, on closer examination, the people he was originally gunning for seem so reasonable. Years later, looking back on the glowing write-up that resulted, he winces at how easily he allowed himself to be manipulated, shrugs his shoulders and blames it on a heavy lunch. But it is hard to argue that Martini's keen intelligence was momentarily befuddled by the justifications presented by the colonial officials he met. In later life, he never showed any sign of regretting his role as co-author of the vital report. What puzzled contemporaries described as Martini's 'conversion' to the colonial cause was to be a permanent change of heart.

Did the quest for self-advancement play a role? Here, the picture becomes more murky. Martini was undoubtedly vain and hugely ambitious. It seems unlikely that he could already have had his eye on the post of Eritrea's governorship, which would only be created 10 years into the future. But once granted a place on a high-profile royal inquiry, investigating a topic known to be particularly close to King Umberto's heart, Martini must have been aware that a bland finding would mean political rewards somewhere down the line.

Martini's own explanation for his U-turn – however nigglingly unsatisfying – probably lies implicit in the pages of *Nell'Affrica Italiana*, a highly personalized account of the Eritrea trip published after his return. Written in the self-consciously literary language of the day, but blessed with the author's characteristic sharp eye for detail, it became a runaway best-seller, appearing in 10 editions and remaining in print for 40 years. Reaching a far wider audience than a dry government report ever could, *Nell'Affrica Italiana*, it could be argued, played a more crucial role in shaping public opinion towards Eritrea than anything else Martini wrote.

In it, Martini pulls no punches about the Italian-made horrors he witnessed in Eritrea. He describes the notorious 'Field of Hunger' – a desolate plain outside Massawa where the town governor had ordered destitute natives to be taken and left to die. 'Corpses lay here and there, their faces covered in rags; one, a horrible sight, so swarmed with insects, which snaked their way through limbs twisted and melted by the rays of the sun, he actually seemed to be moving. The dead were waiting for the hyenas, the living were waiting for death.'

Martini takes to his heels after glimpsing a group of young Eritrean girls sifting through mounds of camel dung in search of undigested grain, fighting for mouthfuls from a horse's rotting corpse. 'I fled, horrified, stupefied, mortified by my own impotence, hiding my watch chain, ashamed of the breakfast I had eaten and the lunch that awaited me.'

He winces at the use to which the *curbash* is put, on both sides of the recently-established border. 'Across the whole of Abyssinia, not excluding our own Eritrean colony, the *curbash* is an institution. Native policemen and guards are issued with it and when needs must (and it seems, from what I saw, that needs must rather often) they flog without mercy.'

Visiting an orphanage, he is repulsed by the sight of the Eritrean sons of Eritrean rebels, shot 'for the sole crime of not wanting Europeans and not wanting to take orders', being taught to sing Rome's praises. 'Conquest always comes with its own sad, sometimes dishonest, demands. Yet this seemed, and still seems, an outrage against human nature. Even now, remembering it, I feel a rush of blood to my head.'

Elsewhere, he bitterly ruminates on the hypocrisy of the Italian colonial project. 'We are liars. We say we want to spread civilization in Abyssinia, but it is not true ... Far from being barbaric and idolatrous, these people have been Christians for centuries ... We claim to want to end the fratricidal wars that have crushed any sprig of human industry in those regions, yet each day we sign up Abyssinians in our forces and pay them to butcher other Abyssinians.'

Yet having supped full of such horrors, having grasped the extent of his government's hypocrisy, Martini comes to what might seem a counterintuitive conclusion. It is now too late, he argues, for Italy to pull out of Eritrea. By embarking in Africa, Italy has set in motion an unstoppable process of racial extermination which, however distasteful, must be allowed to run its course. Any other policy would be shameful. Rather than wasting time fretting over the legal niceties of land confiscation, he argues, Italy should be dispatching farmers to start work. 'Let me repeat it for the 10th time: I would have preferred us never to have gone to Africa: I did what little I could, when there was still time, to get us to return home: but now that that time has passed ... it is neither wise nor honest to keep spreading exaggerated stories.' One can hear a sardonic disdain in Martini's voice as he imagines the eventual fate of Africa's indigenous tribes. 'We have started the job. Succeeding generations will continue to depopulate Africa of its ancient inhabitants, down to the last but one. Not quite the last – he will be trained at college to sing our praises, celebrating how, by destroying the negro race, we finally succeeded in wiping out the slave trade!'

The white race is ordained to supplant the African. 'One race must replace another, it's that or nothing ... The native is a hindrance; whether we like it or not, we will have to hunt him down and encourage him to disappear, just as has been done elsewhere with the Redskins, using all the methods civilization – which the native instinctively hates – can provide: gunfire and a daily dose of firewater.'

His language is staggeringly blunt, but it is meant to shock. Martini's main message to his Italian readers, to paraphrase it in crude modern terms, was: 'Let's cut the crap.' A genocide is already under way in Eritrea, he tells his audience, a genocide that is the expression of ineluctable historical forces. 'We have invaded Abyssinia without provocation, violently and unjustly. We excuse ourselves saying that the English, Russians, French, Germans and Spaniards have done the same elsewhere. So be it . . . injustice and violence will be necessary, sooner or later, and the greater our success, the more vital it will be not to allow trivial details or human rights to hold us up.' Moral squeamishness cannot be allowed to stand in the way of a glorious master project. Let us not shrink from what is necessary, however distasteful. But let us, at the very least, have the decency to admit what we are doing.[8]

In modern-day Eritrea, popular memory tends to divide the Italian colonial era into two halves; the Martini years, time of benign paternalism, when Eritreans and Italians muddled along together well enough; and the Fascist years, when the Italians introduced a series of racial laws as callous as anything seen in apartheid South Africa. But as *Nell'Affrica Italiana* shows, the assumptions of biological determinism that came to form the bedrock of both Fascism and Nazism were present from the first days of the Italian presence in Africa. The thread runs strong and clear through half a century of occupation. If men of Martini's generation, in contrast with their successors, felt no need to enshrine every aspect of their racial superiority in a specific set of laws, it was only because they took their supremacy utterly for granted.

Martini is a fascinating example of how it is possible for a man to be both painfully sensitive and chillingly mechanistic. The views he expressed were the notions of his day, an era in which Darwin's theories of Natural Selection and survival of the fittest were used to justify the slaughter of Congo's tribes by Belgian King Leopold's mercenaries, the German massacre of the Herero tribesmen in South West Africa and the British eradication of Tasmania's natives. Like the rabbits a British landowner introduced to Australia, like the rampant European weeds overrunning the New World, the intellectually and technologically superior white races would push aboriginal tribes into extinction. British Prime Minister, Lord Salisbury summarized the philosophy in a famous 1898 speech. 'You may roughly divide the nations of the world as the living and the dying. The weak states are becoming weaker and the strong states are becoming stronger . . . the living nations will gradually encroach on the territory of the dying.'9

Nor was Martini alone in finding the process distressing to watch. A strange kind of benevolent ruthlessness has always been the hallmark of the colonial conqueror. From H. Rider Haggard's fictional hero Allan Quatermain muttering 'poor wretch' as he puts a bullet through yet another Zulu warrior's heart, to the real-life Winston Churchill, shuddering with excitement and horror as shellfire rips through Mahdi lines at Omdurman, the literature of the day is peppered with compassionate exterminators. Martini was too intelligent not to grasp the humanity of the wretched Eritreans he met. Their plight, he told his readers, haunted his dreams. But at the end of the day, despite all his anti-establishment posturing and elegant irony, nothing mattered more to this Italian patriot than the greater glory of the Motherland.

Nell'Affrica Italiana contains one last clue as to why Martini changed his mind on Eritrea, though it is hard to distinguish authentic feeling from the rhetoric considered appropriate to the closing paragraphs of a 19th-century memoir. Sailing out of Massawa, Martini launches into a high-octane paean to Africa, the continent where, he says, 'the mind purifies, the spirit repairs itself and we find God'. 'Oh vast silence, oh nights spent in the open air, how you invigorate the body and strengthen the soul!' he raves. Adopting the pose of jaundiced Westerner weighed down by the burdens of civilization, he envies the nomads of Africa. In their 'happy ignorance', he says, they never think to ask the moon why it moves across the sky or interrogate their flocks on the meaning of life. 'How sweet it is to dream, amongst sands untouched from one month to another by a human footprint, of a society without sickness or strife, without wars or tail-coats, without *coups d'état* and visiting cards!' It is a vision of the Noble Savage that owes everything to Rousseau and Romantic poetry and nothing to reality. Like so many travellers to Africa before and after him, Martini confused the absence of a set of rules recognized by a European with personal freedom. Plagued by outbreaks of cholera and the raids of local warlords, bound by their own community's conservative codes of behaviour, Eritrea's nomads had far more reason to feel hemmed in than an effete Italian aristocrat on a government expense account.

But underneath all the hyperbole, one catches a glint of sincere emotion. For Martini, it had been easy enough to argue for Eritrea's abandonment from the distance of Rome. But criss-crossing the Hamasien plateau by mule, watching flying fish skipping over the Red Sea, basted by Eritrea's harsh light, Martini had blossomed. Part of him had fallen in love with the place, a love affair that would last the rest of his life and bring him back. He was not about to pronounce the death sentence on a land that had touched his heart. Perhaps this was the true reason why, with typical sophistry, he managed to convince himself that a doomed and destructive colonial project was, in fact, the soundest of investments.

* * *

Driving back to Asmara in the evening light, I decided to take up Cicoria's suggestion. The old Italian cemetery sprawls in rococo magnificence on the edge of town, next to its strangely anonymous modern Eritrean equivalent. Bougainvillea billows around weeping angels, stone fingers tear stone hair in grief. Between the cypresses, separated by a yellow scrub rustling with crickets and lizards, the old family mausoleums stand proud. In the more recent section, the gravestones bear Tigrinya lettering and photographs of Eritreans in graduation robes, instead of portraits of stolid Italian matrons in black. But the old mausoleums are exclusively the white man's province. Serenaded by cooing doves, I strolled between the mini mansions, reading the names which must have once featured in local newspaper articles and taken pride of place on government committees. 'Famiglia Ricupito d'Amico', 'Famiglia Giannavola', 'Famiglia Antonio Ponzio'. Asmara's burghers had not stinted when it came to their final resting places. With their gothic turrets and marbled doorways, the chapels were more substantial than many Eritrean homes. This was a cemetery built by a conquering power, established by people so sure they were in Africa to stay they had laid down vaults for the great-grandchildren they knew would succeed them.

As one of the colony's earliest settlers, the Cicoria family had claimed a prime site near the entrance. The chapel next door was being used as a storeroom by the elderly graveyard workers, paint tins resting on the floor. Undoing a rusty wire securing the door, I slipped inside. All Souls' Day had just been and someone had left flowers, an old family friend, perhaps, able to grant the Cicorias the forgiveness they seemed incapable of offering one another. Water dripped from the cut blooms, gathering in a small rivulet that ran along the floor. Looking at the black-and-white photographs marking each resting place, I was struck by the hardness of the expressions. No smiles or tenderness here. The face of Antonio Cicoria, Filippo's bridge-building grandfather, bore the deep grooves of a life in which nothing had come easy. Flinty and implacable, he looked a paterfamilias who would wield the belt with enthusiasm when disciplining wife and children. Another white-haired Cicoria stared from the slab above, chin jutting aggressively. Was this the hated father? There was no inscription, but he bore a passing resemblance to Filippo. The Italian equivalent of 'What's it to you?' seemed to hover on his lips. With relatives like this, I thought, no wonder Cicoria had run away.

As I headed for the gates, I noticed a pile of splintered gravestones stacked in a corner. Every man tries to leave his mark upon the earth, but even stones eventually wear out. When these headstones had cracked, no solicitous Italian descendant was left in Eritrea to order a replacement. Would Cicoria's body be brought here when his straining breath finally ran out? It seemed unlikely. And once the family friend stopped visiting and the rusty wire dropped off the door, this chapel, too, might end up serving as a workman's shed. I was to visit the cemetery many times after that, but only once overlapped with a relative fussing over a tomb, a young *meticcio* based in Rome. On his rare visits to Eritrea, he said, he fought a losing battle against the weeds slowly obliterating his parents' grave. Burial grounds, like hospitals, need fresh clientele to stay alive. In Asmara's cemetery, you could feel the Italian story coming to a stop.

Clever as he was, Martini could not have got it more wrong. He never faltered in his belief in a future white Eritrea, a little Italy in Africa. Amid the bombastic self-confidence of the late 19th century, it seemed a foregone conclusion, so certain that only the methodology remained to be discussed. But Martini's 'doomed' native proved more resilient than expected. Across Africa, the supposedly unstoppable flood of European settlers was easily dammed and reversed. Earning a living in Eritrea proved too tough for even the hardy peasants of Sicily and Calabria. Italy's African colonies would never absorb more than one per cent of the country's émigrés, compared to the 40 per cent that headed to America. In the 1940s, ridiculing Italy's pretensions to empire, the British – who had so many of their own – started sending Italian settlers back from the Horn. When Ethiopia's regime turned Marxist and nationalized Italian businesses in the 1970s, those who had clung on registered sadly for repatriation. Today the breed facing imminent extinction in Eritrea is white, not black. Less than 120 Italian families remain, liver-spotted men and women in their seventies and eighties who came back after independence in 1993 to die in the only place that felt like home. Not a single country estate lies in Italian hands and each year Vittorio Volpicelli, manager of the Casa degli Italiani, the Italian Club, is called upon to organize yet another medical evacuation, yet another funeral mass at the Church of Our Lady of the Rosary.

With each disappearance, the dwindling community grows a little more mournful, a little more inward-looking. Martini's descendants, dubbed 'soft Fascists' by some Eritreans, have none of his brash confidence. If they still meet friends for an espresso at the Casa, where the Fascist party insignia – a bundle of rods symbolizing 'strength through unity' – graces the main gate, the Italians rarely allow the 'F' word to pass their lips. 'You know, when they're annoyed with us they like to throw Fascism in our faces. But if you look at the origins of the word, it actually stands for something rather beautiful,' a faded Italian beauty told me as we sat having our hair done in Gino and Gina's. Gino was Asmara's first Italian hairdresser and his salon's walls are decorated with photographs of heavily made-up European models, showing off the latest in 1960s styles. Now he potters around in a confusion of Alzheimer's, collecting towels and taking orders from his wife. 'This used to be such a beautiful, beautiful city,' the signora reminisced. 'Every day, a plane would fly in from Rome with fresh orchids for the flower shops. But now . . .' There was no point going on. Asmara's Italians may purse their lips, remembering days bathed in the golden light of memory, but they know better than to voice such views in public. They stay out of politics, keep themselves to themselves. Having experienced one nationalization, they know what angry African governments can do to unwanted white communities. Masters of yesteryear, they are now here on sufferance.

CHAPTER 3

The Steel Snake

'Truly I could say that I built a colony and gave it to Italy.'
Ferdinando Martini

I noticed the scar on the first trip I made to Eritrea. It was impossible to miss: a thin white line that traced a winding route through the clumps of fig cactus and clusters of spiky aloes, lying like upturned octopi on the bottom of a fisherman's boat. At times, the overgrown track ran alongside the road. At others, it veered off, plunging through a tunnel into the bowels of the mountain, only to resurface, gulping for air, a few minutes later. Lurching from side to side as the car took the hairpin bends on the road from Asmara to Massawa, I caught glimpses of terracotta brick buttressing hugging the cliff face, viaducts rearing high above the valley, bridges hurled recklessly across gorges. 'That? It's the old Italian railway,' a friend explained. 'A railway? Up here? Surely that's not possible.' 'Oh yes. They were good builders, those Italians. They understood the mountain.'

It had been closed by the Ethiopians when the guerrilla war began to bite in the 1970s, its sleepers ripped up by soldiers and rebel fighters who used them to line the trenches. The elegant Italian arches now supported nothing at all, the track was just a convenient shortcut for Eritreans strolling to the nearest hamlet in the position they found so comfortable: walking stick slung across the shoulders, hands flopped, prisoner-of-war-like, from the pole. While structurally intact, the tunnels had followed the inexorable rule governing all dark places near human dwellings, it was doused in the acrid aroma of urine. But this *al fresco* toilet would have won the admiration of Brunel himself.

Only a people that had already thrown railroads across the Alps and Dolomites would have dared take on the Eritrean escarpment. Trains, which cannot shift into lower gear or roar round hairpin bends when the gradient begins to bite, are not really designed to go up mountains. Between Massawa and Asmara the land soars from sea level to 2,300 metres in just 70 km. The engineers of the 19th century considered a 1 in 100 gradient to be 'heavy', a gradient of 1 in 16 represented the physical limit a railroad could tackle without cog or cable. At its steepest, on the vertiginous climb between the town of Ghinda and Asmara, the Eritrean railway would touch 1 in 28. And that gradient was only achieved by sending the narrow-gauge track looping for 45 km through the mountains, a sinuous, fiendishly-clever itinerary that won it the sobriquet '*serpente d'acciaio*' – 'steel snake'. The key Massawa–Asmara section alone, I later discovered, boasted 30 tunnels, 35 bridges, 14 arches and 667 curves.

Fastidious in their choice of route, the Italians were equally ingenious when it came to choice of hardware. The techniques adopted, whose idiosyncrasies have turned Eritrea into a place of pilgrimage for modern trainspotters, ranged from the childishly simple to the sophisticated. The Italians imported French steam locomotives, specially designed for mountain transport, whose engines boasted twice the grip of ordinary models, thanks to a system that recycled steam from the main cylinders to a powered front bogie. The locomotives' normally rigid blast pipe was designed to be flexible, allowing the trains to take the tightest of curves. And wagons were fitted with individual hand brakes, which railwaymen spun to prevent the train picking up too much speed on the downhill run and released on the flat. It all made for a very slow, if spectacular ride: 10 hours from coast to capital.[1]

Even to the untutored eye, the Eritrean railway was clearly something of an engineering masterpiece. And the man responsible for this gravity-defying marvel, which would take 30 years to complete, was none other than Ferdinando Martini, epigram-loving politico and raconteur.

Why did Martini return to Eritrea? When the royal inquiry team packed its bags and set sail from Massawa in June 1891, the parliamentarian had every reason to believe that, thanks in part to his own efforts, the colony's future was now assured. But Martini could never have predicted the blow Rome would be dealt five years later, a humiliation so profound it would leave its public feeling heartily sick of all things colonial and ready to throw in the towel on his beloved Eritrea.

Well before Menelik II succeeded Yohannes as Emperor of Abyssinia, it had been clear that two expansionist forces which had been rubbing up against one another – resurgent Abyssinian nationalism and embryonic Italian colonialism – must one day clash head on. Having stamped its mark on Eritrea and signed a series of treaties with sultans on the Somali coast, Italy continued to circle the Horn of Africa's real prize: Abyssinia. The eventual trigger for this shuddering collision was to be the Treaty of Uccialli, an agreement Menelik II signed with the Italians in the belief he was trading recognition of an Eritrean border encompassing the kebessa highlands for the right to import arms through Rome's new colony to his landlocked empire. While Menelik had agreed certain terms in the treaty's Amharic version, he gradually came to the outraged realization that he had put his name to very different undertakings in the Italian translation, which contained a sly clause turning his nation into a protectorate of Rome – effectively a vassal state. When Italy refused to reverse what must qualify as one of the crudest sleights of hand in diplomatic history, war became inevitable.

The battle that followed, staged outside the Tigrayan town of Adua in 1896, pitted 19,000 Italian-led troops against 100,000 Abyssinians, many of them equipped, ironically, with Remington rifles obligingly supplied by Italian emissaries trying to ingratiate themselves with Menelik.[2] Outwitted and outmanoeuvred, some 6,000 Italians and their Eritrean *ascari* recruits were slaughtered by the Abyssinians, more men dying in one day than throughout the whole of Italy's war of independence. To ensure they never fought again, the Abyssinians amputated the right arms and left legs of surviving *ascaris*, a harsh lesson to those who took the white man's silver. It was the first time a Western army of such a size had been bested by an African force, the most shocking setback experienced on the continent by a 19th-century colonial power. Stunned by a defeat that was in part attributable to the automatic assumption that primitive black warriors would stand no chance against modern white troops, in part to Italian Prime Minister Francesco Crispi's disastrous habit of second-guessing his generals, Rome sued for peace.

Menelik could have capitalized on this stunning victory and attempted to eject the Italians altogether from the Horn. Instead, while insisting on Uccialli's abrogation, he accepted the principle of an enlarged Italian Eritrea. But such concessions did little to dilute Adua's devastating impact back in Italy. It was not for mountainous Eritrea and arid Somalia that the Italian public had supported the government's expensive colonial project. Its eyes had always been locked on the green pastures further to the south, the fertile, farmable Abyssinian lands Menelik II had now decreed forever out of reach. Chanting 'Viva Menelik', furious crowds demonstrated against the Italian government, while socialist members of parliament renewed calls for Italy to pull out of Africa. Some already heard Eritrea's death knell tolling: 'The colony no longer lives, it breathes its last,' pronounced Eteocle Cagnassi, the official who had so deftly escaped punishment for the Massawa atrocities. 'The ministry is demolishing, not running it; it no longer has a governor, very soon there will be no settlers either. Even in its most difficult and dangerous moments, Eritrea never went through a more inauspicious and painful time.'[3]

It was at this delicate juncture that the government called in Martini, offering him the post of Eritrea's first civilian governor. He turned it down, hesitated, then accepted. For conservative leader Antonio Di Rudini, who had taken over from the disgraced Crispi as prime minister, Martini was a canny choice. Although Di Rudini had made huge political capital out of criticizing the government's handling of Adua, he was hardly a passionate colonialist. He had decided to hang on to Eritrea, but only after playing briefly with the idea of handing the colony to Belgium's King Leopold, master of the Congo. He realized that he could only successfully defy public opinion if the colony, focus of so much controversy, assumed the lowest of profiles. It must be removed from the control of a profligate military, its shifting border needed to be fixed and, above all, its demands on Italy's exchequer must be drastically reduced. By recruiting Martini, who had somehow managed to survive the Massawa debacle with his reputation for feistiness intact, Di Rudini could appear to be responding to public sentiment. In fact, both he and King Umberto knew they were placing the colony in the safest of hands.

When Martini set to sea, there was talk in Rome of pruning Eritrea down to a triangle linking Massawa on the coast with Asmara and the highlands town of Keren, or something even more modest. Many of Martini's colleagues actually expected the new governor to waste no time in winding up Eritrea's affairs. But the establishment was intent on consolidation, not dissolution. 'I have made quite enough sacrifices to public opinion on this African issue,' the King confided to Martini before he left. 'I will not make the ultimate sacrifice: we must and will not descend from the plateau.'[4] And Martini, the ever-equivocal Martini, was on exactly the same wavelength. 'If I can stop Africa being a thorn in our flesh ... if I can pacify the colony, raise it to a point where it is self-supporting, allow it to become, so to speak, forgotten, wouldn't I be doing the country a major service?' he mused.[5] He spelled out his position in a letter to a friend: 'I will not return a single inch of territory ... the day the government asks me to descend to Massawa is the day I land in Brindisi.'

For Martini, this represented a risky career move. By the time he left for Eritrea in December 1897, he was 56 years old, an age where the delights of African travel, with its malarial bouts, month-long mule treks and most basic of amenities, begin to pall. The job, which meant leaving behind his family, was no sinecure, and others had rejected it. He had already done well for himself, rising briefly to the post of Education Minister. By going to Eritrea, he would be removing himself from the buzz and chit-chat of Montecitorio, with all the opportunities it represented. But at his age, with so much already achieved, such things mattered less than they once had. There were times, indeed, when he felt nothing but disgust for politics, sorry he had ever entered the game. 'When I look back on my 23 years in parliament, I mourn all that wasted time,' he told a friend. 'If I stay here, what will I do? Make speeches to the chamber: Sibylline words, scattered by the wind.'[6] The clear skies he had lauded in *Nell'Affrica Italiana* were calling. Eritrea's first civilian governor, he knew, would be a huge fish in a tiny pool, always a cheering position to hold. It must have been enormously flattering to think that, once again, the future of Italy's 'first-born' rested largely on his shoulders. Who else, after all, knew more about Eritrean affairs? Who else could be trusted to do the right thing?

His nine-year stint as governor is recorded in *Il Diario Eritreo*, 7,000 pages of handwritten entries which constitute a priceless resource of the Italian colonial era. Although he indexed each of its 26 volumes, Martini never seems to have had publication in mind, referring to the work only as a collection of 'notes'. At most, he probably intended the diary to serve as source material for an African memoir he never, in the end, got round to writing. Had it not been for Italy's Ministry of African Affairs, which ordered it published in 1946 – nearly 20 years after Martini's death – the diary would have remained locked away in the family's archives. Why did he put so much care into what was meant as no more than a personal aide-mémoire? Because, one has to conclude, Martini simply could not do otherwise. A man with his inquiring mind, with his lifelong habit of capturing impressions on paper, simply had to record the intense sensations that came with his return to Eritrea. To write something down was to endow it with value, to allot it its proper meaning – the habit came to him as naturally as breathing. 'There is more satisfaction to be won from writing what seems a stylish page than in overturning a ministry,' he once remarked. Whether at sea, on the road, or at home, he faithfully kept his diary, rarely skipping a day. And the fact that publication was never on the agenda makes the diary far fresher, funnier and more accessible than the flowery *Nell'Affrica Italiana*. Martini himself never understood this. 'In

Africa, one writes rather badly,' he says at one point. 'This is certainly not a good page.'[7] In fact, to modern eyes, he writes far better. A sustained ironic conversation with himself, the diary's very lack of artifice brings 19th-century Eritrea to life in a way his more laboured writing never could.

Here is Martini the amused sociologist, fascinated at the goings-on in the stretch of open ground outside his Asmara villa, which serves, he discovers, as a communal latrine. 'This wretched valley is the debating society for those who feel the need to shed excess body weight . . . One man comes along and squats. The effect is contagious. Another comes along, measures the distance and squats a dozen metres from the first, in the same position and with the same aim in mind. And then a third, a fourth; sometimes a fifth and a sixth. And the conversation starts . . . Simultaneous, contemporaneous, in parallel . . . Words are not the only thing to emerge, but they last longer than the rest.'[8]

And here is Martini the urban sophisticate, despairing, as Eritrea's attorney-general reads out a report, at his colleagues' pitiable level of education. 'My God! What a business! It was the most laughable thing imaginable: logic, dignity of expression, grammar, were never so badly mangled. And to think these are the magistrates the government sends to civilize Africa!'[9]

Everything interests him, from the awed reaction of Massawa's residents to his governor's regalia of plumed hat and gold braid, to the flavour of the turtle soup and ostrich steak ('like veal', he notes) he is served at a welcome ceremony. The sexual mores of Eritrea's tribes, the way in which a visiting chieftain falls in love with his reflection in a mirror, the staggering ugliness of a group of Englishwomen spotted in a Cairo hotel, the gossip in Asmara's expatriate community, all are recorded with Martini's characteristic impish sense of humour. The task he had been set, he soon realized, was immense.

Nearly 30 years after its arrival in the Horn, Italy had pitifully little to show for its investment. The Eritrea depicted in his diary is Italy's version of the Wild West, swept by locust swarms and cholera outbreaks, braced for outbreaks of the plague; a land in which villages are raided by hostile tribes and shipping attacked by pirates. It is a frontier country in which slaves are still traded, shady European businessmen mingle with known spies and where government officials still fight – and die – in duels staged over adulterous wives.

Just as he had been warned in Rome, the military administration had careered out of control, spending Italian taxpayers' money as though it would never be held to account. 'Either idiots or criminals', the dregs of the soldiering profession were drawn to Eritrea, he noted, men who believed 'that colonizing Africa and screwing the Italian government are one and the same thing'. 'Dirty, out of uniform, they frequent the brothels until late, while the officers divide their time between prostitutes and the gaming table.'[10] He was appalled to see how the military had lavished government funds on officers' villas instead of investing in the roads, bridges and sewerage the colony so clearly needed. 'Even the best soldiers feel they are only doing their duty when they throw money out of the window,' he lamented after discovering, rotting in Massawa's storerooms, 60,000 men's shoes, enough spurs to equip an army, 40,000 mattocks, 9 years' supply of salt, 3 years' of wine, 2 years' of jam, 52 months' worth of coffee and 22 months' of sugar.

His Eritrean subjects, who appeared to have accepted Italian rule as a necessary evil, were the least of his problems. 'They do not love us, but understand the benefits that come with our rule,' remarked Martini, noting that local administrators regarded the Italians as 'good but stupid'.[11] The settlers were the real disappointment. Far from serving as an alternative destination for the tens of thousands of Italians heading for the

Americas, Eritrea held less than 4,000 'Europeans', and that tally actually included hundreds of Egyptians, Syrians, Turks and Indians judged civilized enough to count as 'white'. Land had been confiscated and experimental agricultural projects launched, but the going had proved so tough many Italian families begged to be sent home. Martini was none too impressed by those who remained, noting that their Greek colleagues seemed less prone to frittering away their profits. 'The Greek does not buy horses and does not keep mistresses, the Italian keeps both horse and mistress.'[12] The constant complaints by the hard core that remained drove him wild. 'I've always said that governing 20 Italians in the colony requires more patience, courage, and skill than governing 400,000 natives,' he fumed. When Rome had the temerity to inquire whether an Eritrean display should feature in the Paris Exhibition's colonial section, an exasperated Martini lost his temper: 'All we can send are dead men's bones, bungled battle plans and columns of wasted money. Up till now these are the only fruits of our colonial harvest.'[13]

Moving the capital from Massawa to cool Asmara, he set about his work with characteristic briskness. A series of decrees created a new civilian administration, placing the army firmly under its control. Strict limits were set to the number of civil servants employed in Eritrea, a move that slashed Rome's expenditure. The worst soldiers and officers were simply expelled. 'These steps will cause a great deal of ill feeling, but I know I am doing my duty. Order, discipline, justice and thrift: without these the colony can neither be governed nor saved,' Martini pronounced.[14] The colony was divided up into nine provinces, each with its own capital, and Martini established the building blocks of a modern society: an independent judiciary, a telegraph system and departments of finance, health and education.

The man who had calmly predicted the disappearance of Eritrea's indigenous peoples quickly changed his tone. It was all very well airily discussing the elimination of local tribes as a passing visitor. Now that he was actually running Eritrea and could see for himself the damage – both political and commercial – done by military confrontation, Martini turned accommodating pacifier. Determined to shore up the Eritrean border, he became the perfect neighbour, putting an end to Rome's long tradition of double-dealing. When rebel chiefs on the other side of the frontier challenged Menelik's rule, Martini turned a deaf ear to their pleas for weapons. Instead of fantasizing, like so many Italian contemporaries, about avenging Adua, he cooperated with Menelik's attempts to check the lawlessness on their mutual frontier, stabilizing the region in the process. As for emigration, Martini quickly realized how poorly judged the royal inquiry report had been. The colony was simply not ready for a flood of Italian labourers, who risked clashing with locals and would, in any case, be undercut by Eritreans willing to accept a fraction of what a European considered an honest wage. He scrapped legislation authorizing further land confiscation and pushed employers to narrow the huge differential between the wages paid Italians and Eritreans.

But while righting certain blatant injustices, Martini was never a soft touch. If Eritrea was to survive, the locals must be taught a lesson in the pitiless consistency of colonial law, the merest hint of insubordination ruthlessly crushed. Mutinous *ascaris* were shackled or whipped and the sweltering coastal jails filled with prisoners who often paid the ultimate price. 'I've never had a bloodthirsty reputation and I really don't deserve one,' Martini wrote, after refusing to pardon a condemned bandit. 'But here, without a death penalty, you cannot govern.'[15] He was building a state, virtually from scratch, and often he felt as though he was doing the work single-handed.

'There is not a dog here with whom one can hold an *intellectual* discussion,' he complained in a letter to his daughter.[16] It was a lonely, heady experience, bound to encourage delusions of grandeur. 'At times, unfortunately,' he confessed to his diary, 'I feel it would not be too arrogant to say, adapting the words of Louis 14th, "I *am* the colony".'[17]

The longer he stayed, the more convinced he became that the success of this monumental project hinged on one key element. He knew Eritrea had gold, fish stocks in abundance and river valleys capable of producing coffee and grain, cotton and sisal. But as long as a rickety mule track was the only way of scaling the mountains separating hinterland from sea, Eritrea would remain forever cut off from the African continent, its ports idle, its administration reliant on government subsidies. Only a railroad could unlock the riches of the plateau and – beyond it – the markets of Abyssinia and Sudan. It was the one explicit undertaking Martini had sought in exchange for his loyal service during his final conversation with King Umberto. 'Without a railway joining Massawa with the highlands, nothing good, lasting or productive will ever come from Eritrea,' he told the monarch. 'Rest assured,' the King had promised. 'The railway will be built.'[18]

The close of the 19th century was the golden era of African railways. Flinging their sleepers and coal-eating locomotives across savannah and jungle, the colonial powers sent a blunt message to the locals: progress was unstoppable. The railroad was both an instrument of war, depositing troops armed with machine guns within range of their spear-carrying enemies, and an instrument of commercial penetration, bringing the ivory, minerals and spices at the continent's heart to market, opening the interior to land-hungry farmers and hopeful miners. Cecil Rhodes dreamt of one that would run from Cape to Cairo, the explorer Henry Stanley, nicknamed 'Breaker of

Rocks', was building one which would link Leopoldville to the sea, the British were braving man-eating lions to connect Uganda with the Swahili coast. Railways were the equivalent of today's national airlines – no African colony worth its salt could be without one.

Martini did not intend to be left out, although he knew Eritrea's topography made this a uniquely demanding challenge. When Martini arrived, the Italian army had already laid 28 km of track to the town of Saati, carrying troops to fight Ras Alula. But the work had been carried out in such haste, it all needed to be redone. There were drawings to be sketched, sites visited, contracts put out to tender and strikes to be settled. It all fell to Martini, acutely aware that Italy's colonial rivals were establishing their own trade routes into the interior, with France and Britain vying for control of a railway that would link Djibouti with Addis Ababa. 'The railway means peace, both inside and outside our borders, and huge savings on the budget,' he told his diary, time and again. Despite the King's promise, winning the funding did not prove easy. Having sent Martini out with orders to cut spending, Rome did not take kindly to constant requests for money. He would waste months peppering the Foreign Ministry with telegrams, winning his bosses round to the railroad's merits, only to see the government fall and a new set of ministers take office, who all had to be persuaded afresh. The railway, fretted Martini, 'would be the only really effective remedy to many – perhaps all – of the colony's ills. But in Rome they do not want to know.'[19]

He assembled a small army of 1,100 Eritrean labourers and 200 Italian overseers for the backbreaking and dangerous work, hacking and blasting through the rock, building stations and water-storage vaults as the railroad inched forwards. Struggling to master the technical minutiae of rail engineering, Martini found himself acting as peacemaker between irate private contractors and his abrasive head of works, Francesco Schupfer, a stickler for detail capable of forcing a company caught using sub-standard materials to knock down a stretch of earthworks and start again. 'Perhaps he is too rough, but he is a gentleman,' Martini pondered, intervening yet again to smooth ruffled feathers. 'He is hated by everyone, but very dear to me.'[20] When Britain raised the possibility of connecting Sudan's rail network to the Eritrean line – a move that would have turned Massawa into eastern Sudan's conduit to the sea – Martini was almost beside himself with excitement. 'This is a matter of life and death, either the railway reaches as far as Sabderat or we must leave Eritrea,' he pronounced.[21] Just when his plans looked set in concrete, Rome began wondering – in a reflection of the changing technological times – whether it might not be better off investing in a highway to Gonder and Addis Ababa instead.

Despite all the telegrams and discussions, the stops and starts, the track slowly edged its way up to Asmara. By 1904, the crews had reached Ghinda, by 1911, four years after Martini had returned to Italy, it had reached Asmara. The final heave up the mountain proved the trickiest. Even today, old men living in Shegriny ('the difficult place'), remember the dispute that lent their hamlet its name, as a father-and-son engineering team squabbled over the best route to take, each retiring to sulk in his tent before the precipitous route along 'Devil's Gate' – little more than a narrow cliff ledge looking out over nothingness – was finally agreed.

The single most expensive public project undertaken by the Italians in Eritrea, Martini's railway was emblematic of his rule. Its construction marked the time when Eritrea, exposed to Western influences and endowed with the infrastructure of a modern industrial state, started down a path that would lead its citizens further and further away from their neighbours in feudal Abyssinia. Yet, as far as Martini was concerned, this gathering sense of national identity was almost an accidental by-product. Like so many colonial Big Men, he was haunted by the need to tame the landscape, to carve his initials into Eritrea's very rocks. Literally hammering the nuts and bolts of a nation into place, he was more interested in the mechanical structures taking shape than what was going on in the heads of his African subjects. This colony was being created for Italy's sake and if much of what he did improved life for Eritreans, it was motivated by an understanding of what was in Rome's long-term interests, not altruism. No one could accuse Martini of remaining aloof – he toured constantly, setting up his white marquee under the trees and receiving subjects whose customs and traditions he recorded in his diary. But these were more the contacts of a deity with his worshippers than a parliamentarian with his constituents. This was the interest a lepidopterist shows in his butterfly collection – cool, distant and with a touch of deadly chloroform.

The approach is at its clearest when Martini writes about the two areas in which intimate contact between the races was possible: sex and education. Racial segregation had been practised in the colony since its inception. In Asmara, Eritreans were confined to the stinking warren of dwellings around the markets, while the Europeans, whose most prominent members donned white tie and tails to attend Martini's balls, lived in villas on the south side of the main street. Public transport was also segregated: Eritreans would have to wait another half-century to share the novel experience of using a bus's front door. But the races still mingled far more than the prudish Martini felt comfortable with. He disapproved of prostitutes, but was also repelled by the widespread phenomenon of *madamismo*, in which Italian officials took Eritrean women as concubines, setting up house together. The practice, he warned, raised a truly ghastly prospect. 'A black man must not cuckold a white man. So a white man must not place himself in a position where he can be cuckolded by a native.'[22] If the offspring of such unsavoury unions were abandoned, it would bring shame upon 'the dominant race'; if decently reared, it could ruin the Italian official concerned. Either outcome was to be deplored, so the entire situation was best avoided. It was an attempt at social engineering that enjoyed almost no success. By 1935, Asmara's 3,500 Italians had produced 1,000 *meticci*, evidence of a healthy level of interbreeding.[23]

But it is for his stance on education that Martini is chiefly resented by Eritreans today. The former education minister violently rejected – 'No, no and once again, no' – any notion of mixed-race schooling. His justification was characteristically quixotic, the opposite of what one might expect from a man who had embraced the credo of racial superiority. 'In my view, the blacks are more quick-witted than us,' he remarked, noticing how swiftly Eritrean pupils picked up foreign languages.[24] This posed a problem at school, he said, where 'the white man's superiority, the basis of every colonial regime, is undermined'. No mixed-race schooling meant there would be no opportunity for bright young Eritreans to form subversive views on their dim future masters. 'Let us avoid making comparisons.' The natives must be kept in their place, taught only what they need to fulfil the subservient roles for which Rome thought them best suited. It was a variation of the philosophy Belgium would apply to the Congolese in the field of education: *'Pas d'élites, pas d'ennemis'* ('No elites, no enemies').

In 1907, Martini asked to be recalled. He had pulled off a final diplomatic coup, travelling to Addis to pay his respects to an ailing Menelik II – 'one of the ugliest men I have ever seen, but with a very sweet smile'. It was a nightmarish journey during which the mules plunged up to their stomachs in mud and Martini, vain as ever, fussed constantly over the size of the

ceremonial guard each provincial ruler sent to meet him.[25] His
work on the railway was not complete. It would never, in
fact, be completed to his satisfaction, for Italy's invasion of
Ethiopia in the 1930s would interrupt construction of a final
section intended to link the Eritrean line to Sudan's network.
But Rome's procrastination had fatigued him. Being lord of all
he surveyed had been enjoyable, but the small-mindedness
of colonial life depressed him and he was fed up with army
intrigues. As he prepared to embark aboard a P&O liner,
with Eritrea's notables – both black and white – mustered in
Massawa to say goodbye, the man of letters was, for once, lost
for words. 'I feel such emotion that I have neither the strength
nor ability to express it.'

His farewell message to the Eritrean people reveals just
how far the anti-colonialist of yesteryear had travelled, how
heady the role of Lord Jim, sustained over nearly a decade,
had proved. It reads more like a prayer penned by an Old
Testament patriarch ascending to his rightful place at God's
side, than an Italian politician returning to his Tuscan constituency and, eventually, the top job at a newly-created
Ministry for Colonies.

'People from the Mareb to the sea, hear me! His Majesty
the King of Italy desired that I should come amongst you and
govern in his name. And for ten years I listened and I judged,
I rewarded and I punished, in the King's name. And for ten
years I travelled the lands of the Christian and the Moslem, the
plains and the mountain, and I said "go forth and trade" to
the merchants and "go forth and cultivate" to the farmers, in
the King's name. And peace was with you, and the roads were
opened to trade, and the harvests were safe in the fields. Hear
me! His Majesty the King learnt that his will had been done, by
the Grace of God, and has permitted me to return to my own
country. I bid farewell to great and small, rich and poor. May

your trade prosper and your lands remain fertile. May God give you peace!'[26]

With this portentous salutation, the Martini era came to a close.

He left behind a society transformed, but one – as far as its Eritrean majority was concerned – that held him in awe rather than affection. Today, when most Eritreans learn English at school, Martini has become little more than a name, his thoughts and achievements obscured by the barrier of language. Asmara holds not a single monument to this seminal figure. But older, Italian-speaking Eritreans remember, and their assessment of Martini is as ambivalent as the man himself. 'His legacy has been enormous, yet his aim was always to keep Eritrea in chains,' says Dr Aba Isaak, a local historian. 'He was a number one racist, but a superb statesman. I admire him, even while I regard him as my enemy.'[27]

When Martini left, there was no doubt in his mind that his government owed him thanks beyond measure. By his own immodest assessment, he had shored up a bankrupt enterprise and 'saved' an entire colony from abandonment, transforming a military garrison into a modern nation-state. But Martini had also laid the groundwork – quite literally, in the case of the railway – for the sour years of Fascism, when the implicit racism of his generation of administrators was turned into explicit law, and a colonial regime that had seemed a necessary irritation began to feel to Eritreans like an intolerable burden.

In the years that followed, the colony would serve as little more than a supplier of cannon fodder for Italy's campaign in Libya, sending its *ascaris* to seize Tripolitania and Cyrenaica from the Turks in 1911. Italy's African pretensions were largely forgotten as the country was plunged into the horrors of the

First World War. The Allied carve-up of foreign territories following that conflict left Italians bruised. Right-wingers who still quietly pined for an African empire felt their country had been promised a great deal while the fighting raged, only to be palmed off with very little by the Allies when the danger of German victory passed. It was an anger that played perfectly into the hands of the bully who was about to seize control of Italy.

As a youthful Socialist, Benito Mussolini had railed against liberals such as Martini for frittering away funds he felt would have been better spent tackling Italy's underdeveloped south, actually going to prison for opposing Italy's invasion of Libya. But once he assumed office in 1922 as prime minister, Mussolini's attitude to empire changed. Hardline Fascist commanders were dispatched to Libya and Somalia, where they ruthlessly crushed local resistance and expropriated the most fertile land. The extreme nationalism at Fascism's core required a rallying cause and Mussolini was a great believer in the purifying power of battle. 'To remain healthy, a nation should wage war every 25 years,' he maintained. He was determined to prove to other European powers that Il Duce deserved a seat at the negotiating table. Nursing expansionist plans for Europe, he needed a quick war that could be decisively won, giving the public morale a boost before it faced more formidable challenges closer to home. Abyssinia, which many Italians continued to regard, in defiance of all logic, as rightfully theirs, seemed the perfect choice. France had Algeria, Britain had Kenya. It was only fair Italy should have her 'place in the sun'.

As the official propaganda machine cranked into action, Italians were once again sold the idea of Abyssinia as an El Dorado of gold, platinum, oil and coal, a land ready to soak up Italian settlers – Mussolini put the number at a blatantly absurd 10 million. Once again, one of Africa's oldest civilizations was portrayed as a land of barbarians, who needed to be 'liberated' for their own good. Italian officials were not alone in nursing a vision of Abyssinia that could have sprung from the pages of *Gulliver's Travels*. 'There human slavery still flourishes,' *Time* magazine told its readers in August 1926. 'There the most trifling jubilation provides an excuse for tearing out the entrails of a living cow, that they may be gorged raw by old and young.' Itching for a pretext to declare war on Ras Tafari, the former Abyssinian regent who had been crowned Emperor Haile Selassie in 1930, Mussolini finally seized on a clash between Italian and Abyssinian troops at an oasis in Wal Wal as a pre-text. Retribution had been a long time coming, but the battle of Adua was about to be avenged.

For Eritrea, the obvious location for Italy's logistical base, the forthcoming invasion meant boom times. *Ca Custa Lon Ca Custa* ('Whatever it costs') reads the slogan, written in Piedmontese dialect, carved into the cement of the ugly Fascist bridge which fords the river at Dogali. It epitomized Mussolini's entire approach to the war he launched in the autumn of 1935, ordering a mixed force of Italian soldiers and Eritrean *ascaris* to cross the Mareb river dividing Eritrea from Abyssinia. 'There will be no lack of money,' he had promised the general in charge of operations, Emilio de Bono, and the ensuing campaign would be characterized by massive over-supply.[28] When de Bono asked for three divisions, Mussolini sent him 10, explain-ing: 'For the lack of a few thousand men, we lost the day at Adua. We shall never make that mistake. I am willing to commit a sin of excess but never a sin of deficiency.'[29] Some 650,000 men, including tens of thousands of Blackshirt volunteers, were eventually sent to the region and with them went 2m tonnes of material, probably 10 times as much as was actually needed. Flooded with supplies – much of it would sit rotting on the Massawa quayside, only, eventually, to be dumped in the sea – Eritrea's facilities suddenly looked in dire need of modernization.

A 50,000-strong workcrew was dispatched to do the necessary: widening Massawa port, building hangars, warehouses, barracks and a brand-new hospital. The road to Asmara was resurfaced, airports built, bridges constructed. Martini's heart would have thrilled with pride, as his beloved railway finally came into its own. Trains shuttled between Massawa and Asmara nearly 40 times a day, laden with supplies for the front. Even this was not considered sufficient, however, and, in 1936, work started on another miracle of engineering, the longest, highest freight-carrying cableway in the world. The 72-km ropeway erected by the Italian company of Ceretti and Tanfani, strung like a steel necklace across the mountain ranges, was as much about demonstrating the white man's mastery over the landscape as meeting any practical need. It was exactly the kind of high-profile, macho project Mussolini loved.

Asmara blossomed. New offices and arsenals, car parks and laboratories sprang up, traffic queues for the first time formed on the city's streets. The most modern city in Africa boasted more traffic lights than Rome itself. Soon the simple one-storey houses of the 19th century were dwarfed by Modernist palazzi. In the space of three frenzied years, Italy's avant-garde architects, presented with a nearly blank canvas and generous state sponsorship, created a new city. A mere five years before Mussolini's new Roman empire was to crumble into dust, Eritrea's designers dug foundations and poured cement, never doubting, it seems, that this empire was destined to endure.

It was a short military campaign. By May 2, 1936, Italy's tactic of bombing Abyssinian hospitals and its widespread use of mustard gas, which poisoned water sources and brought the skin out in leprous, festering blisters, had had the desired effect.

With his army in tatters and Italian troops marching on Addis, Haile Selassie fled the country. He made one last poignant appeal for help before the League of Nations in Geneva, where, jeered by right-wing Italian journalists, he warned member states that their failure to stop Mussolini would destroy the principle of collective security that had been the organization's *raison d'être*. 'International morality is at stake,' he said, 'what answer am I to take back to my people?'[30] European powers, who had already decided to take no more than token action, listened in silent embarrassment to this Cassandra-like warning. Riding a wave of popular rejoicing, Mussolini set about dividing Haile Selassie's territory on ethnic lines. Abyssinia was swallowed up in Italian East Africa, a vast new Roman empire which embraced Eritrea and Somalia and covered 1.7 million sq km, stretching from the Indian Ocean to the borders of Kenya, Uganda and Sudan.

In Eritrea, this should have been a golden age, for white and black alike. But while the economy thrived, relations between Eritreans and Italians had never been worse. The new Italians, Eritreans quickly noticed, were different from the old. They came from the same modest backgrounds as their predecessors, but they seemed, like Il Duce himself, to feel a swaggering need to demonstrate constantly who was boss. There was little danger of these new arrivals, convinced of their Aryan superiority, becoming *insabbiati*: they despised the locals too thoroughly to mix. 'Every hour of the day, the native should view the Italian as his master, sure of himself and his future, with clear and defined objectives,' explained an Italian writer of the day.[31] To that end, a raft of increasingly oppressive racial laws was introduced across Italian East Africa between 1936 and 1940. Part and parcel of the anti-Semitic legislation being adopted in mainland Italy, they aimed at keeping the black man firmly in his place.[32]

Asmarinos today still refer to the city as '*piglo Roma*' and the centre of town as the '*combishtato*', bastardizations of the '*piccolo Roma*' Italy recreated on the Hamasien plateau and the *campo cintato* ('enclosed area'), ruled off-limits for Eritreans outside working hours 'for reasons of public order and hygiene'. Eritrean merchants with premises on prime shopping streets were forced to surrender their leases to Italian entrepreneurs. Consigned to the public gallery at the cinema, Eritreans were barred from restaurants, bars and hotels and made to form separate queues at post offices and banks. Africans actually *preferred* to keep their distance, claimed the Italian Ministry for African Affairs in justification.[33] Once, Eritreans and their white compatriots had greeted each other as '*arku*' ('friend'). In future, Fascism decreed, Italians would address Eritreans with the peremptory '*atta*' and '*atti*' ('you'), while the Eritrean was expected to use the respectful '*goitana*' ('master') towards his white superior.

The new legislation enshrined the principle of separate education Martini had first embraced. And no matter how talented or well-heeled, an Eritrean could not stay longer than four years at his all-black school. Italy needed obedient translators, respectful artisans and disciplined *ascaris*, not trouble-making intellectuals. 'The Eritrean student should be able to speak our language moderately well; he should know the four arithmetical operations within normal limits; he should be a convinced propagandist of the principles of hygiene, and of history, he should know only the names of those who have made Italy great,' announced the colony's Director of Education.[34] The subject of Italy's Risorgimento was dropped entirely from the syllabus, for fear it might spark inappropriate ideas.

If the Fascist administrators disliked the notion of uppity natives, the prospect of an expanding 'breed of hybrids' positively appalled.[35] Young Italian soldiers, whose tendency to acquire female camp followers was noticed by reporters covering the conflict, marched into Abyssinia singing the popular hit 'Facetta Nera' ('Little Black Face'), in which a black Abyssinian beauty is saved from slavery, taken to Rome by her lover and dressed in Fascism's black shirt. A year later, the authorities were attempting to suppress the song as Italian newspapers warned that the Fascist Empire was in danger of becoming an 'empire of mulattos'.[36] The new laws betrayed a vindictive determination to wipe out any vestige of affection, loyalty and love between the races. 'Conjugal relations' between Italians and colonial subjects were prohibited, marriages declared null and void. Italians who visited places reserved for 'natives' were liable to imprisonment and it was ruled that an Italian parent could neither recognize, adopt or give his surname to a *meticcio*. With a stroke of the pen, Rome turned a generation of mixed-race Eritreans into bastards. '*Figlio di N*' was the mocking playground cry that greeted the mixed-race child, officially stripped of inheritance, citizenship and name: 'son of X'.

Long before racial segregation was adopted as an official credo in South Africa, Eritrea had already tasted the delights of apartheid. In its day, it was the most racist regime in Africa. Eritreans no longer regarded their Italian administrators as 'good but stupid'. Every Eritrean who lived through that era nurses in his memory a moment of humiliation he can today shake his head over with the bitter satisfaction that comes from knowing history has had the last laugh, a deep guffaw that comes from the belly. 'When a white man walked along the street, you always followed a couple of steps behind, never alongside,' an old railwayman told me. 'The white man always walked alone.' 'You could be dying of thirst, but the cafés in the town centre would still refuse to serve you so much as a glass of water,' said another. A pastor remembered how, as a boy, he once made the mistake of crossing the road in front of an

Italian policeman on a motorbike. 'He was quite a long way off, but as far as he was concerned I should have waited for him to pass. He caught up with me and slapped me round the face.' The pastor recalled an old Italian lawyer expostulating at his failure to step into the gutter as he passed. 'He said "Hey, can't you see that a white man is coming?",' he chuckled. 'It was his way of saying "Get off the pavement".'

For Italy, the conquest of Abyssinia and racial subjugation of Eritrea would prove Pyrrhic victories. Resistance by Haile Selassie's followers meant much of the Abyssinian countryside remained unsafe, and the number of settlers never rose above the disappointing. The extravagantly-funded war plunged Rome into debt and while other European powers milked fortunes from their territories abroad, Italy, embarrassingly, never managed to make colonialism work for her financially. Pouring investment into both Eritrea and Ethiopia, her empire cost her more than she gained and the government was juggling ballooning budget deficits when the Second World War began. Thanks to the predictably easy victory in Abyssinia, Italians would enter that campaign with a dangerously unrealistic belief in their military might, a confidence which shattered at huge national cost. And while the European powers agreed temporarily to turn a blind eye to Mussolini's bullying, the Abyssinian campaign also marked the moment when Il Duce's eventual destiny as Hitler's patsy began to take shape. Having thoroughly alienated the liberal democracies with his behaviour towards Abyssinia, Mussolini's natural place, increasingly, would seem alongside the Nazi leader.

The war's most dramatic long-term outcome was to effectively kill off Italian colonialism. Crude and invasive, often loathed by Italian settlers whose presence in Eritrea predated Fascism, the racial legislation made daily life so ghastly for Eritreans that when history finally granted Italy's African subjects an opportunity to decide their fate, they would turn and spit in the face of their former masters.

Looking back, many Eritrean intellectuals view the 1930s as the period in which the characteristics now regarded as quintessentially Eritrean began to take recognizable form. Every country which experiences colonialism is defined by how it digests humiliation. In many African states, the experience corroded a community's sense of self-worth, dripping through the generations like acid. But in Eritrea's disciplined, tight-knit communities, sure in the knowledge of their ancient traditions and religious faith, subjugation ate into the soul in a different way. In the *kebessa*, families had always prided themselves on settling their disputes without recourse to outsiders. Nothing was considered more undignified than being seen to lose control, to let go. The brutality of the racial laws was met with tight-lipped self-restraint. Turning inwards, the Eritreans bottled up their emotions and waited to see what the future would bring. 'Whatever sun rises in the morning is our sun, and whichever king sits on the throne is our king,' runs a Tigrinya proverb which summarizes the bittersweet philosophy of a people accustomed to having things done to them. Since there was no point standing up to a mightier adversary, silence seemed the only way of salvaging self-respect. 'You knew what the consequences would be if you openly revolted, so you were advised to bide your time, to be patient. That was the advice my father always gave me when I was growing up: "just be patient",' says Dawit Mesfin, an Eritrean intellectual living in London.[37] He calls this preternatural calm, the apparent passivity cultivated in that era, 'quietism'. Like all superficial passivity, it was a lid on a pressure cooker, clamped over a storm of hurt pride and a longing for retribution.

CHAPTER 4

This Horrible Escarpment

'It was contrary to every book that had ever been written, but it came off.'

Lieutenant-General Sir William Platt

If you drive two hours north-west of Asmara, on what was once Mussolini's Imperial Way, you eventually reach the town of Keren. The road takes you through eucalyptus groves, whose leaves, in the early morning, give off a heady medicinal perfume, before crossing a plateau of almost unimaginable bleakness. When it came to power in 1993, Eritrea's new government launched an ambitious reforestation campaign and brave green saplings, their bark protected from nibbling goats by little iron wigwams, have been planted at neatly-spaced intervals along the way. But it will be decades before their handkerchiefs of shade make any impression on this glaring, denuded landscape. Much of the Hamasien highlands looks as though it was scooped up at a quarry and deposited from the tail-end of a dumper-truck. With so little standing in its path, the wind is free to harry the cappuccino-coloured dust, so soft it could have been poured from a lady's powder compact, so fine a trailing fingertip registers no contact at all. Donkeys stand motionless in the middle of the road, as though stunned into imbecility by the heat and sheer unfairness of being expected to graze off an expanse of rubble.

Keren itself is a bustling crossroads of a town, a natural meeting point for Eritrea's various tribes. White-robed nomads bring their animals to the weekly livestock market, a moaning, bleating, piebald medley of camels, sheep and goats. City dwellers make weekend jaunts to a Holy Shrine hidden in the womb of an ancient baobab and stroll along the colonnaded street where silversmiths crouch over bracelets and rings. But Keren only makes historical sense if you keep going, through the centre and out the other side. On the edge of town, by the side of the road that points towards the border with Sudan, lies a British cemetery from the Second World War. Jump over the low wall – visitors here are too occasional for anyone to man the gates permanently – and you can pace the rows of neat white British graves, each with its own crinkly bush of red geranium. The birthday-card triteness of the mottoes carved into the stones betrays a terrible anguish for young lives abruptly ended. 'He gave his life, forget him never, for in giving he lives for ever,' plead the relatives of 22-year-old W Hollings, who served with the West Yorkshire Regiment. 'With many sad regrets, we shall remember when the rest of the world forgets,' promises the family of PFC Mapes, 26. 'In memory's garden, we meet every day,' is the only consolation the mother of AV Simmons, just 18 when his life came to a sudden stop, can offer herself.

Further on, by the wayside, a grieving Madonna inclines her head in an expression of infinite tenderness. Her location is not accidental. A moment later, the road plunges, weaving and winding its tortured way through a ragged gash in the mountains which the British, mangling the local name, dubbed Dongolaas Gorge. Cliffs of bulbous brown rock crowd in on all sides, blocking out the sun, while the sheer slopes below are scattered with the rusting debris of vehicles that missed the curves. There is something oppressive about this pass, and it is a relief to emerge finally in the valley, a scrubby floodplain so dry it seems impossible water should ever run across these yellow sands. It is only when you hit the bottom and look back to where you came, that Keren's geography suddenly becomes clear. Between you and the town a knobbly range of barren peaks now stretches, a giant amphitheatre coddling Keren in its lap. With Dongolaas Gorge offering the only easy access route across this rock wall, Keren is one of the most formidable fortresses ever thrown up by nature.

This was the sight that met a force of around 30,000 British troops as it chased Italy's retreating army from Kassala, on the Sudanese border, east across Eritrea. It was February 2, 1941 and the first dents were being knocked into what had until then seemed like an unstoppable Nazi and Fascist war machine. In the preceding two years, Hitler had swallowed up Czechoslovakia and Poland, invaded Norway and Denmark, toppled the French government and forced the humiliating evacuation of British forces from continental Europe. Hanging on to the German dictator's coat-tails, Mussolini had invaded Albania and Greece and used his colonies in north and east Africa to attack Egypt and eject the British from Somaliland. Afraid of losing control of the vital supply route through the Suez Canal to the Indian Ocean, Britain was hitting back. It had trounced Italian forces in Libyan Cyrenaica and was now sending three separate forces – one bearing with it the deposed Haile Selassie – in a vast pincer movement aimed at squeezing the life out of the marooned Italian administration in the Horn. Before the Red Sea could be made safe for Allied shipping, both Asmara and Massawa had to be captured. But between the advancing British forces and their target lay Keren.

No one remembers the battle of Keren now, except the

men who fought there. Perhaps that is only to be expected. The Normandy landings were staged on beaches within reach of modern European holidaymakers, the Blitz left permanent scars on the map of London, every time Berliners gaze at their city's skyline, they are reminded of the devastation of Allied bombing. Keren took place in a country whose name was unfamiliar even to the soldiers dispatched there to fight, in a part of the world so alien as to be virtually unimaginable to those back home, against an enemy regarded as a joke. El Alamein and Tobruk were to be the African campaigns of the Second World War remembered by British audiences, not Keren. Even Eritreans, who live amongst memorials and cemeteries spawned by the battle, are distinctly fuzzy about the details, too obsessed with their own still raw military history to show much interest in an episode logged in the general category of 'neo-colonialist adventures'.

Yet it was a linchpin episode, on which turned a long sequence of events stretching to eventual Allied victory. The BBC's decision to send its star reporter Richard Dimbleby to cover the battle shows it recognized that fact. But as far as the BBC was concerned, Dimbleby was covering an early bout in a mighty strategic contest, not a liberation campaign. Not for the first time, Eritrea would play unwitting host to a battle in which its citizens would be slaughtered, yet their own aspirations remained almost immaterial, of interest purely for their passing propaganda potential. The battle of Keren might have been fought on Eritrean soil, but it was plotted, planned and ultimately capitalized on in the capitals of Axis and Allied powers.

It was to prove a grinding, infinitely testing campaign in which victory against the Italians, viewed until then with amused condescension, never looked assured. 'It was a ding-dong battle, a soldiers' battle, fought against an enemy infinitely superior in numbers, on ground of his own choosing,' General William Platt, head of the armed forces in Sudan and mastermind behind the Eritrean campaign, later said. 'We got down very nearly to bedrock, very nearly.'[1] Those who took part and went on to fight in the deserts of North Africa, the streets of European cities and the jungles of Burma were to recall the fighting at Keren as the most dreadful they ever experienced. 'Physically, by World War Two standards, it was sheer hell,' remembered Major John Searight, of the Royal Fusiliers, in a letter written after the war. 'NOTHING I met in nine months as a company commander in NW Europe compared with it.'[2]

It is interesting to log the process by which the mind gives form and shape to landscape. Colonial explorers in Africa had a knack for it, baptizing waterfalls and lakes whose existence had been known to locals for centuries after childhood sweethearts and royal patrons. When the British forces arrived on the floodplain they looked at the scenery with the outsider's lazy eye. The soldiers, recruited not only from Britain but from India, Sudan and Palestine – these were the days of the British empire, after all – saw a craggy range of mountains, rising to 7,000 ft, pierced by an occasional curious nipple of rock. Foothills formed rucks in a ribbon of brown land that stretched across the horizon. From a distance, what little vegetation there was – sprays of candelabra cactus, stunted thorn bushes here and there – resembled a light dusting of pepper sprinkled by a giant hand. To the far east crouched a curious promontory resembling a squatting cat, to the far west rose a peak like a ripped-out molar.

Fear changes one's perspective, lending what was of purely abstract interest a sudden urgent relevance. By the time the men had moved on, 53 days later, every inch of the terrain had acquired a dreadful, unwanted intimacy, more familiar to them than the soft folds of their Yorkshire valleys or the alleyways

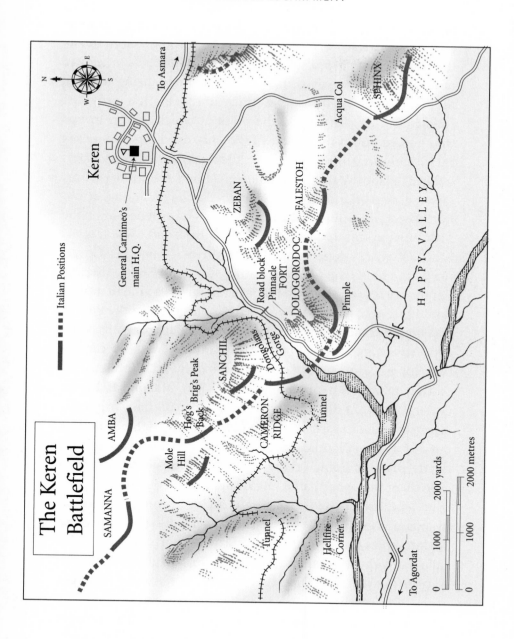

The Keren Battlefield

---- Italian Positions

To Asmara

Keren

General Carnimeo's
main H.Q.

ZEBAN

Road block
Pinnacle
FORT
DOLOGORODOC

Dongolaas
Gorge

SANCHIL

Hog's Brig's Peak
Back

AMBA

Mole
Hill

SAMANNA

CAMERON
RIDGE

Tunnel

Tunnel

Hellfire
Corner

To Agordat

FALESTOH

Acqua Col

SPHINX

Pimple

HAPPY VALLEY

2000 yards

2000 metres

1000

1000

0

0

of their villages in the Punjab. Peaks and hillocks had been captured and lost, recaptured only to be surrendered again, as the front line juddered forwards and backwards. Each feature of the land had acquired its own distinct, tantalizing identity in the bloody baptisms that tracked the battle's progress. Hellfire Corner, they called the point at which British troops were forced out of the shelter of the hills, exposing themselves to withering enemy fire. Above the horribly exposed plain, dubbed 'Happy Valley' with the squaddie's traditional irony, towered razor-sharp Mount Sanchil. The highest peak, it ran north-west, stretching into Brig's Peak – christened after the 11th Brigade sent to conquer it – Hog's Back and Mole Hill. At Sanchil's base, overlooking the mouth of Dongolaas Gorge, reared the bluff that would be named Cameron Ridge, after the Cameron Highlanders who managed to scrape a vital first position there. Across the pass, above the foothills called the Pimple and the Pinnacle, rose Fort Dologorodoc, a cluster of trenches and cement parapets enjoying panoramic views. The cat-like formation became, with a certain inevitability, the Sphinx.

The Italians had never meant to make a stand at Keren. General Luigi Frusci, governor of Eritrea, had originally planned to dig in further up the road to Asmara. Since pulling out of Kassala in January, the Italians had been consistently on the defensive, holding up the British advance when the terrain gave them the upper hand, but always pulling out before losses rose. The British soldiers had almost been caught off-guard by an ambush at a place called Keru, where an Italian officer on a white horse had led a troop of Eritrean horsemen in a wild gallop straight into the mouths of the British guns.[3] Taken by surprise by what must have been one of the last cavalry charges of the 20th century, British troops had recovered just in time, tipping their cannon so that they pointed into the ground in front of the horses' hooves. 'It was suicide, really, but it was gallantry one wouldn't expect of the Italians. They were totally destroyed,' remembered Colin Kerr, an intelligence officer with the Cameron Highlanders.[4]

As his troops withdrew ever further up the Imperial Way, Frusci belatedly realized the huge advantages Keren offered and ordered his men to stop and turn. There would have been no subsequent battle, had the British forces succeeded in keeping up their momentum. But the Italians had blown up a bridge behind them; and mined the river bed. By the time the obstacle had been cleared and the British convoy trundled into Happy Valley, the dull thud of explosions was audible ahead. General Orlando Lorenzini, the 'Lion of the Sahara' to his men, was dynamiting the walls of Dongolaas Gorge, triggering a massive rock fall that closed off the pass. If Keren could be compared to a medieval fortress, the Italians had just rushed across the moat and pulled up the drawbridge.

Hoping to catch the Italians while they were still out of breath, British commanders ordered an immediate assault on Mount Sanchil. The Cameron Highlanders and Punjab Regiment stormed the slopes, capturing Cameron Ridge and Brig's Peak, only to be forced off the latter by an Italian counterblast. 'I saw a certain amount of the war after that, in various places, but the first 24 hours of Keren were about as unpleasant as any I saw anywhere,' recalled Kerr. 'It was incessant shelling and we were unable to reply to it because we were under observation.' The first attack ground to a halt as a sober understanding of the enormity of the task ahead dawned. The Italians, whose ranks were being swelled by reinforcements from the rear, were now dug in. They had set up machine guns and heavy mortars on all the loftiest points in the mountain range and designated them 'last man, last round' positions. There was to be no surrender, only death.

In contrast with what many civilians assume, military action does not always require extraordinary degrees of personal courage. Usually, the landscape offers so many possibilities for ambushes and outflanking movements, careful advance and sensible retreat, that a soldier can convince himself he stands a fair chance of surviving. Not so in Eritrea, where the campaign would essentially be fought along one road. On that road, Keren represented the head-on confrontation that brings with it the oppressive awareness of likely death. 'It was the first set-piece battle I'd been near. The others were skirmishes where one could dodge out of the way. Here there was no avoiding it,' said one former artilleryman.[5] Reconnaissance revealed that there was going to be no way of sidestepping this particular showdown: the mountain range stretched for 150 miles. 'It was a question,' recalled Platt, 'of fighting this horrible escarpment somewhere or coming somewhat back. We could not live continuously under the escarpment.'[6]

When, in 2002, a British general took command of a UN peacekeeping force posted to Eritrea, he decided to use the battle of Keren as a training exercise for his under-employed troops. His aide-de-camp was ordered to dig out the 60-year-old military records, and groups of blue-capped UN officers were taken on tours of the battlefield to discuss how, facing the same logistical challenge but equipped with today's equipment, they would plan their attack.

I joined one of the tours, tagging along as the peacekeepers scrambled part way up Cameron Ridge to survey the terrain. None of us was carrying heavy loads, but we were soon stumbling and staggering, rasping for breath. The sand churned beneath our feet, making it hard to get a grip, rocks used for leverage had a nasty habit of rolling back onto you. It was only 10.00 in the morning, but it was already 41° C. The water in our plastic bottles had turned bathwater warm. Within a few minutes, I realized, to my embarrassment, that I was on the brink of fainting. Dots were dancing before my eyes. I couldn't seem to get enough oxygen into my lungs. The men in the group, I noticed with relief, looked no better: their faces had flushed an unbecoming gammon-pink.

'Look around you,' trumpeted the boisterous British general. 'No shelter anywhere from the sniper fire. Just imagine what it must have been like.' Our faces coated in a shiny slick of sweat, we gazed from the bare brown boulders on which the British had once crouched up to the unforgiving heights of Sanchil, where the Italians had sat waiting with their machine guns. 'Horrid terrain to fight in,' muttered the general. 'Horrid.'

Few of the soldiers mustering in Happy Valley had any real idea of the ordeal awaiting them. Their advance across Eritrea had gone swimmingly up until that point, with victory tumbling upon victory and few casualties suffered on the way. 'We were very confident. Morale was grand,' recalled Patrick Winchester, an artillery officer with the 4th Indian Division, just 21 at the time. 'We'd been up in the Western Desert, which was the first good thing to happen in the war. We thought "We'll go and sort them out over there."'[7]

Such optimism was compounded by the low regard in which the Italians were held as fighting men. The sight of thousands of Italian prisoners-of-war straggling along in dejected columns had registered on the British troops gathering in Happy Valley. Your average Itie, it was felt, was essentially a fun-loving, happy-go-lucky sort, who packed up fairly easily, showing nothing approaching the steely focus of the German. Italian equipment was lightweight and insubstantial because, in his rush to arm a ballooning military, Mussolini had cut corners and gone for the cheapest options. Often it lay unused at the depot, because the munitions from Rome did not match with hardware on the battlefield. If the Italians nominally boasted 250,000 men in eastern Africa, Eritrean and Ethiopian *ascaris* accounted for 75 per cent of these forces, and their loyalty to their masters – who were only paying their wages spasmodically – was shaky. What was more, the Italians, cut off in the Horn, would inevitably face tremendous problems of resupply once they came under pressure. In trying to protect an African empire stretching across 1 million square miles, Italian forces were dangerously over-extended.

What the British troops initially failed to appreciate, however, was that those facing them were no ordinary Italian troops. The regiments and battalions sent to fight in Keren – the Savoia Grenadiers, the Bersaglieri and the Alpini – were the best Italy could muster. The moment they crossed the border with Sudan and stepped onto Eritrean soil, these men were fighting on what they considered home turf, defending Italy's oldest colony from a foreign invader. In the case of the Alpini, they had grown up in a northern version of the mountainous terrain now confronting them. 'What are those goats running up there for?' a British colonel asked one of his officers on arrival in Happy Valley, spotting movement on the heights. 'Those are not goats, sir,' was the ominous reply. 'They are Alpini.'[8]

The Italians had the confidence that comes with knowing you enjoy a nearly impregnable position. They held the high ground. The slopes were so rocky that the British could never dig trenches, sheltering instead behind sangars, piles of rocks that provided little protection against a direct hit. What was more, however cheap Italian hardware might be, British equipment – much of which dated back to the First World War – was ill-suited to Keren. The lumbering 25-pound guns were too cumbersome to be taken up the slopes and their long, low trajectories were most effective on flat terrain. Used in mountains, the shells either screamed harmlessly over the heads of Italians crouched on the far sides of the ridges or clipped the tops of the crests on which British troops moved, subjecting soldiers to terrifying 'friendly' fire. 'To be shot up the back passage by your own guns when your rickety breast work is designed to give you protection from the front creates a paralysing terror,' remarked Peter Cochrane, who served with the Cameron Highlanders.[9] In contrast, the nimble Italians had pack guns which could be moved about on the backs of mules, small red grenades which were simply dropped onto advancing troops and mortars which lobbed their missiles high in the air and neatly onto British positions.

The terrain in itself was challenge enough, but there was also the climate to contend with. On the plains, temperatures rose so high that convoys would sometimes grind to a halt within minutes of departure; radiators boiling, fit young soldiers keeling over with sunstroke. Today, the UN issues its forces in Eritrea with four and a half litres of drinking water a day. The soldiers at Keren were expected to march, run and fight on a pint a day, dysentery or no dysentery. No one wanted to waste a precious drop on washing, so teeth grew black, faces so caked with dirt soldiers often struggled to recognize one another. Rations were monotonous: dried biscuits and bully beef served in metal tins which became too hot to hold and whose instruction labels – 'chill before serving' – were a source of bleak amusement. In other campaigns, British forces supplemented their diet with fruits and vegetables collected on the way. Here, with the exception of the occasional guinea fowl, the landscape had almost nothing to offer. The result was a low-vitamin diet that undermined the body's immune system. Scratch your arm on a thorn tree and within a few days it would go septic. Then the 'desert sore' would spread until much of the limb was a suppurating mass of pus requiring hospital treatment. The baggy shorts and kilts worn by the troops were a menace, the soldiers soon concluded: in daytime they failed to protect the legs from sun and scratches, at night the men shuddered with cold on the hillsides.

This was not to be a sophisticated campaign. With mechanized transport ruled out by the gradient, fighting often took on an almost medieval directness. When the big guns fell silent, the battle of Keren was reduced to a low-tech war in which the readiness to emerge from cover, stagger up a mountain slope and simply bludgeon a way through mattered far more than weaponry. When Italian positions were overrun, it was a question of hand-to-hand fighting, with few prisoners taken and killing done by bayonet. At one stage, an Indian brigade actually made itself shields of corrugated iron. Held over the head in true storming-the-ramparts tradition, they proved surprisingly effective in warding off the pepperpot grenades raining down from above.

Three days after the first assault on Sanchil and Brig's Peak had failed, leaving only Cameron Ridge in British hands, Platt ordered an outflanking attempt to be made at Acqua Col, the mountain pass over which the Sphinx held watch. But commanders had only a vague grasp of how the land lay and the Rajputana Rifles were obliged to grope their way forward in the dark. The Italians, enjoying sweeping views, realized what the British were attempting well ahead of time and prepared a devastating response. After four days of fighting, both sides pulled back, exhausted.

Platt was coming under enormous pressure to wrap the Keren campaign up. In London, Winston Churchill was desperate to move British forces back to the Western Desert, to combat the growing menace posed by Rommel's Afrika Korps. The rains were due to start in a few weeks' time – another good reason to get Keren over with. But the first two attempts had shown the town could neither be taken by surprise nor from the side. Frontal assault, the great taboo of military strategists, was going to be the only way in. British forces would have to take the mountain range the hard way: feature by individual feature. It was not a task, Platt knew, to be undertaken without a great deal of logistical preparation.

The next month was spent in a frantic whirr of activity as the British stockpiled ammunition, fuel, rations and water. The old Italian railway was extended, every available wagon, truck and mule and camel press-ganged into service as the British set up a supply chain that stretched from Port Sudan, via Kassala, all the way to Happy Valley. The troops would not attempt another push, Platt decided, until he had enough supplies to last two divisions for 14 days of sustained assault.

While the stockpiling continued, soldiers up on Cameron Ridge gritted their teeth and hung on to their precious bluff of captured land. They had never experienced conditions like this before. Italian fire was killing or wounding between 25 and 30 men on the ridge every day. Adopting a technique that would later become a way of life for the Eritrean guerrilla movement, the British forces learnt it was wisest to move at night, covering the ground in scampering crawls. Wounded soldiers would lie stifling their groans during daylight, waiting for the darkness that made evacuation possible. It took 12 men to carry one man down the mountain. There was no way of safely burying the dead, so bodies lay where they had fallen or were pushed into nearby ravines, where they blew up in the heat, attracting dark clouds of flies. Rotten odours seem to travel further in dry heat, there was no avoiding the smell. Digging latrines was out of the question, so the stench of human excrement was soon added to the sweet stink of rotting flesh. The men smoked hard to drown out the nauseous odour and when the tobacco ran out, they smoked tea leaves. At night they could hear jackals tearing at the corpses of their former friends, a sound to turn the stomach. In daytime, another local species made its unwelcome appearance. 'There'd be a scrambling around and a rush of stones. Everyone would get in a state of tension and think this was the start of an attack, only to see the coloured behind of a baboon passing on the skyline,' said Kerr.[10]

Keeping the Cameron Highlanders up on the ridge supplied with water presented Platt with a major operational challenge in itself. A train of mules had been brought in from Cyprus, but even these surefooted animals regularly lost their footing, their bodies bouncing down the slopes. In any case, there were never enough mules for the task, so the troops were enrolled as pack animals, carrying water and bullets to those on the front line.

But the Italians were also suffering. Normally, attacking forces sustain the lion's share of injuries. At Keren, however, the hard rock surface was responsible for hundreds of deaths by indirect fire, as each landing shell sent shards of shattered stone flying in all directions. 'My troops are exhausted both physically and morally,' General Nicola Carnimeo, who repeatedly demanded reinforcements, warned Frusci. 'We have had very heavy casualties and the length of the perimeter is such that the defence will necessarily be thin on the ground. If the British concentrate a sufficiently strong force, they may well smash a gap in the ring.'[11] Italian forces had also been thinned by desertions. The RAF had been dropping leaflets over enemy lines announcing that Emperor Haile Selassie was returning to his rightful throne in Ethiopia and calling upon the *ascaris* to rise up against their Italian masters. Whether the propaganda had any effect or the British bombardment was simply too ghastly to endure, hundreds of Eritrean and Ethiopian recruits were defecting.

In mid-March, Platt decided he was ready and summoned

his officers. 'Do not let anybody think this is going to be a walk-over,' he told them. 'It is not. It is going to be a bloody battle, a bloody battle against both enemy and ground. It will be won by the side which lasts longest. I know you will last longer than they do. And I promise you I will last longer than my opposite number.'[12] As far as Platt was concerned, this represented the last throw of the dice, for he had run out of alternative schemes. 'What will you do if it doesn't come off?' General Archibald Wavell, Commander-in-Chief of the Middle East, asked him in the days leading up to the attack. 'I'm damned if I know, sir,' he replied. March the 15th was designated as the day of the assault, a week earlier than Platt would have liked, but head-quarters were pushing. Platt quickly regretted his scheduling: the day dawned oppressively muggy. 'I could not have chosen a worse date. Some of the efforts of the troops that day were defeated almost as much by the heat and heat exhaustion as by hostile opposition.'

As 96 British big guns opened fire and the Italians responded with a barrage of machine-gun fire and a hail of mortars, the 4th Indian Division launched a multi-pronged assault on Hog's Back, Brig's Peak and Sanchil, while the 5th Indian Division followed up with an attack on Fort Dologorodoc. This was the rationale behind the long weeks of stockpiling. Over the next 12 days, British artillery worked its way through a stockpile of 110,000 shells, representing, Platt later calculated, the equivalent of 1,000 lorry-loads of ammunition. Firing continually, the British were inflicting a highly effective form of psychological torture: sleep deprivation. 'We'd be given a list of pre-arranged targets and we would go through them in chronological order through the night, keeping them on their toes,' recalled Winchester.[13] In that rocky terrain the shock waves bounced from cliff to cliff, numbing the brain and shattering the senses. Cochrane experienced the intolerable effect of this constant bombardment from the other side after being taken prisoner by the Italians. 'By the second night,' he recorded in his memoirs, 'I would have given anything to get off the hill.'[14]

The West Yorkshires finally broke through on March 16, seizing first the Pimple, then the Pinnacle and finally Fort Dologorodoc. For the first time, the British could venture close enough to inspect the pile of rocks blocking Dongolaas Gorge. It proved to be nothing like as solid as feared – sappers estimated they could clear it in two days. Suddenly Keren no longer seemed quite so impregnable. 'Keren is ours!' declared one British commander.

He was a trifle premature. Realizing that Dologorodoc represented a fatal breach, Carnimeo tried repeatedly to re-capture the fort, sending wave upon wave of Savoia Grenadiers and native troops unsuccessfully against the position. As bodies piled up at the fort perimeter, Italian morale began to waver. It nearly buckled entirely when the popular Lorenzini, said by the *ascaris* to be immortal, was killed reconnoitring the ground for a seventh counterattack. Italian units were now down to two-thirds of their original strength. But British nerves were also starting to go. On the crests, young men who knew they were about to die penned farewell letters to their parents. Casualties were running so high that drivers, orderlies and mess sergeants were being mustered to form new companies thrown into the fray. 'There was a nasty moment when one or two people got so-called shell shock and we had to take a very firm line,' remembered Kerr. 'It started spreading, demoralization is very infectious. There are moments when you simply have to say "Go back where you came from and don't come running down here." At heart we all wanted to turn away. It was only pride or shame or a sense of responsibility that kept you going.'[15]

In the early hours of March 25, the British played their final card. Two brigades attacked on either side of Dongolaas Gorge, one unit working its way silently along what proved to be a poorly-barricaded railway tunnel high on the ridge, the other moving up from Fort Dologorodoc. By the morning of March 27, a last desperate Italian counterattack had been repulsed and sappers had cleared the pass. The guns fell silent as white flags shot up from Italian soldiers on the peaks. With his units in tatters, Frusci had ordered a withdrawal, congratulating his soldiers for their heroism in a florid declaration. 'Our many dead, who include one general and five senior officers, remain in Keren as armed guards and a warning to the enemy. We have left Keren only temporarily,' he promised, unconvincingly. 'We will soon return there and the sacred flag of our country will once again flutter in the light of our future glory.'[16]

If Platt had fulfilled his pledge to last longer than his opposite number, it had only been by a hair's breadth. The British had come within a whisper of calling off the assault. The general later confessed that in the last three days of the battle, his reserves had shrunk to just three tanks. 'A company commander said to me when he heard that, "Was that quite sound sir?" No, it was contrary to every book that had ever been written, but it came off.'[17]

The battle of Keren was over and with it, Italy's most spirited military performance of the Second World War. The official Italian tally was 3,120 dead – a total that characteristically omitted around 9,000 Eritrean and Ethiopian *ascaris* who had fallen alongside their European comrades.[18] British forces, which had pulled off what was as much a quartermaster's as a soldier's victory, had lost between 4,000 and 5,000 men.[19] Added together, both sides had probably sustained more than 50,000 casualties, averaging out at around 1,000 dead and wounded each day. 'It was incredibly tough, and it is a source of wonder how we ever succeeded,' an officer in the West

Yorkshire Regiment later recalled. 'It will never, like some battlefields of the First World War, look small and insignificant, but will stand always, huge and rugged, the gateway to Eritrea.'[20]

The feared escarpment had gone from insurmountable threat to just another geological feature. The Pimple, the Pinnacle and the Sphinx were no longer of any interest to the British soldiers who had crouched in the dust, trying to guess what hidden gullies and unexpected ridges – the dips and bumps that held the key to survival or destruction – lay ahead. As instructions were shouted, equipment packed and trucks and tanks jostled, nose to tail, for their place in the grey-green caterpillar working its way up Dongolaas Gorge, past the inevitable anti-climax that was Keren itself, the men were already forgetting a landscape they would never see again. For many, there was a dreamlike quality to the sudden telescopic shift in focus. 'It always surprised me, in any battle, how limited one's life was while the battle was going on,' remarked Kerr. 'You knew every stone for the next 50 yards. It always struck me as extraordinary how when a battle ended, like in Keren, how the next day the birds were there, peace reigned, the place was in a bit of a mess, suddenly there were trees and everyone walking about and standing up in daylight and one wondered at how different it was from yesterday, a different world entirely – what had we been doing all those weeks? At one moment somewhere is a battlefield and life is being lost right and left. And the next day, total peace and silence.'[21] The convoy roared through Keren – 'a pathetic little town', commented Richard Dimbleby, before putting it out of his mind forever – and ground on to Asmara.

Frusci was to stage a series of rearguard actions further up the Imperial Way, but his men had lost their stomach for the fight. The trouble with Maginot lines, as military strategists know, is the symbolic significance they come to acquire in the eyes of those who shelter behind them. When they collapse, so does the notion of further resistance. The Italians knew that they were not going to stumble on a better position than Keren, and Keren had gone. A few days later, Dimbleby, who had given his radio listeners a crisply eloquent account of the campaign, watched open-mouthed as a small touring car loaded with Italian officers and dignitaries, waving a large white flag, drove past him. They had come to negotiate a surrender. On April 1, Asmara was declared an open city, saving its elegant boulevards from the ravages of British artillery. Massawa fell a week later. After half a century of occupation, the Italians had lost their first-born colony, and with that defeat the surrender of Ethiopia to the south became a matter of time.

Mussolini's new Roman Empire was imploding, and Eritrea's surrender freed up the troops Wavell desperately needed. They were allowed only the briefest of breathing spaces before being whipped away to fight Rommel. Had Keren not fallen when it did, British morale, bruised by Dunkirk and the Blitz, might never have recovered. Its conquest was a small but crucial part in turning the tide of the Second World War, from a position where a vast Nazi empire seemed a certainty to a point where Allied victory was for the first time conceivable.

Strikingly absent from this whole strategic picture – staggeringly absent, indeed, from all the vivid veterans' memories and detailed military reports on Keren – is any mention of the people most immediately concerned by the events of 1941: the Eritreans themselves. The British soldiers who fought in Keren struggle to recall a single encounter with a local, an unsurprising lacuna, perhaps, given that until Asmara, Eritrean towns had either been bombed or marched through by the Allies, but never occupied or administered. Asked about the Eritrean countryside, one officer mused, 'There was no countryside, really,' as if he had been marching across a blank vista. What the invaders retained was the impression of a landscape bereft of people, stripped of vegetation, a moonscape so desolate it seemed the ideal setting for a war. As for the Italians, the words they ordered to be carved on the white gravestones erected over the tombs of every Eritrean and Ethiopian who fell at Keren say it all. 'Ascaro Ignoto' – 'Unknown Ascaro'. The Italians didn't even know the names of the natives who died for them.

The post-independence Kenyan politician Tom Mboya used to recount how a white customer once poked her head into the office where he was sitting working, looked around, and said: 'Ah, nobody here,' as an example of how colonial assumptions about authority rendered blacks effectively invisible. There are echoes of the Tom Mboya experience for the Eritreans at this juncture in their history: to the outside world they seemed as insubstantial and transparent as the chill mountain air. Despite all the promises made in the leaflets sprinkled by the RAF, Britain had not invaded Eritrea to free the natives from colonial rule. It had fought the battle of Keren for strategic reasons that stretched far beyond Eritrea's borders and bore no connection to local wishes, a matter of supreme indifference, at this stage, to London.

It is a view of the world reflected in the story that has passed into Eritrean history concerning Keren. In a way, it doesn't really matter whether the tale is apocryphal or not, because it says so much about the gathering cynicism of a people who had come to understand their country was no more than a proxy location for a war, this merely a dress rehearsal of the great fight between Fascism and Liberal Democracy that would be concluded elsewhere.

Popular legend has it that a British captain leading his weary men on the march from Keren into Asmara was met on the road by an old Eritrean woman, wrapped in the ghostly white

shroud of the highlands. She was ululating in traditional greeting, celebrating her country's liberation from Italian Fascist
rule and the start of a new era of hoped-for prosperity. Perhaps
that high-pitched shrilling irritated the captain, extenuated by
a campaign he thought he might not survive. In any case, he
is said to have stopped her in mid-flow with one throwaway
line designed to crush any illusions about why he and his men
were fighting in Eritrea. 'I didn't do it for you, nigger,' he said,
before striding on towards Asmara.

CHAPTER 5

The Curse of the Queen of Sheba

'All people come from God, but the Ethiopians more than most.'

A Geez scholar at Addis Ababa University

There was something very English about the new administration set up in Asmara. Italy's military collapse in the Horn left Britain with a vast area to administer at a time when, with the German army still to defeat in Africa, it could barely spare the men or muster the energy. There could be only one solution: co-opt the defeated enemy, assign a skeleton crew as overseers and move on. Eritrea's first British civilian administration consisted of just nine Sudanese policemen, eight British officers and a former governor from Sudan.[1] This team issued orders to thousands of Italian officials and civil servants who could theoretically have risen up and overwhelmed it at any time. But most Italians were in no mood for insurrection, and the mutually-agreed fiction of British power held good.[2]

The colony did offer some strategic advantages, the new occupiers swiftly realized, isolation coming top of the list. Eritrea's distance from the main battlegrounds of the Second World War meant it was safe from direct Nazi attack. In late 1941, when America was still nominally neutral in the war,

London and Washington struck a secret deal, arranging for RAF aircraft damaged by Rommel's Afrika Korps to be repaired at an American-staffed base in Gura, south-east of Asmara, before being sent into battle once again. Britain also hit upon the idea of using Eritrea as a prison. In 1944, 251 hardcore members of the Jewish underground, which had been assassinating British VIPs in its campaign for a Jewish state in Palestine, were deposited in a camp on the capital's outskirts. Cut off from friends and family in this 1940s version of Guantanamo Bay, the prisoners, who included future Israeli prime minister Yitzhak Shamir, were thought unlikely to escape. In fact, they organized a dozen breakouts, keeping Asmara's police force constantly on its toes.[3]

A repair shop, a penal dumping ground: such small benefits could not cancel out the overall British perception of Eritrea as an unwanted responsibility, little more than a nuisance, to be shouldered with grumpy bad grace. British officials who wrote histories of this period would later congratulate themselves for liberating Eritreans, opening up secondary education, allowing political parties and trade unions and encouraging the birth of a free press. But there was always a pinched, parsimonious quality to their relationship with this inherited colony. The Italians had rolled up their sleeves and set to work, falling in love with the country, its golden prospects and – despite all the strictures against inter-racial contact – its women. The British, aware they would not be staying long, content to leave day-to-day administration in the hands of Italian intermediaries, kept themselves to themselves, a fact that could be measured by one statistic – or rather, the lack of it. 'By the time Italy's rule had come to an end, there were thousands of *meticci* in Eritrea. I don't know of a single recorded incident of a British official fathering a mixed-race child. Not one,' says Eritrean historian Alemseged Tesfai.[4]

This aloofness was exemplified in British attitudes to the hated racial laws, which continued to be applied despite protests from Italian community leaders in Asmara who felt their time had passed. 'This is a very tedious, not an important subject,' was the airy observation of Colonel EJ Maxwell, legal adviser to British headquarters in the Middle East, when the idea of scrapping them was raised.[5] The truth was that while London had scoffed at the myth of Aryan superiority and publicly decried Fascism's record on human rights, it had long enforced a system of racial discrimination in its own African colonies. Repealing Eritrea's legislation, much of which seemed pretty unexceptional to British eyes – Brigadier Stephen Longrigg, Eritrea's military administrator, actually considered *strengthening*, rather than diluting it – might set an embarrassing precedent. The foot-dragging ensured that when the post-Mussolini government in Rome itself scrapped the laws in 1943, British officials in Eritrea found themselves in the surreal position of maintaining a set of laws rejected as morally repugnant even by the country that had dreamt them up. Legally-sanctioned racial discrimination only ended in Eritrea a full four years after the battle of Keren. Truly, the British had not 'done it' for ordinary Eritreans.

One of the reasons the British hung on so long to these obnoxious laws was that they saw the usefulness of being able to ban seditious meetings and deport subversives. They felt they needed such powers, for after the decades of rigid Italian control, Eritrea was in a state of fizzing political ferment. Who would run Eritrea once the British – there purely as caretakers – withdrew? The newly-created UN, set up to replace the discredited League of Nations? Ethiopia – as Abyssinia would in future be known – where Haile Selassie was painfully clawing back sovereign powers from grudging British occupiers? Disgraced Italy? Or the Eritreans themselves? The question obsessed Eritrea's brand-new political parties, whose manifestos roughly reflected the country's fundamental fault line, its half–half religious divide. Christian highlanders, determined to break the link with Rome and fearful of Arab encroachment, looked to Addis Ababa for their salvation. In contrast, Arabic-speaking Moslems from the lowlands, worried about religious persecution if Eritrea became part of Christian Ethiopia, leaned towards independence.[6] The political uncertainty underpinned the activities of the gangs of *shifta* bandits terrorizing the countryside, whose numbers had been swelled by out-of-work *ascaris*, still carrying the weapons issued by the Italians.

In fact, the Eritreans were never going to be left to their own devices when it came to deciding the future. Italy, Britain, the US, the Soviet Union and Egypt: they all thought they knew what was best. But no foreign leader felt a keener interest, no ruler was more convinced he deserved a final say than Ethiopia's Emperor. The reason for that intense focus could be traced to a rubble of ancient masonry, obscured by straggling shrubs and submerged in the sand, lying a couple of miles from the Gulf of Zula.

Mention to a highlander that you are planning a trip along the gnarled strip of land that forms Eritrea's southern coastline and he will whistle under his breath, uneasy at the thought of venturing into a mysterious region where the maps may tell one story, but the foolhardy traveller discovers quite another. This is where the Danakil Depression starts, a huge geographical basin scooped into the earth's crust, dipping in places 125 m below sea level. It is one of the earth's hottest zones, a land of volcanic lakes, cracked salt flats and treacherous mangrove swamps, which slurp the cars of unwary drivers down into their sucking embrace. The light here plays strange tricks with the mind. Plum-coloured peaks float suspended above the plains, rootless as clouds. At times, sheets of water, dotted with islands, seem to flood the lemon-yellow plains, only to evaporate on approach, reverting to sand and scrub. At other times, the horizon dissolves completely, melting away into a throbbing blue glow.

In summer, when the sky turns migraine-white and the heat shimmers ahead, treacly as oil on water, the dunes are left to the ostrich and white-backed antelope. Afraid their tyres will melt on the tarmac, Eritrea's truckers wait for darkness before starting their engines and hitting the long, lonely road that runs to the port of Assab. Even the Afars abandon the stifling inland settlements, moving to bleached little fishing villages in search of fresher air. But the sea offers no real relief: the miles of postcard-pretty golden sands and aquamarine waters are a cheat. Heavy with salt, the Red Sea is as hot as a bath – 'fish soup', they call these waters – and that sticky syrup breeds curious creatures. Abandoned on the beach, like monsters from some medieval nightmare, grin the decapitated heads of fish large enough to swallow Jonah. Pyramids of wet sand, piled above dark burrows, line the waters' edge, dug not by rats or rabbits, but giant crabs. These muggy waters, churning with sleek brown dolphins, turtles and dugong – the beasts sailors of yore mistook for mermaids – are what all Africa's warm seas must have been like before commercial trawlers vacuumed up the shoals and tourists' feet pounded the coral reefs. Just as Eritrea's 30-year war preserved Asmara's architecture, it has protected the Danakil coast. One day, perhaps, cement mixers will set to work on high-rise hotels. Dark Afar warriors will service middle-aged women travellers and empty Coke bottles will litter the sands. But we are not there yet. Assab is a ghost town – its Ethiopian residents fled in the 1998 conflict – and on the once-busy coastal road the loudest noise is usually the scream of sea birds flocking above a submerged shoal.

Wilfred Thesiger called these deserted areas, dreaming in the torpid heat, 'blanks in time'.[7] But they have known their moments of historical glory. It was from this coastline, anthropologists say, that homo sapiens first left Africa 80,000 years ago, swimming across the Red Sea to what is now Yemen, the start of a long wandering that would scatter mankind around the world. It is also along this stretch, historians believe, that a Semitic people – the Sabaeans – journeyed in the opposite direction in the first millennium BC, crossing from the south Arabian peninsula into Africa and bringing their sophisticated language, metal-working and stone-cutting skills to the indigenous Hamitic dwellers on the plateau. Half an hour's drive north of the modern Eritrean village of Foro lies the evidence of what that Sabaean influx eventually gave rise to: the partially-excavated ruins of Adulis, port of one of the greatest trading nations the world has ever seen.

The Axumite empire spilled over into what is today Yemen and Saudi Arabia, embracing Eritrea, Djibouti, northern Ethiopia and stretching into Sudan and Somalia. It lasted around 1,000 years, although dates are incredibly sketchy: some historians place its birth in the first century AD, some in 300 BC, others as far back as 600 BC. In its day, it was considered one of the most powerful kingdoms on the globe, ranked alongside those of Persia, China and Rome. Sailing from Adulis, its merchants exported myrrh, frankincense, gold dust, ivory and slaves as far as India. The descendants of the Sabaeans developed Geez, precursor of the languages used today in the highlands of Eritrea and Ethiopia; they manufactured glass, minted coins and carved the vast, strangely modern-looking obelisks that still tower over the northern Ethiopian town of Axum. But, for Ethiopian patriots, the Axumite empire has always been more than a merely terrestrial wonder. It was a kingdom blessed by God himself. They know this because the book which provides Ethiopia with its extra-ordinary founding myth – the *Kebra Negast* ('Glory of the Kings') tells them so.

Written in Geez, the *Kebra Negast* was compiled in the 14th century by a group of scribes in Axum, who attempted to lend their work credibility by claiming it was a translation of a pre-existing Coptic text. Their motivation seems fairly clear. Fifty years before they wrote the *Kebra Negast*, Ethiopia's Zagwe royal dynasty had been overturned. Justifying his claim to the throne, the Amhara usurper, Emperor Yekuno Amlak, claimed to be restoring a Semitic bloodline running back to King Solomon of Israel. Just as Virgil wrote the Aeneid to provide the Romans with a made-to-measure legend estab-lishing a link with the heroes of Troy, the Axum clerics provided the King and his descendants with a national epic that both legitimized his rule and united the nation around a glorious destiny. Drawing on the Old and New Testaments, the Apocrypha, Talmud and Koran, weaving in Ethiopian legends handed down by word of mouth, they came up with an exotic composite story. It was the tale of a royal date rape, a radical rewrite of the Bible that substituted Ethiopia for Israel as God's Chosen People. In the process, it offered an answer to the mystery that continues to obsess scholars of the Biblical world: whatever happened to the Ark of the Covenant, the holy vessel holding the tablets inscribed with the Ten Commandments?[8]

The *Kebra Negast* tells the story of the Queen of Sheba, also known as Makeda, a beautiful, pagan monarch who reigned over Ethiopia. Uncertain of her own fitness to rule, she is impressed when word reaches her of Solomon, King of Israel, a man revered for his knowledge. 'The light of his heart was like a lamp in the darkness and his wisdom was abundant as sand ... Of the speech of the beasts and the birds there was nothing hidden from him and he forced the devils to obey him.'[9] She decides to meet this marvel face-to-face, setting off on a long and arduous journey to Jerusalem.

Described as 'a lover of women', Solomon already has 400 queens and 600 concubines by the time he meets the Queen of Sheba. But she makes an immediate impression, 'for she was vigorous in strength and beauty of form and she was un-defiled in her virginity'. As for the Queen, she is swept away by Solomon's intellect. As 'she marvelled in her heart and was utterly astonished in her mind', Solomon sets about converting the queen to Judaism. He is soon successful. 'From this moment,' declares Makeda, 'I will not worship the sun, but will worship the creator of the sun, the God of Israel.'

With rather more than religious education on his mind, Solomon invites the Queen to complete her instruction. Planning meticulously for the forthcoming seduction, he orders a chamber to be draped in carpets and festooned in purple hangings, perfumed with incense and sprinkled with myrrh. He then arranges for the Queen to be brought 'meats which would make her thirsty and drinks that were mingled with vinegar, and fish and dishes made with pepper', and once she has eaten, invites her to sleep by his side.

'Swear to me by thy God, the God of Israel, that thou wilt not take me by force,' demands the anxious Queen, sensing something is afoot. Solomon agrees, but on one condition: she, too, must promise not to take anything of his by force. When Makeda wakes in the night with a raging thirst and reaches for a jug, Solomon grabs her hand: she is about to break her promise, he claims, for what could be more precious to him than water? 'He permitted her to drink water, and after she had drunk water he worked his will with her,' records the *Kebra Negast*. The strip cartoon paintings that tell the story show first one female head on the pillow, then a male and a female head together: the relationship has been consummated. If she has been tricked into sex, the Queen of Sheba is usually portrayed as accepting the fact with philosophical resignation – the paintings often show the faintest of post-coital smirks.

This is not to be a lasting union, for Makeda must return to her kingdom. But on the long trip home she gives birth to a son, Menelik, the great founding leader from whom Ethiopia's long line of emperors will be descended. When the boy comes of age, he goes to Jerusalem to meet his father, who showers him in gold and silver and offers him the throne of Israel. But Menelik, like his mother, decides to return to Ethiopia, taking a retinue of noblemen's sons. It is only when his convoy reaches the Red Sea that these young Israelis confess what they did before leaving. Stealing into Jerusalem's Temple, they have removed the Ark of the Covenant. Menelik, they say, must take the holy vessel, possessed of terrifying divine powers, into Ethiopia to guide him and his royal successors. The theft appears to have God's blessing, for when a furious Solomon sends troops to retrieve the Ark, the Archangel Michael marches before Menelik and his followers, allowing them to walk across the water. 'Everyone travelled in the wagons like a ship on the sea when the wind bloweth, like a bat in the air and like an eagle when his body glideth above the wind.'

Although Solomon was to live another 11 years, 'his heart turned aside from the love of God, and he forgot his wisdom, through his excessive love of women'. Both he and his nation had paid a high price for his libido. For in allowing the Ark to leave Jerusalem, the *Kebra Negast* made clear, God removed his special favour from Israel and transferred it to Ethiopia, a kingdom which could trace its lineage, via Solomon, back to Adam himself. 'And now God hath chosen thee to be the servant of the holy and heavenly Zion, the Tabernacle of the Law of God,'

Menelik was told, 'and it shall be a guide to thee for ever, thee and to thy seed after thee.' Axum was the new Zion. If Axum's empire flourished, it was thanks to the divine favour made manifest in the Ark, and when word of Jesus Christ was brought to the city in the fourth century by two young boys, it was hardly surprising that the Ethiopians embraced it as the word of God, while the Jews, who had lost their way, foolishly rejected the Messiah.

The myth of the Queen of Sheba was a gift to any prospective ruler. If he could only establish a blood link with the great Menelik I, then his claim to supreme power was theoretically safe from challenge. His reign was not of this world, but divinely ordained. No wonder so many of Ethiopia's leaders were ready to draw the most contorted of family trees, seeking to establish a connection with the royal Solomonic dynasty that sprang from Menelik's loins.

The Ethiopian monarchs were hardly the first to try and pull off the trick of heavenly endorsement. Egypt's pharaohs presented themselves as incarnations of the Sun-God; Charles Stuart, King of England, famously proclaimed the Divine Right of Kings. But the time of pharaohs expired with Cleopatra in 30 BC. King Charles I was executed in the mid-17th century, his claim to special status rejected by parliament. What is unnerving about the Queen of Sheba legend is that it is not a quaint historical tale. The claim that Ethiopia's rulers were descended from Sheba and Solomon would be enshrined in the country's constitution as late as 1955. Haile Selassie, routinely addressed as 'Emperor of Ethiopia, King of Kings, Lord of Lords, Conquering Lion of the Tribe of Judah, Elect of God, Light of the World', 225th descendant of Sheba and Solomon, was still insisting on his direct link with the supreme deity in the middle of the 20th century, era of television, the combustion engine and the jet plane.

No one knows for certain whether the Queen of Sheba ever existed or, if she did, where her kingdom actually stretched. Yemen, unimpressed by Ethiopia's supposed bond, has claimed her as its own. While the Old Testament certainly mentions a love affair between Sheba and Solomon, celebrated in the erotic Song of Songs, it makes no reference to a son of theirs stealing the Ark, which simply disappears, without explanation, from the narrative. The Axumite empire itself, so intimately associated with Makeda, was probably founded hundreds of years after her supposed reign. And the notion that Axum's sophisticated skills can be attributed to a Semitic exodus from across the Red Sea has been shaken by excavations revealing the existence in Eritrea of settled communities dating back to 800 BC.

But none of this really matters. Tourists visiting Axum today are told by their guides, speaking with a quiet certainty, that inside the church of St Mary of Zion, only ever glimpsed by the official guardian, lies the Ark of the Covenant. In each of Ethiopia's 35,000 Orthodox churches rests at least one *tabot*, a wooden replica of this most feared and venerated of objects, constant reminder of a nation's special relationship with God. Told and retold by Ethiopia's 500,000-strong clergy, the Solomonic myth seeped into the national psyche, justifying a deep-rooted sense of racial and cultural superiority. It is possible to know in one's head that something is a fable, while feeling in one's heart it holds an essential truth. I once asked the manager of a computer school in the northern town of Mekelle whether he really believed the Ark was in Axum, expecting a sceptical response from a young Ethiopian at ease with Microsoft Word and Google. 'Where else would it be?' he replied, with genuine puzzlement. His contemporaries in Addis are more prone to dismiss the Queen of Sheba story as pictur-esque mumbo jumbo, believed only by their fathers' church-going generation. But ask them if they feel Ethiopia is part of Africa and they respond with what only seems a non-sequitur: 'We are an ancient civilization, we go back more than 3,000 years.' Featured on coffee packets and tea towels, painted on restaurant ceilings and shop walls, the story captures a sense of uniqueness. 'These legends lie at the heart of the Ethiopian sense of national identity,' says Shiferaw Bekele, associate pro-fessor of history at Addis Ababa University. 'It is a complicated, zig-zag line, but the feeling that we are somehow special, can be traced back to them.'[10]

Through the centuries, Ethiopians held on to the convic-tion that their nation, a Christian outpost in a sea of Islam and pagan belief, was different from the rest of Africa: black, tribal, subjugated. Ethiopia had once possessed a great empire, Ethiopia had never been colonized, Ethiopia, in fact, traced its ancestry across the sea, back to Biblical lands: the light skins and fine features of the ruling Amhara and Tigrayan elite proved the Semitic link. 'You are knocking at the wrong door,' Menelik II once politely informed a West Indian asking him to lead a campaign for the 'Amelioration of the Negro Race'. 'I am not a Negro at all, I am a Caucasian.'[11] As for Haile Selassie, although he would come to be hailed by Rastafarians as a living God, he was no keener than Menelik to establish a connection with the negro race, associated with the humiliations of both the slave trade and colonialism. Ethiopians, he said, were 'a mixed Hamito-Semitic people'. A popular fable captured this sense of racial superiority: in creating man, it was said, God the baker had put three trays of dough in the oven. The whites were removed too early, the blacks too late, but with the Ethiopians, God got the timing just right.

Nationalism and religion, woven so tight the individual

threads cannot be unpicked: there can be no more seductive or dangerous a combination. History's most implacable wars have always been fought by peoples convinced they enjoyed a direct line to God. For Eritreans, the myth of the Queen of Sheba and all that went with it was to prove the most malign of curses. For when Ethiopia's 20th-century rulers dreamt of restoring the glory that was Greater Ethiopia, they glossed over the complex tumble of events separating the present from the fall of the Axumite kingdom, seeing instead one long continuous chain, interrupted only by the aberrant blip of European colonialism. Thanks to the *Kebra Negast*, the irredentists pined for the great nation that had gobbled up a swathe of Africa and the Middle East, its proud sailing vessels bringing the riches of the Orient to the bustling port of Adulis. 'There is definitely a religious element to Ethiopian feelings towards Eritrea,' an expert on Geez at the university told me. 'For an Ethiopian, losing Eritrea has always felt like losing part of our body.'

No matter how dramatic, Ethiopia's ideological chops and changes barely made a dent in this sense of frustrated geographical oneness. 'For 4,000 years Eritrea and Ethiopia have been identical: identical in their origins, identical in their historical development, identical in their defence of the Ethiopian and Eritrean region,' Haile Selassie's Foreign Minister Aklilou Habte Wold told the UN in 1949.[12] The Marxist regime that took over in 1974 executed Aklilou, but embraced his vision of history. 'Until the second half of the 19th century, the strong link between the Eritrean region and the central government has never been severed,' claimed the Derg.[13] Why, you only had to look at the faces, clothing and hairstyles of the Tigrinya-speaking communities on both sides of the Eritrean border to see the link. Italy had messed things up, poking its nose where it didn't belong. But by rights, the Ethiopian empire came with coast attached. The Eritrean coast.

Sifting through the library shelves of writings on Ethiopia, what is striking is the extent to which the myth of the Queen of Sheba, while nominally designated a legend, has cast its romantic spell over scholars and historians, Ethiopian and Western alike. Discussing Ethiopia's ambitions for Eritrea, most writers use 'restoration', 'recovery' and 'return' – words which take for granted the notion of modern Ethiopia as a direct descendant of ancient Axum. Most of the Western politicians who debated Eritrea's future in the post-war years did likewise. Confronted by such widespread acceptance of a reinvented past, the Eritrean liberation movement would produce their own, alternative history, in which a fierce sense of separate Eritrean identity existed long before the first Italian jumped ashore at Assab. Bent on challenging a historical makeover, the EPLF swerved too far in the opposite direction. Nonetheless, on examination the notion of a smooth continuum, with Eritrea bound to Ethiopia by a 3,000-year common history, looks decidedly ropey.

In fact, the history of the Christian empire in the Horn was a stop-start, messy affair, in which one ruling dynasty's area of influence rarely overlapped with its predecessor's and 'control', in any case, hardly ever meant more than a frightened community's payment of tribute to escape looting by an angry warlord. As Moslem power grew between 800 and 1,000 AD, Axum lost its hold on the Red Sea and Adulis withered away. With the coast subjected to repeated Arab invasions, what remained of Axum's Christian civilization shifted inland. The Zagwe dynasty that eventually followed was centred on Lalibela, a good 300 km south of Eritrea's current border, while north-western Eritrea was ruled by the Bejas, a Cushitic people who had migrated from Sudan. After the Solomonic empire was 'restored' in the 13th century, the Bahr Negash, a ruler who controlled the Eritrean highlands, did pay annual tribute to the

Ethiopian kings, then based in Gonder. But that tenuous link was destroyed in the 17th century when fighting between regional princes stripped the king of most of his power. 'There was no sense of the peoples of Eritrea being a constituent part of a territorial state with clear boundaries,' argues David Pool, a British expert on Eritrea. 'Indeed, it is not at all clear after the sixteenth century, and until the late nineteenth century, that Ethiopia existed as a unified state with a recognised ruler.'[14]

It was only in the mid-19th century, just when European colonialism was making its first inroads into the Horn, that strong Ethiopian emperors such as Tewodros II and Yohannes IV showed themselves capable of exercising anything approaching centralized control and even that rarely extended further north than Eritrea's *kebessa*. As for the coast, it was seized by the Ottomans at the start of the 16th century and remained in Turkish control until Egypt took over in the early 19th century, eventually ceding to Italy. Viewed from this angle, Ethiopia's 'historical' claim to the whole of Eritrea as an estranged northern province makes about as much sense as Britain arguing that the French region of Normandy is rightfully hers today on the basis of the marriage between Henry II and Eleanor of Aquitaine in 1152.

But claim it, Haile Selassie did, just as his fathers had done. Underlying any sense of religious predestination was a strong element of strategic pragmatism. Ethiopia's rulers had initially congratulated themselves on the natural protection provided by their geographical inaccessibility beyond the mountains. European visitors to the 15th-century court were warmly welcomed, but told that they would never be allowed to return home to spread tales about the land of Prester John, the legendary Christian empire in the East. But Ethiopia's leaders had come to register the price paid for such isolation. With no port of their own, they depended on their neighbours for contact with the outside world and their neighbours, happy for this regional giant to remain locked in the Middle Ages, had done their best to block trade, and the weapons that always made up the most valuable part of trade. Emperors had been reduced to begging the few Italian explorers and British adventurers who ventured south to bring muskets and ammunition, and their pleas often went ignored. Access to the sea: it was the only way of relieving this sensation of choking, the claustrophobia of the humiliatingly landlocked. 'Let me tell you a secret,' Menelik II confided in Ferdinando Martini at their one and only meeting. 'Every other nation has a port, Abyssinia has none. You must ask the King to give me one, too. My access to the sea harms no one.'[15] The reinstated Haile Selassie could never forget how, in the run-up to the Fascist invasion, the French had banned weapons shipments along the railway that ran from Djibouti to Addis Ababa, making it impossible for his country to arm itself. Ethiopia must have its own ports, operating free from European meddling. The myth of the Queen of Sheba, his country's awareness of past greatness, merely sanctified what was, in the Emperor's view, a practical necessity.

In the years to come, winning Eritrea was to become a near-obsession, a craving that demanded to be satisfied. But while Haile Selassie waited for an opportunity to stake his claim, the European nation entrusted with Eritrea's care was not sitting idle. When the Emperor finally got his hands on what he regarded as his rightful inheritance, he would find that while he had certainly won access to the sea, the rest of the legacy had been plundered.

CHAPTER 6

The Feminist Fuzzy-Wuzzy

'She is an unscrupulous woman whose information is inaccurate, views distorted and influence practically nil.'

A Foreign Office employee gives his
assessment of Sylvia Pankhurst

On April 17, 1950, Count Gherardo Cornaggia Medici Castiglioni, the Italian government's somewhat extravagantly-named representative in London, was ushered into the African Department at the Foreign Office. The Italian nobleman was not happy, but he was careful to keep his tone of voice level, his manner deferential. This was something more than the diplomat's professional self-control. As the envoy of a defeated nation forced to accept the occupation of its African colonies, the count knew he enjoyed no leverage over the British official sitting before him. Going on the offensive would gain him nothing.

Presenting his compliments, he confessed that Rome had been disappointed by a letter the British government had sent in response to concerns raised, he admitted, on a purely unofficial basis. Italy, he said, did not want to argue the legal issues. It felt, however, that what the British were doing in

Eritrea went against basic economic good sense. To support his comments, he handed over a wad of black-and-white photographs. They showed what had so outraged Cornaggia and his colleagues back in Italy.

The naval base in the port of Massawa, built by the Fascists to hold 1,000 sailors, had been bulldozed to the ground. Navy headquarters, a 500-bed hospital with its air-conditioning plant, the oil storage tankers, main water supply tank, electricity unit, naval warehouses, customs offices – more than 75 buildings in all – had been reduced to an expanse of rubble. The photographs tracked the relentless demolition from start to finish. The first showed the neoclassical colonnaded buildings standing tall. In the next, sweating Eritrean labourers with pickaxes stood perched on collapsed piles of masonry that had been toppled by explosives. The last showed the steel rods and blocks of timber extracted from the reinforced concrete, neatly crated up on the quayside and waiting to be shipped abroad.

R Scrivener, the British official who received the count, was not unsympathetic. 'The Italians have no official standing in this matter,' he acknowledged in a note to his superiors. Nonetheless, he said, 'it would be helpful if we were to show that our action was economically sound and not just wanton. I think the Italians feel we are just being spiteful over this. This is foolish of them, but we ought to try and smooth the ruffled feathers even if we are under no obligation to do so.'[1] A month later, the African Department drafted a lofty response. If the Italians raised the issue in future, it decreed, they could be told that buildings decayed rapidly in the humidity of the Red Sea and were liable to be looted by 'local natives'. Since the British taxpayer could hardly be expected to shell out for the police force needed to guard the site, destruction had been the only option. The possibility that grubby mercantile motives might have played a role was not even deemed worthy of mention.

Nowadays, anyone who cares to can leaf through the same photographs and read the furious comments scrawled on their backs by some Italian hand. They lie in individual waxed paper bags at the Public Record Office in Kew, south London.[2] Only made public in 1981 under the 30-year rule, they record the final, small-minded episode in a sequence of breathtakingly petty British actions in the Horn of Africa. Both Eritrea and Ethiopia would share a strikingly similar fate at Britain's hands, a fate that would have gone totally unnoticed in the West had it not been for a pigheaded iconoclast called Sylvia Pankhurst and her equally stubborn son.

Miss Pankhurst – for she was always to be a 'Miss', despite the eventual acquisition of both partner and child – could be described as one of Ethiopia's earliest True Believers.

She discovered the country late in life, but like all converts, she made up in enthusiasm for what she lacked in experience. 'This confounded Pankhurst woman,' a Conservative member of parliament once complained in a letter to Anthony Eden, the then Foreign Secretary, '[is] *plus fuzzy-wuzzie que les fuzzy-wuzzies*.'[3] If the language grates today, he had nonetheless put a finger on the characteristic that constituted both Sylvia's most admirable quality and her biggest weakness. She would always be her own government's most implacable critic, showing an almost uncanny instinct for the baser motives that lay behind the British establishment's obfuscations. But when it came to Ethiopia – the land in which she chose to end her days – everything would always be viewed through an indulgent golden haze.

By the time she latched on to what was to be the last of her many all-consuming crusades – a passion that would span over a quarter of a century – Sylvia was already in her early fifties and had learned all there was to know about being an effective lobbyist and relentless agitator. An understanding of how to achieve a maximum of political impact at a minimum of expense had virtually been imbibed at her mother's breast, for Sylvia came from the least ordinary of families.

During the 1900s and 1910s Sylvia, her more glamorous sister Christabel, the quiet Adela and their widowed mother Emmeline had all braved violence on the political podium in their campaign to win British women the vote. The suffragettes, as they came to be known, had been manhandled by male hecklers, arrested by police and – when they went on hunger strike to protest their imprisonment – been subjected to the horrors of force-feeding, an experience the women regarded as akin to rape. But as the suffragette cause had gradually triumphed, the family had split asunder, riven by ideological differences and long-simmering jealousies.

Emmeline had swung firmly to the right. Christabel had found God and lectured on the Second Coming. Adela had sailed for Australia and become a pacifist. Sylvia – now not on speaking terms with Christabel and Emmeline – had set herself up in London's deprived East End, publishing the socialist paper *Women's Dreadnought* through the First World War. She had embraced Bolshevism – an enthusiasm that waned with time – and travelled to Moscow to meet Lenin, moves that brought her under the scrutiny of the British police's Special Branch.

Of solid middle-class stock, she had led a Spartan existence, the modest sums earned by journalism and political writing, combined with the odd contribution from a former patron of the suffragette cause, barely saving her from real poverty. For the Pankhursts, money matters always came a careless second. Like every member of the family, Sylvia could not imagine life without a cause, preferably one that seemed, at least at its inception, doomed to failure. Once the slogan 'Votes for Women' had lost its crowd-pulling force, she dabbled with Indian history, took up Interlingua – an alternative to Esperanto – and grew interested in the Romanian poet Mihael Eminescu. But Mussolini's coming to power was to provide her with the fresh intellectual focus she craved.

To her credit she grasped, more clearly than most of her contemporaries, that Mussolini's brand of thuggish expansionism posed a threat extending far beyond the frontiers of Italy itself. Encouraging her to concentrate on events in Rome was Silvio Corio, an Italian anarchist seven years her senior. Exiled to London, he had worked with Sylvia on the various political journals she produced, becoming her lover and father of her only child. They never wed, a fact Sylvia – very much the dominant partner – made no attempt to conceal. A vocal exponent of free love, who had in her youth conducted a doomed relationship with the married Labour politician Keir Hardie, she brazenly sold the story of her 'eugenic' baby – 'eugenic' because he was born to two intelligent adults free from hereditary disease and untrammelled by social convention – to the sensationalist *News of the World*. 'I suppose you think I am awfully silly, don't you?' the 45-year-old unmarried mother asked a journalist in an uncharacteristic moment of girlishness. Her own mother thought her not so much silly as shameless. When Emmeline died six months after the birth of Richard Keir Pethick Pankhurst – named after Sylvia's adored father – many friends blamed it on the shock caused by her estranged daughter's outrageous behaviour.

With Corio as self-effacing editorial helpmate, providing her with insights and news from his native country, Sylvia turned her attention to exposing the horrors of Mussolini's Italy. Campaigning against Fascism led ineluctably to campaigning on behalf of the African nation Italy was preparing to attack. To the British public, the invasion of Ethiopia was little more than an exotic distant adventure, immortalized in Evelyn Waugh's semi-fictional *Scoop*, which allowed a generation of foreign correspondents to win their spurs. Glossing over the cudgellings, castor-oil treatments and murders that had accompanied Il Duce on his rise to the top, many Britons dismissed Mussolini as a grandstanding buffoon, all comic-opera wind and bluster. His African escapade was soon sidelined, as public attention focused on Hitler's ascension and the threat of a second, devastating world war.

For Sylvia, the two developments could not be separated. Ethiopia, she sensed, represented a dry run for the Fascists; Rome's readiness to use chemical weapons was likely to be a grim precursor of atrocities to come in Europe. With remarkable prescience, she warned that appeasement in Ethiopia would merely encourage Europe's bully boys to make ever more presumptuous territorial claims. 'REMEMBER – Everywhere, Always, Fascism means War',[4] she was to tell her readers, long before many could bear to think she might be right.

She had always found it difficult to share the limelight or work alongside fellow activists. Now, realizing she had found a neglected niche in which she could flourish virtually alone, she set about making the Ethiopian cause her own. On May 5, 1936, the day the Italian army entered Addis, she launched the *New Times and Ethiopia News*, a weekly newspaper dedicated to keeping the Ethiopian war in the public eye. If the concept of collective security was to have any meaning, the League of Nations should impose sanctions on Italy, she argued. 'As I view it,' she explained to a friend, 'both Fascist Italy and Nazi Germany are going to make war whenever they find a sufficiently good chance of success, and I am perfectly convinced that if the [Italian] armies come back from Africa with a victory, a new enterprise will be entered upon before long . . . You cannot buy these super-militarist governments off by treating them well.'[5]

History was to prove her correct, but it was not a message the British government was disposed to hear. The timing was not right: frantically scrabbling to rearm ahead of the coming storm, London wanted Mussolini indulged and appeased as long as possible rather than punished and pushed into Hitler's arms. The invasion of Ethiopia seemed a small sacrifice to make in the context of the greater game of gearing up for another world war, hence the eventual decision by both the United Kingdom and France to recognize the Italian conquest. This was exactly the kind of short-term, pragmatic calculation Sylvia, by her very nature, was incapable of making.

By the age of 50, George Orwell once wrote, everyone has the face they deserve. In Sylvia's case, his maxim did not quite ring true. Her true face crept up on her through the decades, the gap between personality and appearance narrowing with the years. But a vital element somehow remained missing: nothing in the placid portraits left behind explains why quite so many British officials came to regret coming into contact with Miss Pankhurst, or what element in her make-up prompted the splenetic comments spluttered across government paperwork of the day.

Photographs of the young Sylvia show a gloomy child, just the kind of oversensitive introvert likely, as contemporaries remembered, to indulge in fits of copious weeping. The artistically-gifted daughter of a left-wing, politically-active Manchester lawyer, she was brought up in an agnostic, anti-monarchist, cash-strapped, decidedly eccentric household in which the children were often left to educate themselves,

Emmeline Pankhurst having developed a deep suspicion of formal schooling. Sylvia registered early on that she was not her mother's favourite – that role went to Christabel – and the realization cast a plaintive shadow across her life. As the years passed and the puppy fat of youth disappeared, it became clear that the girl had not inherited the aristocratic cheekbones and high forehead of Emmeline, vivacious beauty of the family. The adult outline that emerges is that of a somewhat doleful young woman. From her hooded eyelids to her protruding top lip and receding chin, Sylvia's features seem to have surrendered early to gravity's pull: they drooped more dramatically than was necessary, a mournful effect heightened by the severe hairstyle of the bluestocking. Here was a woman, the photographs suggest, resigned to suffering, capable of self-sacrifice and nobility of purpose. A woman commanding respect, but not someone you would necessarily want sitting next to you at a dinner party. Somehow, it is impossible to imagine her in fits of laughter. 'She had an overwhelming personality,' confirms Eritrean academic Dr Bereket Habte Selassie, one of the many students Sylvia was to take under her maternal wing. 'When she was in the room she filled it. But she had absolutely no sense of humour.'[6]

What the photographs fail to capture, at any age, is the element of steel. She had a talent for self-flagellation, setting suffragette records in Holloway prison for the refusal of food, water and sleep. On one occasion, she was left so debilitated she had to be carried in a stretcher along a demonstration's route. Her long life saw a vast outpouring of physical and mental energy, funnelled into her grittily realistic paintings, impassioned speeches and relentless lobbying work. All the while, she produced a ceaseless stream of books, articles, poems and translations. The *New Times and Ethiopia News*, for which she penned a 2–3,000 word weekly editorial, was largely made up of her own contributions. While juggling that commitment, she pumped out a flow of importuning letters, political pamphlets and appeals. Even in old age, the kindly grand-mother's face that gazes out at us, its contours softened by trailing wisps of grey hair, gives little hint of this tremendous capacity for hard work. 'A biographer once asked us what we used to do in the holidays,' her daughter-in-law Rita Pankhurst later recalled. 'It was only then we realized that there were no holidays. With Sylvia it was always work, work, work.'[7]

Ironically, given her early Marxist convictions, Sylvia's own life defied the deterministic vision of history as a vast, economic process whose grinding wheels render free will a fantasy, reducing individuals to destiny's pawns. She always acted on the assumption that a single, driven person could exert a level of influence out of all proportion to their political backing or social standing. Her formal education might have been shaky, but she knew exactly how to go about drumming up international sympathy for Ethiopia. 'To get your answer into the hands of people who count has been and remains terribly important, and should be done through the Press, by pamph-lets, circularisation and so forth,' she lectured Ethiopia's representative to Britain. 'I strongly urge propaganda in every possible way.'[8]

Her previous campaigns had taught her the value of headed notepaper. She knew the classic lobbyist's trick: that a letter sent from an 'institute' or 'international council' will always have more impact than the individual appeal, even when the institute is run from a front living room and its staff consists of a few relatives and friends. She set up the International Ethiopian Council, in whose name she wrote letters to *The Times* and impertinently peppered such notables as Winston Churchill, Clement Attlee, the Archbishop of Canterbury, King George VI and President Franklin D Roosevelt with advice.

Her home in Woodford Green, a north-east London suburb, became a cosmopolitan gathering place for Italian anti-fascists, lonely Ethiopian students and African dissidents, where she arranged for friendly MPs to raise questions in parliament, organized bazaars to raise funds, and pulled together demonstrations.

The *New Times and Ethiopia News*, sent out to MPs, journalists, churches and ambassadors, had a weekly print run of only 10,000 copies, but its influence spread far further than this figure suggests. Around the world, black resentment of white rule was stirring. Kwame Nkrumah, Ghana's future president, was in Britain at the time of Mussolini's invasion and he later articulated the outrage of a generation of Africans who heard the news. 'It was almost as if the whole of London had suddenly declared war on me personally,' Nkrumah recalled in his autobiography. 'For the next few minutes I could do nothing but glare at each impassive face wondering if those people could possibly realise the wickedness of colonialism.'[9] For an embryonic black consciousness movement in search of heroes, Ethiopia – hitherto untainted by European colonialism – served as an inspiration and rallying cause. Its beleaguered Emperor, nominated 1936 'Man of the Year' by *Time* magazine, seemed nothing short of a saint.[10] Nkrumah, Jomo Kenyatta, Hastings Banda, ITA Wallace-Johnson: London in those years was a gathering place for the educated young Africans who would go on to become the continent's post-independence leaders. The *New Times and Ethiopia News* fed their appetite for news from Ethiopia and disseminated Pan-African notions to the restless colonies. It was sent to the West Indies, read as far afield as South Africa and Sierra Leone, and its articles were reproduced in Nigerian and Ghanaian newspapers.

While the uneven war in Ethiopia still raged, Sylvia published chilling accounts of the effects of mustard gas and shocking photographs showing grinning Italian soldiers brandishing the severed heads of Ethiopian warriors. When Haile Selassie arrived in Britain to start a five-year exile in Bath, she organized a noisy welcome at Waterloo Station, knowing this would embarrass her government, then busied herself trying to find him a residence. At a time when British officials were assiduously avoiding contact with their unwelcome guest, the *New Times and Ethiopia News* published long, admiring interviews with HIM (His Imperial Majesty), who had the sense to recognize a valuable mouthpiece in this dowdy, garrulous Englishwoman. Hoovering up information from Ethiopians in London and informants in the Horn, Sylvia tracked the Ethiopian resistance movement as it made a mockery of Mussolini's conquest.

It made things decidedly awkward for the British government, which had already recognized Italy's Victor Emmanuel as King of Ethiopia. So awkward, that during the nine-month hiatus between the outbreak of the Second World War and Mussolini's entry on Hitler's side, Sylvia's newspaper was placed on a 'Stop List' of publications whose export was deemed likely to damage Anglo–Italian relations. Then, in June 1940, everything changed. 'There were a large number of Italian refugees at home and I remember that we were having pasta and someone turned the radio on,' recalls Richard Pankhurst.[11] 'We heard a newscaster saying "Signor Mussolini, in his declaration of war . . ." There was great rejoicing.' As the diners knew, the keystone of Britain's hypocritical foreign policy – the need to placate Il Duce – had just been removed.

One argument – the threat Fascism posed to world peace – had been won, but new ones remained to be fought for someone fretting over Ethiopia's future status in the world. Despite the fact that Mussolini was now the enemy and British soldiers would soon be confronting Italian Alpini in the Horn, London refused to rescind its recognition of the dictator's Ethiopian conquest. The RAF flew the Emperor to Sudan, but British generals proved strangely reluctant to use him as a rallying figure for Ethiopia's resistance, ensuring that when Addis was liberated on April 6, 1941, it was by South African troops, not Ethiopian warriors under imperial command.

The truth was that Britain still viewed Africa through colonialism's cold lens. Wars had always presented European powers with treasured opportunities for redrawing the world map and by spelling the *de jure* end of the Ethiopian nation-state, Mussolini's invasion had made a new carve-up possible. Looking at the Horn with the calculating eyes of a land surveyor, London sensed a chance to tie up some colonial loose ends to the benefit of its own African territories. The region, British policymakers suggested, would be more stable if the various areas inhabited by Somalis – including Ethiopia's eastern Ogaden – were merged to form a 'Greater Somalia'. Borana, in Ethiopia's south, should go to Kenya and as for Eritrea, it could be sliced in two, with the Moslem west attached to Sudan and the Christian highlands and ports given to Ethiopia. No wonder the government was in no hurry to recognize Ethiopia as an ally, or grant the reinstated Emperor too long a leash. 'We are free of any obligation not to disturb the existing legal position and have our hands free to make such settlement of the future of Abyssinia as we may think fit,' one Foreign Office official confidentially noted, when Sylvia sought clarification of Britain's stance. 'Let us remember,' a colleague added, 'that what we are doing in Abyssinia is for our own benefit, not for that of the Abyssinians, and it is possible to imagine circumstances in which it might suit us to throw them over.'[12]

In public, London insisted it had no intention of tinkering with Ethiopia's frontiers. But Sylvia had a nose for colonial machinations. Haile Selassie, she felt, had not in the past been sufficiently alert to the possibility that Britain might simply replace Italy as his country's new master. 'With all his ability and dignity, I often felt ... as though I was talking to a sick child who did not know how to deal with the politicians about him,' she told a sympathetic member of the House of Lords.[13] She herself wrote so many letters denouncing what she described as 'the predatory subterranean policy of the Foreign Office', the ministry in question opened a special file labelled, 'How to deal with letters from Miss Sylvia Pankhurst', whose entries log the increasing testiness of the officials assigned to answer her. 'Unbalanced and fanatical', 'Busybody Miss Pankhurst', the Foreign Office men called her in their internal correspondence, giving expression to a fury they could not voice in public. 'Miss Pankhurst only wants to be tiresome,' one man has scrawled. 'Does it matter what Miss P. thinks?' another asks. 'She's a crashing bore who will deserve to be snubbed if in fact it were possible to snub her.'[14]

In 1944, her world opened up. Travelling by air for the first time in her life, she flew to British-administered Eritrea, then on to Addis, where a grateful Haile Selassie, who had already named a street in the capital after her, awarded her a Patriot's medal and the Order of the Queen of Sheba, an honour usually reserved for foreign queens. After so many years spent writing about an imagined country, she found the reality of what she dubbed 'fairyland' exhilarating and overwhelming: 'I was enchanted and bewildered; I seem to be living in a dream.'[15] The Eritrean leg of her trip outraged the Foreign Office, which sniped at her failure to show proper respect for British administrators on the ground. Real anxiety ran below their complaints, for while Sylvia was in Asmara she met prominent Eritreans campaigning for unification with Ethiopia – encounters London deemed too dangerous to be repeated. Sylvia's hatred of British imperialism had, ironically, turned her into an enthusiastic campaigner for an African empire: Greater Ethiopia, encompassing the entire Horn. She had swallowed the myth of the Queen of Sheba in one enthusiastic gulp, touting a vision of the future in her newspaper editorials that London had little wish to see take shape. 'I beg you to believe that we here had no desire whatever to promote the lady's journey,' a sorrowing Foreign Office employee assured a former diplomat who raged against Sylvia being granted a visa. 'She is in fact one of our most persistent and unscrupulous persecutors.'[16] In light of her 'egregious behaviour and scurrilous attacks on HMG', the Foreign Office briefly played with the idea of using the Order of Sheba episode as an opportunity for delivering 'a sharp rebuff' – British subjects could supposedly only accept foreign decorations with the King's prior approval. 'The lady is a blister, and deserves a rap,' fulminated one official. But the suggestion was pursued no further.

Sylvia's worst suspicions were confirmed in April 1946, when Britain, without consulting Haile Selassie, publicly unveiled its plans for a United Somalia at a Foreign Ministers' meeting in Paris. Her denunciations so infuriated the British legation in Addis Ababa, they called for legal action. Cooler heads prevailed in London, which could all too easily imagine to what effective propaganda use Sylvia would put a court case. 'Infamous as her slanders are, it is a fact . . . that public probing into our record since 1941 in the Ogaden would produce quite a lot that would embarrass us,' warned one Foreign Office man. Nonetheless, he assured the legation, his colleagues agreed 'wholeheartedly with you in your evident wish that this horrid old harridan should be choked to death with her own pamphlets'.[17] The partition plan met with such intense international opposition it had to be abandoned, but the British would only withdraw from the Ogaden a full 13 years after Addis was liberated.

Reading the Foreign Office files, one savours Sylvia's ability to rub the British government up the wrong way. The voice of the establishment comes across clearly, and it is not a pleasant one: sexist, patrician, pompous, utterly convinced of its own superiority – it's hard to say whether the men airing these views held Sylvia or the 'local natives' they constantly refer to in greater contempt. Either way, they knew they were right. But setting aside a certain instinctive empathy for a woman content to play the role of establishment scourge, it is difficult to escape the conclusion that Sylvia, at this stage in her life, had lost her political way. The woman who had spent her career battling Western imperialism was, it seems, blind to the dangers of African colonialism. The radical who had challenged the bourgeoisie was now accepting honours from a monarch claiming a direct link with God. Perhaps her biggest mistake was the failure to recognize the ethnic loyalties and historical experiences that differentiated Moslem lowlander from Christian highlander, Eritrean from Ethiopian, Amhara from Somali.[18] To her they were all HIM's rightful subjects, longing to return to the Motherland after years of Fascist exploitation. These were easy mistakes to make when she was tracking events from the distance of Woodford Green, harder to justify once she began visiting the Horn and saw for herself the complex realities on the ground.

Modern Eritreans find it hard to forgive her these failings. They hold a particular grudge against Sylvia for the aggressive role her newspaper played in trumpeting union with Ethiopia. When the suffragette's name comes up in conversation, it is usually with a nudge and a wink, the suggestion being that her relationship with Haile Selassie was more than merely platonic. Nonetheless, Eritreans know they owe her a reluctant debt of gratitude. Were it not for the campaign she launched after her second trip to the Horn, a key episode in the moulding of the Eritrean psyche would have passed unrecorded. On the issue of British asset-stripping, this True Believer, however inadvertently, was to prove a real Friend of Eritrea.

According to Sylvia's own account, she came across the destruction ordered by the British authorities in Asmara – by then in its final phases – on a 1952 trip to Massawa with an Ethiopian acquaintance and an Eritrean merchant. 'The port was in the process of being dismantled; all its installations were being systematically destroyed or removed,' she wrote in *Eritrea on the Eve*, the book she rushed into publication. 'It is a disgrace to British civilisation,' the merchant told her, and she felt ashamed. 'His words affected me painfully, like blows, so just they were in my opinion. I was grieved and downhearted.' Her book was illustrated with the same photographs Count Cornaggia had handed to the Foreign Office two years earlier – how they ended up in Sylvia's possession is not clear. But the damage she logged was far more extensive than that the Italian aristocrat had cited, and she made a new, embarrassing allegation: dismantled industrial equipment and stripped wood, iron and steel had gone to benefit British-run territories in Africa, the Middle East and Asia.

In Massawa itself, apart from the levelled naval base and hospital, a cement factory had been removed and sent to Sudan, an aerodrome had been sold to Egypt for demolition, the only dredger working the Red Sea had been sold and a floating dock and two giant cranes had been dispatched to Pakistan, Egypt and Malta, while 500 oil reservoirs had been razed and 20 radio stations dismantled. 'As a result of the demolitions in the port its capacity has been reduced by three-fourths,' she reported. Further along the coast at the town of Zulla, 400 Italian officers' houses had been demolished, in Fatma Dari a potash factory sold. In Assab, a radio station had been transported to Kenya, a salt factory dispatched to Aden and the motors of scuttled ships removed for sale. Three hundred railway wagons, plus rail construction material imported by the Italians to connect Agordat in Eritrea's lowlands with the Ethiopian town of Gonder, had been sold off, as had the motors running Eritrea's remarkable ropeway. In Gura, the American aerodrome had been demolished and exported to India. While some of this property had certainly been brought to Eritrea by the Allies as part of the war effort, the lion's share, she alleged, had originally been erected by Italy.

In railing against what she described as an act of 'hideous sabotage', Sylvia's concern was never for the Italians, effectively robbed of a sizeable state investment, or for the Eritreans most directly affected. She was outraged that her beloved Ethiopian government, whose right to the coastline she took as read, had been deprived of a priceless industrial resource just as it was seeking to rebuild and modernize. Leaving the Italian installations intact, she argued, would have served as 'essential, though scanty reparation' for the wrongs suffered by the occupied Ethiopian people – wrongs experienced, she for once refrained from mentioning, during a Fascist invasion the West had not lifted a finger to prevent. Keeping the Allied hardware in place 'would have been a gracious act'. Astutely, she questioned whether Britain had acted legally. Appointed by the UN temporary 'caretaker' of Eritrea, the British Administration had a duty to 'preserve and cherish' assets *in situ*. It seemed it had abused that trust. 'The right of the British Administration to destroy or dismantle installations in Eritrea is exceedingly doubtful.' And she poured appropriate scorn on the ludicrous Foreign Office claim that the buildings had been flattened because they represented a potential threat to life and limb. Left unguarded, the buildings might conceivably have been stripped of their roofs by looters, but 'the strong reinforced concrete structures erected by the Italians would have weathered the test', she said. Querying whether Britain's secret agenda had been to render Ethiopia militarily defenceless, she went on to raise another interesting possibility as to the motives behind the asset-stripping: 'Americans who have viewed the demolitions declare the British have effected them in order to handicap an economic rival.'

The deeper she delved the more she found to fuel her fury. Many Eritrean assets, she claimed, had been dumped on the market at a fraction of their genuine value. Thousands of metres of armoured cable had been sold for as little as 5s 3d a metre, when the real market price was £2 10s. Not only had the Addis government never received a penny in compensation, the British government actually forced Ethiopia to hand over the sum of £950,000 (a hefty £15.6m at today's prices) when it finally ceded control of Eritrea, on the grounds that Her Majesty's Government had lost revenue during a decade spent running the territory. This last financial transaction was not one the government was eager to publicize. The Foreign Office noted in January 1954 that Peter Freeman, an MP who often posed awkward questions in the House of Commons at Sylvia's prompting, might raise the issue of the £950,000 payment. 'The Department recommend that as little as possible should be said about this separate matter.'[19]

Calling for a public inquiry, Sylvia pursued the British government over the episode for a relentless three years, firing off letter after letter to her usual targets: the Foreign Office, Churchill, Attlee and Eden. She sent details of the British dismantling to John Spencer, the Emperor's American adviser, who used the information to block the removal of a consignment of rails the British proposed to sell as scrap – one small victory won from what was, overall, a crushing defeat.[20]

At her instigation, letters were published in *The Times*, the issue repeatedly raised during parliamentary Question Time. The official denials and dismissive slap-downs came thick and fast, the British government never wavering in its response: it had done nothing to be ashamed of.

The establishment denial has lasted to this day. Veteran historian Edward Ullendorff, who worked for the British Administration in Eritrea in his twenties, becomes apoplectic when Sylvia's name comes up. 'The British dismantling never happened. It is all a myth. Sylvia Pankhurst was a wonderful propagandist but she knew nothing about Eritrea,' he told me over the phone. There was no point in our meeting up to discuss the matter, he said, implying that anyone who treated the allegations seriously was a fool.[21]

The conversation unnerved me enough to return to pore over the files at Kew with special attention. Just how accurate, I wondered, was *Eritrea on the Eve*?

Characteristically, Sylvia did herself no favours by over-stating her case, slipping up on details which Foreign Office staffers then used to undermine her entire thesis. The Assab salt factory she had thought lost, for example, was still operating. The cement factory she mentioned had been requisitioned by the British, but returned to its owners, whose decision to sell to an Englishman who then moved it to Sudan could there-fore be presented as a private matter. Gura's aerodrome had been destroyed by the Americans, not the British, and the Italians had dismantled the radio station in Assab themselves to prevent it falling into enemy hands. Since some of her informa-tion was scrambled, the questions Freeman put were often muddled enough to allow the Secretary of State, by sticking rigidly to the literal terms of the MP's inquiries, to say nothing of interest.

Nonetheless, leafing through the archives, the picture that

emerges is hardly an honourable one. Perhaps the single most valuable document, never made public, is a four-page internal Foreign Office report written by an unnamed British official, probably based in Asmara, who had clearly been asked to assess each of Sylvia's claims for accuracy.[22] Dismissing her findings, his tone is lofty and contemptuous. 'There is really no answer to this utterly ridiculous statement,' he writes at one point, after gleefully highlighting her faulty spelling of local names. Yet even this least critical of employees cannot help but reveal that the British set to with a will in Eritrea.

The demolition of the naval base in Massawa is taken as read. A large dredger, he acknowledges, 'various major vessels' and a floating dock had been 'condemned in prize' – ie, re-quisitioned by the British as war booty – and sold. Equipment salvaged from the bombed aerodrome had been sold, as had oil installations at Ras Dogan. In Zula, 5.7 km of rail track, 850 railway points, 3 tonnes of bolts, 20 turntables and 71 trucks had been classified as 'wasting assets' and sent to Kenya. Another 1,500 m of barbed wire, 3 small cranes, 1 steel jetty, 1 steel signal tower and 1 steel hut had also been sold back to Italy. At Assab, scuttled or sunk vessels had been requisitioned as war booty, as had oil surveying equipment on the Dahlak islands belonging to the Italian company Agip. At Otumlo, 45 km of railway line had been deemed 'susceptible of military use' and sent to the Middle East, a fate also allotted 77 railway wagons and 31 km of railway line the Italians had intended to lay between Agordat and Bisha. In addition, British administra-tors had 'received' 100 train wagons, 2 diesel locomotives and 2 motor trolleys 'from other sources'.

There is enough in this report alone to make the British stance of wounded innocence look decidedly suspect. For those who care to look further, tantalizing clues that something drastic took place in Eritrea lie scattered through the public archives and buried in the histories written by British colonial officers of the day.[23] Perhaps the most damning statistic was published by a commission sent to the Horn by the Allied powers in 1947 to decide the fate of Italy's colonies. Setting out to establish a detailed picture of Eritrea's social, political and economic conditions, investigators asked a representative of the Eritrean Chamber of Commerce to explain what lay behind the country's severe recession. To their obvious amazement, the man immediately cited the requisitioning of Eritrea's best factory equipment – including machinery from its cotton, oil, Lancia and Fiat car plants – as a major factor. As the British Administration in Asmara refused to supply any information, the Four Powers Commission could probe this unexpected disclosure no further. 'No indications are given by the Chamber of Commerce – and no confirmatory information is available – of the manner in which this vast material wealth has been disposed of,' the investigators concluded.[24] But they published the Chamber's estimate of total lost assets anyway: 1,700m East African shillings, the equivalent of £1.85 billion today – a tidy sum for such a small territory.

The British public could not have cared less. Far more terrible things had happened in the war and by the 1950s more pressing domestic concerns – the end of rationing, the coronation of Queen Elizabeth, the conquest of Everest – were filling the newspapers. The question of how Britain had behaved in an obscure African outpost, some sun-bleached leftover from a conflict everyone wanted to put behind them, seemed old news. The steady stream of official denials did its work. To Sylvia's disappointment, the national press never got excited by the story, leaving her to do all the running – a fact commented on with relief by the Foreign Office. The story died a death, and so did the awareness that Sylvia Pankhurst's life had ever involved more than chaining herself to railings in the suffragette cause.

Experienced journalist and seasoned headline-snatcher that she was, Sylvia had nonetheless missed the real story. She had caught only the scraggy tail of a far larger scandal. Looked at in isolation, the asset-stripping in Eritrea might seem a temporary aberration, unworthy of a once-great empire. The episode acquires a different shape, however, when viewed in the context of British policy in the Horn of Africa as a whole. It was to be left to another member of the Pankhurst family – Sylvia's adored son Richard – to uncover a chapter in British history that makes what was done in Eritrea look not just logical, but utterly inevitable.

Richard Pankhurst came to Addis in 1956, a young man of 29 accompanying his mother in the last upheaval of a life that had already seen its share of fresh lodgings and career switches. Haile Selassie, who never forgot the debt he owed his most ardent Western supporter, had invited Sylvia to rebase in the Ethiopian capital. Bearing what many a British spinster would regard as the two essentials of life – a paintbox and a white Persian cat – she installed herself in a one-storey villa in a green glade of garden, looking south-west across the Wuchacha mountains.

To the outsider it might seem extraordinary that a woman who had already suffered one near-fatal heart attack should choose to move so far from home at the age of 73. To Sylvia, not prone to dwelling on the past or looking too far into the future, there was nothing puzzling about it. She was not the kind to campaign from a distance, only to blanch on contact with reality. As a long-term supporter of the Emperor, she was ready to savour life in his country, amongst her fellow fuzzy-wuzzies. One by one, her ties with Britain were fraying. Corio, who would never have agreed to quit the environs of the British Library, had died two years before. Many of her friends from the suffragette days had passed away. Her latest big project, raising funds in London for a modern hospital in Addis, christened after Ethiopia's Princess Tsahai, had been completed. It was time for a new start, and she was confident she would find much to do.

It was a move that determined the course of Richard's life. Now 77, he is recognized as one of the world's leading authorities on Ethiopia. Author of more than a score of ground-breaking histories of the country, founder of the Institute of Ethiopian Studies at Addis Ababa University, his expertise has made him a regular contributor to international symposiums and research journals so obscure, he wryly observes, 'they are probably read by no more than twenty people'.

The former eugenics baby – who only became aware of this quirky claim to fame when he had emerged from childhood and was thankfully immune to teasing – could really be no other woman's son. His face has the same droop as Sylvia's: gravity working overtime. Like her, he seems more at ease in the realm of the concrete than the emotional. Discussing Ethiopian history, he rattles off thoughts at machine-gun speed, but ask him about his mother's frame of mind on moving to Addis and he flounders, suddenly lost, as though the question has no meaning. He has inherited her capacity for the dogged campaign, sustained across the decades: the most recent has been the fight to return to Axum one of the great obelisks Mussolini seized as war booty and used to decorate a Roman square.

There the similarities begin to peter out. Richard, one suspects, is rather more fun than Sylvia can ever have been.

Endowed with the blinking diffidence of the shy Englishman, he has a habit of twitching his lips spasmodically, as though controlling an urge to laugh. The last time we met, he recounted with impish amusement how he and his wife Rita, who divide their time between Ethiopia and Britain, had attended an Italian official reception in London as honoured guests one evening, delicately failing to mention they would be leading a demonstration outside the embassy the following week. 'We thought best not to tell the ambassador we would be manning the picket line.' He lacks Sylvia's stridency. Perhaps the slightly distant gaze of academia militates against his mother's brand of white-hot fury. If a significant part of his work has been dedicated to defending Sylvia's role in history, he is too subtle not to acknowledge some of her failings. 'We were always criticizing her for being too admiring of the Emperor. If she'd lived longer, I think she might have expressed doubts. But it came down to the anti-chauvinism on which her radicalism was based, the rejection of British colonialism that went to the heart of her work. "My job is to criticize my own government," she used to say. "As for criticizing the Ethiopian government, that's up to the Ethiopians."'

It was Richard who, reading systematically through the Ethiopia archives at the Public Record Office in the late 1990s, stumbled across a set of files that cast Britain's behaviour in Eritrea in a completely new light.[25] What had happened in Eritrea had been the tip of the iceberg, this correspondence made clear. A decade before they were to address identical issues in Eritrea, the British officials who occupied Ethiopia in the wake of Mussolini's defeat had already established how to handle Italian assets in the Horn. The approach could best be summarized, to use a colloquial expression, as: 'Gimme, gimme, gimme.'

Just as they had poured investment into Eritrea, the Italians had not stinted during the brief five years their Abyssinian folly lasted, constructing hospitals and hotels, post offices and telephone exchanges, factories and aqueducts. A giant car repair works which could service 6,000 vehicles a year, the finest flour and biscuit factory in Africa, an oxygen plant that could produce 400 cubic metres of gas a day, a state-of-the-art cotton mill: the assets were scattered across Ethiopia. British military officials took one look at the spanking new machinery, rubbed their hands, and decided it was almost all surplus to local requirements.

Just five days after Italy's official surrender on November 27, 1941, General H Wetherall, commander-in-chief for East Africa, sent a telegram to the War Office in London listing the Italian factories he wanted packed up and dispatched to Britain's colonial dependencies. In the following days, the country's 11 most important factories were all neatly categorized under three columns: those 'capable of being moved', those which 'probably' should not be moved because of their contribution to Ethiopia's economy, and those which definitely should 'not be moved' because the effort would not be worth the candle.

The British justified the operation on several grounds. With the war against Nazi expansionism still raging in North Africa and the Middle East, Britain was struggling to cover the cost of its military operations. It was only right and proper, surely, that all surplus assets, especially those paid for by a former enemy, should go to lightening London's financial burden. 'It is essential that we should not waste any possible source of either machinery or labour. Abyssinia represents in this respect a wasting asset,' explained the Intendant-General in Addis Ababa. With skilled Italian workers in Ethiopia scheduled for deportation, the factories would, in any case, swiftly grind to a halt, as Ethiopians had 'by universal report . . . no mechanical aptitude'. Like an acquisitive mother muttering 'oh, he'll only break it' as she snatches a gift from her bawling infant, the British told themselves such munificence would only go to waste in a backward nation.

The argument begins to crumble as soon as you examine the list of items the British authorities asked the Emperor to requisition on their behalf. While shipping crankshaft grinders to Libya made sense – they were used to repair damaged tanks – it is hard to see how sending road-making equipment to India, a brick factory to Nairobi and an oxygen plant to Uganda served any military purpose. Rather than repositioning the Italian plants nearer the North African war zone, the British administrators actually proposed moving them further away, to British Somaliland, Kenya, South Africa and Pakistan.

Soon the original list of 11 factories was being dramatically expanded, irrespective of the likely impact on locals. Removing the oxygen plant, one memorandum from a meeting of the British Military Administration in Addis Ababa made clear, would force Ethiopia's hospitals to go without the life-saving gas. No matter – the plant was slated for removal. One brigadier suggested stationery supplies be spared as the Ethiopian government would find it 'extremely difficult' to locate any writing paper once the city's printing works have been dismantled – he was overruled. By the end of 1941, with the first batch of factories already on their way out of Ethiopia, British officials turned their attention to the CONIEL electricity plant, although their own experts warned its loss 'would be a great blow to Ethiopia'. The inventory of items selected for requisition would eventually fill 16 pages, embracing soap-making equipment and diesel tractors, bridges and fleets of trucks, water-boring works and oil-pressing concerns, saw mills and mining machinery.

Originally, the British had said they would pay the Ethiopian government for the dismantled property. Now, with the wind in their sails, they denied ever making such a promise, although the damning phrase 'acquire by purchase', underlined in red crayon, screams out from the correspondence. As word of the free-for-all spread around the empire, everyone wanted a piece of the action, with competition at its keenest between the military establishments in Cairo and Nairobi. So many different British authorities were at one time homing in on Ethiopia – even the military department in India demanded its share – British officials considered assigning a special officer to deal with 'officers arriving in Addis looking for things they want to take away'.

While Ethiopia's sale of the century was being held, the Emperor did not stand meekly by. But Haile Selassie's leverage was limited. He owed the British – who had granted him exile in Bath, beaten the Italians and, however grudgingly, reinstated him as ruler – a great deal. In exchange for having Ethiopia's independence officially recognized, he had signed a convention obliging him to 'requisition and hand over to the British forces any private property . . . which may be required'. With his country still full of armed British soldiers, open defiance hardly seemed wise. Nonetheless, as the dismantling accelerated, Haile Selassie began to remonstrate with increasing force. 'Emperor and ministers are seriously perturbed and former has now twice spoken to me on the matter of factories,' wrote a British official in the capital, complaining that he was being caused 'serious embarrassment'. Finally, in February 1942, the Emperor took the risky step of ordering his officials in Jimma to prevent a sisal rope factory being dispatched to Kenya. Faced with the strong possibility of an armed clash between their troops and a supposed ally, the British ordered their convoy commander to hold off.

During the diplomatic standoff that ensued, Robert Howe,

the British Minister in Addis Ababa, was ordered to step in and sort the problem out. During an audience with the Emperor, he delivered a list of outstanding British requisition items – a radio transmitting station, road-making equipment, a cement factory – before delivering the ultimate threat. If His Majesty did not comply, Britain was ready to think again about granting Ethiopian independence. The Emperor, Howe noted, 'appeared rather shattered by my plain speaking'.

Although he had behaved like a bullying landlord softening up a tenant who has fallen behind in the rent, Howe was actually sympathetic to the Ethiopian cause. It was time for this to stop, he told his superiors. Once outstanding items had been handed over, he recommended Britain make no further demands. 'Whatever may be the rights and wrongs of this matter there is no doubt that the Ethiopians have got it firmly fixed in their heads that the British army have plundered the country, and I use the expression advisedly,' he told London. 'They estimate that we have removed 80 per cent of the equipment with which the Italians lavishly endowed this country. They point to one item alone of medical stores to the value of £4m which was removed.' His analysis of why the Ethiopians were so upset shed devastating light on establishment prejudices. 'The sight of the removal of all this valuable material from this country has touched them in their most Semitic spot. In this respect, the Emperor is more Semitic than most Ethiopians.' If the Ethiopians balked at being robbed by their liberators it was only, the reader is led to understand, because they had all the money-grabbing instincts of the grasping Jew.

Despite Howe's ultimatum, the Emperor's resistance had had its effect. The best of the Italian equipment had already been removed, in any case, as a result of what British officials now openly referred to in their letters as a 'scorched earth policy'. A telegram to headquarters from a British officer in Yavello, southern Ethiopia, gives some idea of how thoroughly the task had been carried out. 'HAVE WITH DIFFICULTY PREVENTED STRIPPING HOSPITAL ROOF YAVELLO DESPITE OCCUPATION BY ETHIOPIANS,' it reads, before putting in a plea on behalf of a fort housing Ethiopian policemen. 'FEEL DOORS WINDOWS ROOFS SHOULD BE LEFT.' Sizing up the dregs that remained, the British pondered whether it was worth risking another awkward incident and decided against. 'You may write off Ethiopia as a source of supply of any plant or machinery,' a high-ranking British official told the Minister of State in Cairo in August 1942. 'However unfortunate this may be I am sure it is no use pressing any further.'

The eight-month looting spree had ended. During that period, the British had broken international law, come close to clashing militarily with the very regime they had returned to power and removed 80 per cent of the country's Italian assets without paying a penny in compensation to either of the two affected parties. 'Ethiopia didn't get a cent and there is no record of the Italians ever getting a cent either,' says Richard Pankhurst.

London had taken a huge risk, and pulled it off. Miraculously, no Italian or German newspaper ever got hold of the story. As the US consul in Asmara told Washington, news of Ethiopia's treatment would have been enough to make any country think twice about surrendering to the Allies: 'Happily the Axis propaganda experts do not know this, for if they did, they would be shouting it to the heavens,' he wrote. 'Goebbels could now point to Ethiopia and say to the people of Norway, Belgium, Holland, France, Poland, and Greece, "If the United Nations win, they will treat your country as enemy territory, just as Great Britain treated Ethiopia. Look at Ethiopia and be warned! When the British got through with it, what was left?" '26 There was so little adverse comment, a British Ministry of Information publication on the Horn actually felt free, in 1944, to sneer at the enemy on the topic of military appropriation: 'When you occupy a territory, if you are not an Axis power, you do not requisition all you see regardless of humanity and individual rights,' wrote its author, KC Gandar Dower.27

But in the long term, London had scored a devastating own goal. Ethiopia had been treated with humiliating contempt, its plans for a post-war resurgence dealt a shattering blow. 'It clearly didn't help Ethiopia to be deindustrialized at that stage of its development. It essentially had to start again,' comments Richard Pankhurst. Haile Selassie was not the type of leader to forget. His suppressed fury over the affair – just one expression of a wider phenomenon of British highhandedness in Ethiopia – ensured that once he was free to choose his friends, the Emperor looked elsewhere. Britain, which had competed so energetically at the turn of the century with its European rivals for a foothold in the Horn, found the door to Ethiopia slammed in its face.

What explains such ungenerous behaviour? Richard Pankhurst attributes it to a massive failure of imagination by men who brought to Addis the unreflecting arrogance developed while running Britain's African empire. Twenty years before most British colonies on the continent would be granted independence, the notion of an autonomous African state ruled by an Emperor enjoying supreme powers, rather than the limited authority doled out at London's discretion, was anathema. 'The British officials who came here had served before in Sudan, Kenya and Uganda. Many had been colonial governors. They had a whole load of racist assumptions, values so implicit they didn't even need to write them down. The idea of an independent African state was an anachronism, so why give anything to an anachronism?'

145

This was not Britain's traditional zone of influence, and these men felt none of the avuncular – if patronizing – concern they extended towards the subjects of their own African territories. At the back of British officials' minds, one suspects, was the pragmatic awareness that an industrialized Ethiopia would represent a serious commercial challenge to Kenya, Uganda and Tanzania. Grabbing a God-given opportunity to furbish British territories with equipment their own government could not afford, administrators were simultaneously emasculating a potential rival to British-run hubs of economic interest in East Africa.

But they would never be that honest about their own motivations. Instead, they found philosophical justification in the suspect notion of the 'overcapitalized state'. Rather than admiring the Italians for the seriousness of their commitment to their African colonies, the British ticked them off – it was so typical of these flamboyant, hot-headed Mediterranean types – for 'overdeveloping' the Horn. 'Both Eritrea and Somalia are specialised hothouse orchids and parasitism is in their blood,' sniffed Gandar Dower in his leaflet.[28] Dismissing Eritrea as a 'remarkable levitated white elephant', he commiserated with his government for being lumped with 'two over-capitalised, bankrupt semi-deserts, which had never been self-supporting and which had never been intended to be self-supporting'.

'They felt there was too much industry here,' says Richard Pankhurst. 'This was a native state and it didn't need this infrastructure. It could be used more effectively elsewhere, and, coincidentally, "elsewhere" meant elsewhere in British-administered territories.' Argue with sufficient fervour and the most perverse of conclusions begins to sound logical. Rather than crippling Ethiopia and Eritrea for decades to come, the British were actually doing the countries a favour, ridding them of a raft of cumbersome, unwanted technology.

Like a lost piece locking into place on an unfinished puzzle, Richard's research completes Sylvia's work. What seemed bizarre in Eritrea makes perfect sense given what happened in Ethiopia a decade before. Having absorbed the details of British dismantling in Ethiopia, I defy anyone to read Sylvia's *Eritrea on the Eve* with sceptical eyes. When it comes to their dealings in the Horn, the British authorities manifestly do not deserve the benefit of the doubt.

Sylvia herself would never know she had been vindicated. On the afternoon of September 27, 1960, an Australian husband-and-wife medical team working at the Princess Tsahai Memorial Hospital, in which Sylvia continued to take an interest, received a distress call. Richard and Rita had left Addis on a camping trip and in their absence, Sylvia had become ill. When Catherine and Reginald Hamlin arrived at the house, they realized Sylvia had suffered a major heart attack and began administering morphia and oxygen. Soon they were joined by two of the Ethiopian royal princesses, who held vigil as the former suffragette lapsed into unconsciousness. 'As I sat beside her, now and then she squeezed my hand until, after about two hours, she died,' Catherine recorded in her memoirs.[29]

Haile Selassie paid the 78-year-old campaigner a final tribute. The socialist who had in an earlier manifestation argued the finer points of Marxism with Lenin, was given a state funeral at Holy Trinity Cathedral, built to mark the Emperor's triumphant reinstatement. Before thousands of mourners, including members of the royal family, a rebaptized *Walata Kristos* (Child of Christ) was honoured with the kind of grandiose eulogy usually reserved for those killed defending the homeland.

Sylvia's simple grey marble tomb now lies near the

cathedral's front entrance, not far from a bronze memorial dedicated to the Ethiopian ministers shot by the military regime that ousted the Emperor. She died too soon to see Haile Selassie move from admired reformer to despised reactionary, too soon to see the Eritrean merger with Ethiopia she had supported turn into a hateful farce, too soon to see Ethiopian nationalism rival European colonialism for brutality and horror. The Emperor she venerated lies inside the cathedral, in a vast pink granite sarcophagus that lay empty for a quarter of a century until his lost remains were recovered and could be interred next to those of his empress wife. Empty tombs appear to prey on the Ethiopian mind. In recent years, a rumour has spread that the grey marble slab lies above an empty coffin: Richard, it is whispered, stole to the cathedral late one night and quietly removed his mother's body, returning it to Manchester for burial in its native soil. It is a story that shows a profound misunderstanding of the commitment made by both mother and son. Neither Pankhurst would want her buried anywhere else. This is her rightful resting place. As the Ras who delivered Sylvia's eulogy told worshippers: 'Your history will live forever written in blood, with the history of the Ethiopian patriots ... Since by His Imperial Majesty's wish you rest in peace in the earth of Ethiopia, we consider you an Ethiopian.'

Britain's post-war pillaging appears to have vanished from Ethiopia's collective memory. The history books glide over it and, despite the publication of Richard Pankhurst's paper in 1996, I have yet to meet an Ethiopian, no matter how well-read, who is aware of it. Perhaps a country whose sense of identity is rooted in its proud image as an unconquered land, simply refuses to make mental room for such a demeaning interlude.

In Eritrea, in contrast, where British dismantling would form part of a sustained pattern of foreign exploitation, the experience is seared into the public consciousness. As every superpower learns, communities which know themselves to have been victimized develop long memories, able to recite chapter and verse long after the empires concerned have turned senile and benign, snoozing in complacent forgetfulness. When I raised the subject with an elderly history professor at Asmara University he grew suddenly animated and told me how, as a small boy, he had stood with his friends on a bridge looking down on a heavily-laden train pulling out of Asmara. 'We leaned over and spat on it, because we had been told by our parents that the British were taking away our riches, stealing what belonged to us.' For a generation of younger Eritreans, raised during the Struggle, Sylvia's *Eritrea on the Eve* was almost required reading, testimony to how badly not just one, but two European powers behaved during their time at the helm. At independence in 1993, some of these Eritreans, taking it for granted London would recall the episode with the same clarity as they did, could be heard predicting that a shame-faced British government would rush to make amends with generous offers of aid. In fact, of course, the development ministry in London retains no institutional memory of the event, and Britain today has no bilateral aid agreement with Eritrea.

When the angry old men die and *Eritrea on the Eve* is no longer read, the landscape itself will be left to bear witness to what the Foreign Office always denied. A few years ago, a South African businessman thinking of investing in Eritrea learned of an old potash mine the Italians had once operated on the edge of the Danakil Depression and a railway branch line that had ferried output to the coastal village of Mersa Fatuma. The line was still marked on the map – could enough of the original infrastructure remain, Moeletsi Mbeki wondered, to revive the project? The Eritrean government obligingly pro-vided a military helicopter to allow him to survey the area.

Today, anyone who feels strong enough to brave the heat can walk the route he tracked from the air. Criss-crossed by ostriches and gazelles, the trail weaves its way between black table-topped mountains, across the gravel flats, past the rusting debris of a crashed German fighter to a jetty that thrusts into the blue sea. A lot of work clearly went into laying these solid earthworks. But every iron sleeper, every rail, has gone. The railway buildings are a tumble of shattered bricks and the smashed water vats hold the skeletons of goats which lost their way. 'When we landed,' remembers Mbeki, 'all we found was a village elder who laughed, and said the British had taken it all away a long time ago.'[30]

CHAPTER 7

'What do the baboons want?'

'The UN is our conscience. If it succeeds, it is our success, if it fails, it is our failure.'

British sculptress Barbara Hepworth

'The United Nations Resolution on the future of Eritrea constitutes one of the most outstanding and constructive experiments which the United Nations has undertaken throughout the world.'

UN Commissioner Eduardo Anze Matienzo,
October 1951

Arranging my tape recorder on the table, I jolted upright as I felt something warm and wet flicking over the bare toes of my sandalled feet. A sausage dog that had been trotting around the room had just given them an exploratory lick. Losing interest, she waddled to where a paper towel had been laid on the mushroom-coloured carpet. Positioning herself over the fabric, she ground her little bottom frenetically into the floor and had a pee. There was not a word of reproof from the black nurse, who merely picked up the soiled towel and dropped a new one in its place. In the world of John Spencer, longest-serving of

Emperor Haile Selassie's white advisers, Katy the dachshund clearly enjoyed extraordinary privileges.

His mind, in any case, was elsewhere. Lost in the folds of a tweed jacket originally bought for a younger, sturdier man, he sat musing on the past. Around his neck dangled an alarm button. If he pressed it, the white-robed attendants manning the desk at this Long Island 'senior residence' would come running, vaulting the cluster of Zimmer frames wielded by residents heading outdoors to enjoy the spring weather. But despite his physical fragility, Spencer did not really look in need of help. He was hard of hearing – this would be more a shouting match than a conversation – and he jumbled his dates, a weakness he recognized and which embarrassed him. But he rapped out his sentences with the testy peremptoriness of a man who regarded confrontation as a normal mode of discourse. The patrician, high-domed forehead and square jaw, framed by two ruched curtains of skin, were instantly recognizable. I had seen them in an earlier version, in the black-and-white photographs which showed a fresh-faced young American standing respectfully amongst the ministers gathered around an Ethiopian Emperor in his prime. Rather than diluting him, time had boiled Spencer down to his very essence: all bulbous knuckles, clenched jaw line and strong opinions. Given a legal brief and a case to make, one felt, he might still give an adversary a decent run for his money.

I had come to this quiet suburb, where the Stars and Stripes flapped over clipped lawns and motorists drove with the exaggerated care of the old, in a state of incredulity. I could hardly believe that John Spencer, the man who served as Haile Selassie's international legal adviser through four decades – with the occasional interruption for the Italian invasion, the Second World War and family duties – was still alive. The fact that, at the age of 95, he also retained a lucid grasp of the process that led up to Eritrea's incorporation in Ethiopia – a marriage that was to end in the most drawn-out and bitter of divorces – seemed positively miraculous.

He had been recruited in 1935 when Haile Selassie, his territory penetrated by Mussolini's invading troops, had registered the need for an expert who understood the ways of the West, its unfamiliar legal terminology and institutions, and could put Ethiopia's case to an international audience. The Emperor wanted a citizen of a country with no history of colonialism and someone who was at ease in French, the Western language spoken by most of Ethiopia's elite. Spencer, a US national who had spent four years studying law in Paris, met both requirements. Haile Selassie experienced second thoughts when he actually met Spencer, who, at 28, still looked alarmingly young for his age. But by then it was too late, the lawyer had already taken up his post at the Ethiopian Foreign Ministry. There, plagued by fleas, forced to dash into the nearby eucalyptus groves whenever Italian bombers droned overhead, he pounded out press releases on the one and only typewriter, denouncing Italy's use of poison gas and calling for international intervention. Four months later, Spencer's interesting new job sputtered to a premature end. As Italian soldiers advanced on Addis, Haile Selassie quietly boarded an overloaded train to Djibouti with his wife and five children, brushing aside the remonstrations of aghast Ethiopian notables. His latest appointee was left to wander a city that looked as though it had experienced a sudden snowstorm: ripping through mattresses as they systematically trashed the capital, looters had covered the streets in down. After taking part in the armed defence of the US and British legations, under siege from gangs dressed in stolen top hats and tails, Spencer abandoned Addis to its new owners.

This initial encounter with the Lion of Judah gave him an

intriguing foretaste of his employer's ambiguous character. On the front line, Spencer had watched fascinated as Haile Selassie shooed anxious aides away, donned a helmet, climbed into a trench, took the controls of a piece of heavy artillery, and attempted to shoot down an Italian aircraft. The man was undeniably capable of physical bravery. Yet by slipping away into exile, rather than fighting to the death in the tradition of Ethiopia's great warrior-kings, Haile Selassie committed an act of disgraceful cowardice in the eyes of his countrymen, which many never forgave.

Spencer caught up with the Ethiopian delegation in Europe, helping to draft the famous reproach the Emperor delivered to the supine League of Nations. But once that task had been completed, Haile Selassie could no longer afford to keep him on the payroll. Spencer went to work for the State Department and was then caught up in the Second World War. He was called away from his post aboard a US warship as the landings at the Italian port of Salerno got under way in October 1943. The reinstated Emperor needed his international legal adviser and Washington, which had its own agenda in the Horn of Africa, was only too happy to see an American citizen take up the key post.

In the years that followed, Spencer was to demonstrate his enormous value to his employer. He was one of a small team of well-qualified expatriates, dubbed the 'white Ethiopians' by their foreign counterparts, who advised the Emperor. Powerful men hire aggressive lawyers so that they don't have to show aggression themselves – the old 'if you have a dog, don't bark' principle. And Spencer was truly a terrier of a man, worrying his adversaries until they succumbed. If the prickly Ethiopians were always ready to detect signs of condescension amongst foreign colleagues, Spencer was even quicker to take offence on their behalf. Stubborn and astringent, he could – he admits in his memoirs – be legalistic to a fault, often pushing things to a point where even Aklilou Habte Wold, the long-serving Ethiopian Foreign Minister who became a trusted friend, called for compromise.

When he arrived in Addis, Spencer admits in his memoirs, he was 'naively opinionated, deeply suspicious . . . and totally ignorant of the Byzantine arabesques of Ethiopian thought processes, habits, and face-saving devices'. But he learned fast in the imperial court, a nest of jealous rivalries and long-brewed feuds which Haile Selassie dominated by applying the age-old tactic of divide and rule. Those early hair-raising experiences in pillaged Addis left their mark, for Spencer's view of his client was always to be a nuanced one, vacillating between critical exasperation and awe-struck admiration. As he briefed the Emperor, who would invariably sit immobile, silently twisting a ring on his finger, occasionally flashing a dark look of amusement or contempt from the vast throne that dwarfed him, Spencer built up a picture of a complex, contradictory personality.

At this stage of his life, the Emperor possessed a palpable magnetism. 'He was endowed with radiant charisma,' Spencer noted. 'He effortlessly commanded the rapt attention of all who came into his presence.'[1] Having survived the labyrinthine intrigues of Menelik's royal court as a young man and fault-lessly plotted his course from regent to ruler, Haile Selassie had learnt how to create a reverential stillness around him, the shimmering aura of unchallenged power. Lolling ministers leapt to their feet and bowed into space when they heard his voice on the end of the telephone, dignitaries ushered into his presence knelt so low their heads touched the floor. Legend had it that no ordinary mortal could look the Emperor in the eyes, so overwhelming was his gaze. When his motorcade swept through Addis, drivers spotting that aquiline profile and impassive face would brake, get out of their cars and stand respectfully by the roadside.

Yet nature had hardly been generous with Haile Selassie. In Africa, where a leader's path to power is often rooted in the brute fact of physical domination, presidents tend to be built on imposing lines, capable of quieting restless crowds and shouting down mutinying conscripts. As the empty uniforms on display today in Haile Selassie's former palace attest, the Emperor never grew taller than a stunted teenager. With his predilection for sweeping military capes and oversized pith helmets, the effect could be downright comic – from a distance, it looked as though a wilful child had been let loose on his father's wardrobe. A glimpse of the saturnine face on top of the tiny body, framed by craning, stooping courtiers, carried the same disconcerting punch as the wizened features of a music-hall midget. His voice was low and grating, it rasped at the nerves.

But Spencer swiftly registered that the Emperor had developed techniques to combat the disadvantages of his unprepossessing stature and harsh delivery. 'Even when standing, he never directed his gaze upwards towards the taller interlocutor. Unless inspecting a building or following birds or aircraft in flight, he invariably looked out upon the world as from an inner eminence.'[2] Naturally aloof, he kept his words to a minimum – the less said, after all, the more likely that men would hang upon his every utterance, transformed from the mundane into pronouncements from the Oracle. From the measured movement of his hands to the deliberate carriage of the head, the Emperor's deportment was as poised and devoid of spontaneity as the court's elaborate protocol. And when it suited him, the Emperor used his physical vulnerability and contained demeanour as a stratagem. As Ras Tafari, he had risen to the post of regent by convincing powerful ministers he could be manipulated, only to expose the steel in his character once his position was assured. He used the same technique to take control of the Organization of African Unity, walking meekly hand-in-hand with African ministers he had long despised. 'Do not underestimate the power of Tafari. He creeps like a mouse but has jaws like a lion,' a bested Ethiopian warlord once ruefully observed.[3]

The ultimate manipulator, the Emperor had premised his survival on his instinctive suspicion of individuals, nations and policies. He had an extraordinary gift for recall, and his insistence on micro-managing the business of government, centralizing power ever more efficiently in his slender-boned hands, meant he was in possession of a host of potentially embarrassing information. Details of a past dismissal, banishment or indiscretion would be coldly dangled in front of a supplicant who had been counting on official amnesia to get his way. 'At the same time, and in thundering contradiction, he was the constant victim of a confidence which, once placed, he instinctively could not withdraw,' said Spencer, noting how advisers who had failed the Emperor – not once but many times – remained at their posts. Lavishing luxury on himself, the Emperor's love of ceremonial pomp and his greedy determination to secure a personal share of any profitable enterprise established in Ethiopia repelled many a visiting foreign dignitary. Yet he could be surprisingly generous with those around him, anticipating their needs and rendering kindnesses with no thought of return. Courageous and cowardly, remorseless yet forgiving, the Emperor, the legal adviser ultimately decided, was a fascinating split personality. But below all the seeming contradictions, Spencer concluded, lay a bedrock of enormous vanity and overweening egotism. 'He was fundamentally an intensely self-centred person for whom the lives of others counted for little beside his own,' said

Spencer, who never forgot the indiscreet comment that burst from Foreign Minister Aklilou's lips in a moment of fury: 'I swear to you that His Majesty is beyond doubt, even beyond imagination, the most selfish and grasping man I have ever known.'[4]

This, then, was the leader whose interests Spencer championed as Ethiopia set about negotiating an end to Britain's occupation and winning a say in the disposal of defeated Italy's colonies. His painstaking, nitpicking, behind-the-scenes role in that drawn-out operation left Haile Selassie forever in his debt. It also guaranteed him a permanent, less-than-affectionate place in the memories of modern Eritreans. 'Him?' an Eritrean government minister snorted derisively when I mentioned Spencer's name. 'You mean the biggest liar of the lot.'

While British officials smoothly appropriated what took their fancy in the Horn, the diplomatic machinery that would determine Eritrea's messy destiny had been grinding slowly forward. Once Italy signed the 1946 Treaty of Peace, formally renouncing all claims to its African colonies, the fates of Eritrea, Libya and Italian Somaliland, all under temporary British administration, remained to be decided. The four victorious Allied powers sent a team to the Horn to decide what was best, giving themselves a one-year deadline to reach a unanimous decision. They wanted the matter done and dusted. Yet an agonizing six years would pass between Italy's formal surrender and the introduction of a new system of government in Eritrea.

The problem was that no country that expressed a view on Eritrea's future at the international meetings Spencer and his superiors attended in Paris, Geneva and New York came to the issue pure in heart. The stakes seemed too high for that. The war's end was bringing about a new world configuration. Europe's exhausted imperial powers had started their long decline, while the new kids on the block – the United States and Soviet Union – were testing their superpower muscles. Mankind was moving from a world view moulded by European colonial interests to a global arrangement dictated by US–Soviet rivalry, in which the twisted logic of 'my enemy's enemy is my friend' would eventually reign supreme. History's players were elbowing for position, casting suspicious glances at one another as a new pecking order was established. Those responsible for deciding the future of Italy's colonies would vacillate, change their minds and execute the most dramatic of policy U-turns as they tried to work out how to benefit from this complex repositioning.

As a defeated Axis power, Italy initially seemed to stand little chance of pulling off its dearest wish and winning back its colonies under the guise of UN-allocated trusteeships. But Rome found some unexpected allies in its attempt to re-establish a presence on the Horn. Haunted by the fear that its own African colonies would claim independence if Eritrea was allowed to break away, allergic to anything that smacked of an 'Anglo-Saxon' plot, France rallied to the Italian cause. So did the Latin American nations, which felt duty bound to support a fellow Catholic country. Together they accounted for a third of the membership of the newly-established United Nations.

Rome also briefly enjoyed the support of the Soviet Union. Convinced it was owed territorial compensation for the huge losses sustained in the Second World War, Moscow dithered over whether to push for a trusteeship in Eritrea, Tripolitania or the Dodecanese, before plumping for a pro-Italian position when it looked as though Communists were about to seize power in Rome and then, just as abruptly, opting for Eritrean independence after all. There Moscow joined the Middle East states, which viewed anything short of total independence as part of an anti-Islamic conspiracy by a coalition of Christian states.

Initially, Britain viewed Eritrea through the prism of its traditional African and Middle Eastern interests. At one stage, London considered using Eritrea, rather than Palestine, as a haven for Europe's dispossessed Jews. Then it came up with its ingenious Greater Somalia scenario, a plan that depended on splitting Eritrea down the middle and giving half to Ethiopia in exchange for the Ogaden. But by the time Eritrea's partition was definitively rejected by the UN's General Assembly, London too was looking at the world in a new light, registering the need to form a common front with the US against the growling Soviet bear. What was good for Washington was good for London, and for the US Eritrea's highland plateau held – as we will see – enormous Cold War importance. There was a top-secret, hush-hush factor at play that ensured Eritrean independence did not gel with Washington's evolving plans.

Struggling to be heard above the jabber was Ethiopia. Addis Ababa used the powerful weapon provided by collective guilt to whittle away at international indifference. 'You owe us,' was the essence of the message repeated by Spencer and Aklilou. 'Remember how you stood by and did nothing when Italy gassed our villages? Remember how our Emperor spoke so movingly before the League of Nations, yet you ignored him?' Twice in its history, they pointed out, Ethiopia had been invaded from Eritrean soil. It was time to put paid to that threat by 'restoring' Eritrea with what Haile Selassie liked to refer to as 'the Motherland'.

Given such shifting, self-serving agendas, a unanimous Four Powers ruling was never going to be likely. Throwing up their hands in despair, the Four Powers handed the matter over to the UN, which dispatched a second investigative committee to Eritrea. But Ethiopia had already won a major concession. When the Four Powers team went to Eritrea in 1947, its delegates undertook to reach their judgements on the basis of the 'wishes and welfare of the inhabitants and the interests of peace and security'. When the UN's new Commission set out for Asmara two years later, its delegates pledged to defend Eritrean wishes while taking into account 'the rights and claims of Ethiopia based on geographical, historical, ethnic or economic reasons, including in particular Ethiopia's legitimate need for adequate access to the sea'.[5] It seems to have occurred to no one that the two issues – what the Eritreans themselves hoped for and what the Ethiopians wanted – might be mutually exclusive. In theory, the UN was committed to self-determination, a principle enshrined in its founding charter. In practice, Spencer and company had ensured that the Queen of Sheba vision of history was tacitly accepted as valid by the international community, whatever the implications for ordinary Eritreans.

The story of the UN Commission for Eritrea was to prove, in its way, a microcosm of the UN, with what seems at times its systemic incapacity to deliver on well-meaning promises. Democracy is all very well on paper. But as any member of a tenants' association or parents' committee knows, when put into practice it has all the smooth fluidity of a ragged fingernail being scraped along a blackboard.

Clashing egos, personal foibles, carefully-nursed grudges, a tendency to lose sight of the main issue in the pursuit of petty vendettas: every international meeting suffers from them. The UN Commission, however, possessed these characteristics to a degree out of all proportion to its size and task. With only five members and the wishes of less than one million souls –

Eritrea's population at the time – to be established, the extent to which delegates managed to disagree seems, in retrospect, nothing short of extraordinary. The account of this key body's internecine squabbles, published here for the first time, would seem comic, if one could only forget the terrible consequences.

To be fair to the judges and military men who made up the Commission, they were never operating as free agents. Assigning a new body to decide Eritrea's future did nothing to remove the pressures that had originally made agreement by the Four Powers delegation impossible. Each Commission member, under instructions from headquarters to engineer the required result, brought his country's national agenda with him to Asmara. Yet each managed to add a new, personal and vindictive ingredient to the mix. Like recruits for a reality television show picked for their likelihood to rub each other up the wrong way, the delegates moved from indifference to mutual loathing in the course of a few weeks. Resenting the UN bureaucrats assigned to help them, irritated by the Eritreans, loathing one another, they were to demonstrate the eternal truth of Jean Paul Sartre's maxim: 'Hell is other people'.

The new Commission was appointed in November 1949 and it soon became clear this was going to be no walkover. 'I regret to say that the work of the Commission during the first six meetings has been entirely unsatisfactory,' Petrus Schmidt, who headed the 20-man secretariat appointed to smooth the Commission's path, admitted in the first of many candid confidential reports sent back to Trygve Lie, UN Secretary-General of the day.[6] As the group travelled from New York to Cairo and on to Asmara, a potential troublemaker emerged: Guatemalan delegate Carlos Garcia Bauer, who prevented any real work being done with his 'persistent obstructionist tactics'. Despite speaking excellent English, Garcia Bauer had gone in to battle on the language issue, threatening at one point to storm out unless the Secretariat agreed to translate all documents into Spanish. This was a man who thrived on the tedium of protocol, wasting the first five meetings 'with purely formal questions, such as rules of procedure, credentials, points of order, roll-call votes, corrections of summary records, language questions, etc.' The other delegates were already wondering what lay behind such tactics. 'They all agree that the Guatemalan representative is an over-ambitious man, who wants his name on every page of every summary record, that he has at the same time a very obvious and very deep-rooted inferiority complex,' said Schmidt. UN headquarters did its best to soothe ruffled feathers. 'It is clear . . . that you and the other members of the Secretariat are doing an excellent job under very trying circumstances,' replied Andrew Cordier, Executive Assistant to the Secretary-General.[7] The British Foreign Office, being kept abreast of Garcia Bauer's activities by Frank Stafford, its liaison officer in Asmara, was more forthright in its assessment. 'Native dishonesty,' scribbled an Africa Department official on Stafford's report. 'People like Bauer' – presumably he meant shifty, unreliable, greasy Latinos – 'have an ineradicable dislike of facts.'[8]

The Commission landed in a country rent by banditry and bubbling with political skulduggery. By fair means or foul, the countries most directly affected by the Eritrean question were determined to bring about their preferred outcomes. Having surrendered its hopes of winning a UN trusteeship, Italy decided total Eritrean independence was the next best option – anything, so long as Ethiopia remained out of the picture. Rome was pumping thousands of pounds worth of bribes into Eritrea. Its representative in Asmara, Count Gropello, was paying members of the Unionist party, which favoured a merger with Ethiopia, to surrender their cards and join the Independence bloc.[9] Ethiopia, for its part, was subsidizing the

Unionist movement, with the Orthodox Church lending a helping hand by threatening independence supporters with excommunication. There were widespread reports of Ethiopians crossing into Eritrea where, dressed as Moslem lowlanders, they loudly denounced the idea of independence. In addition, British officials were convinced that a dramatic rise in *shifta* activity, which claimed the lives of scores of Italian settlers and members of the separatist Moslem League, was being funded by Addis. Gangs of armed bandits were killing pro-independence campaigners, shooting up buses and raiding Italian farms, then escaping into Ethiopia's Tigray province. The aim was to terrorize the pro-independence vote into silence, but the murders had another useful effect, for those who believed in the joys of Union. The deliberate impression was being created of an ungovernable territory, which would descend into anarchy unless it was placed in Ethiopia's strong, capable hands.[10]

As for Britain, it was busy moulding Eritrea's political scene into a shape that fitted its Greater Somalia partition plan. In a breathtakingly honest letter to the Foreign Office marked 'secret and personal', Frank Stafford gleefully logged the progress he had made, by dint of persuasion and promise, in sabotaging Eritrea's budding independence movement. 'The important thing is that we have now substantially reduced the number of Christian non-unionists on the Plateau. Following this I have returned to the task of persuading the Moslem leaders on the Plateau at least to break away from the Independence bloc.'[11] Stafford had also got to work on leaders in Eritrea's Western Province, which London wanted to attach to Sudan. 'I have set in train movements which will lead to a breakaway of a large proportion of the influential Chiefs in the Western Province from the Muslim League.' He was even franker about his hopes of influencing the UN Commission's findings. 'I have made other contacts with wobblers in the opposition, which are promising. The next step is to ensure that the people in the country who follow the true cause are properly primed in the right answers to give to the Commission when it gets down to the job of ascertaining the wishes of the population.'

Poor Eritrea. With so many clever, ruthless bullies whispering in its citizens' ears, telling them what they wanted and what they felt, it was a wonder they could think at all.

By March 1950, when Schmidt penned his fourth report back to headquarters, 47 people had died in Asmara in six days of rioting between the Moslem League and the Unionist party, brutally highlighting the need for a swift settlement. But relations between Commission members had only deteriorated, as Garcia Bauer flexed his talent for filibustering. 'You should know that the Commission is far from being a happy party. Not only are the delegates at sixes and sevens amongst themselves, but they are all or nearly all at cross purposes with the Secretariat,' British officials in Asmara reported. Schmidt, they noticed, 'does not conceal his contempt for the Commission as a whole'.[12] The Principal Secretary, in his own report, recounted how, while Eritrean organizations waited to put their case, the Commission became locked in a two-hour wrangle over procedure. 'Concessions, politeness, friendly appeals, did not help. On the contrary, it is now obvious that Mr Garcia Bauer explains such gestures as signs of weakness.'[13] Mr Garcia Bauer had not only insulted the Secretariat by accusing officials of tampering with records, he was picking fights with the weak Commission chairman, Norwegian Justice Erling Qvale. 'Mr Qvale is a charming, be it rather vain, old gentleman, with very little knowledge of procedural matters. Consequently, he is an easy prey for Mr Garcia Bauer, who continually shouts his "points of order" at him before he even has a chance to discover what it is all about.' But Mr Qvale himself was no angel.

He could show exasperating obtuseness when the Commission was in the field. 'He often shouts at witnesses and puts questions which are either irrelevant or irritating. At some occasions he acted like a colonial official of the old school shouting at his "native subjects"' reported Schmidt.[14] Qvale had managed to offend the Pakistani representative, Mian Ziaud Din, who had penned a long-winded complaint. As for the Pakistani delegate, while professing particular concern for the rights of Eritrea's fellow Moslems, Mian Ziaud Din seemed to share many of Qvale's racist attitudes. Even British officials had been taken aback by his reaction after seeing the primitive conditions in which Eritrean highlanders lived. 'Why don't we ask the baboons what they want?' he was heard to remark.[15]

Quite apart from all this bad blood, the Commission was wasting huge amounts of time going over ground that had already been covered by the Four Powers Commission that had preceded it. True, the Commissioners had spotted the Unionist party's clumsy ploy of moving its supporters from one public meeting to another to create an inflated impression of support. But they persisted in putting questions to which answers were already known, while failing to probe the only issue that mattered – Eritrea's fast-evolving political situation. 'They work on no system as far as I know . . . Nearly all that they have done in the field could have been done by one man sitting in Guatemala City with a copy of the Four Power Commission Report in front of him,' Stafford sneered.[16] Schmidt's assessment was even more damning: 'The officers of the Secretariat are disappointed and appalled by the shocking waste of United Nations time and money.'[17]

The Commission's mood improved slightly during a series of field trips into rural Eritrea, but when it returned to Asmara in April, it curdled once again. The Guatemalan was now not the only rebel, he had been joined by his Pakistani colleague and the men banded together to launch a putsch, forcing the Norwegian to stand down as chairman. Despite Qvale's sidelining, the atmosphere remained poisonous. 'The feelings of mutual distrust and bitterness have increased rather than diminished,' reported Schmidt.[18] The Commission's quarrels were no secret in Asmara, nor – given its members' indiscretion – was much else. 'Most of the representatives, if not all, talk far too much to any outsider who is willing to listen,' sighed Schmidt.

The Commission moved to Ethiopia, where they were meant to tour the countryside in an attempt to establish whether Haile Selassie's government was up to the task of administering Eritrea. A first-hand glimpse of the terrible hunger in Ethiopia's Welo and Tigray provinces might have stopped them in their tracks. But by arranging a busy schedule of receptions and urging the Commissioners not to waste their time on an arduous road trip, Ethiopian and British officials between them managed to ensure the Commission never ventured beyond Addis. The British Minister in Ethiopia struggled unsuccessfully to silence his conscience. 'I myself have been rather torn between the feeling that the Commission *ought* to see how bad things are in Ethiopia before deciding whether she should acquire any further territory, and the knowledge that if it did so this would tend to diminish the chances for the solution advocated by His Majesty's Government,' Daniel Lascelles privately confessed. 'The fact of the matter is that Ethiopia is not in the least worthy to acquire Eritrea or any part of it.'[19]

The Commission's final report, drafted in June, faithfully reflected all the antagonisms and interference that had dogged its investigations. Its various factions not only failed to agree a solution, their interpretations of the facts on the ground jarred so wildly Commission members might have been visiting entirely different territories.

For the delegates of Norway, South Africa and Burma, swallowing the British version of Eritrea as a 'bankrupt semi-desert', the colony was simply too backward and politically immature to stand alone. 'Political and economic association' with Ethiopia was the only answer. Even in this majority group there was a divide. Burma and South Africa, worried about the rights of Eritrea's Moslem community, favoured federation, while Norway argued for total union. The minority report filed by the troublesome delegates of Guatemala and Pakistan warned that the annexation of Eritrea to Ethiopia would cause 'constant internal friction'. Impressed by Eritrea's economic potential, they recommended complete independence after a 10-year period of UN trusteeship.[20]

Five men had somehow managed to come up with three separate answers to the question of what should be done with Eritrea. The UN General Assembly had travelled full circle, returning to its point of departure. Whatever gloss was put on the situation in public, UN officials did not mince their words in private. 'Without a doubt, the Report of the Eritrean Commission can be considered a failure,' Cordier informed the Secretary-General. 'The level of the Delegations was, to say the least, not equal to their task.'[21]

In light of such fundamental disagreements, what, concretely, was to be done? A middle way must be found. In December 1950, after months of intense behind-the-scenes negotiations between the foreign ministers and ambassadors of interested nations,[22] the UN General Assembly ruled that Eritrea should become 'an autonomous unit federated with Ethiopia under the sovereignty of the Ethiopian Crown' – a phrase that in itself contained a world of potential ambiguity and internal contradiction. Resolution 390 A (V) – a number engraved on the heart of any Eritrean patriot – was passed and a special Commissioner appointed to oversee its implementation.

It was, as the man chosen for the Commissioner's post himself acknowledged, 'essentially a middle-of-the-road formula ... the best possible compromise'.[23] While Eritrea's political parties had campaigned passionately for either annexation or independence, not one had ever called for federation. Nevertheless, that was what the international community, after years of fidgeting, about-turns and second thoughts, had determined they should have.

The new UN Commissioner for Eritrea, Bolivian Eduardo Anze Matienzo, arrived in Asmara in February 1951. A well-meaning, podgy diplomat with receding, curly hair, Matienzo had been appointed in the teeth of fierce opposition from Britain, which had 'a jolly, fat Burmese' judge lined up for the job and feared a Latin American would work in Italy's interests. Eritrea, in the opinion of the British delegation at the UN, was once again hardly getting the cream of the crop. 'He is an amiable but idle South American, quite friendly to us, but lacking in guts, commonsense, any knowledge of the position in Eritrea or indeed any of the required qualifications,' it reported. The British embassy in La Paz was a little kinder. 'Reported honest and competent diplomatist, though conceited' it telegrammed London.[24]

Matienzo immediately plunged into a series of meetings with Eritrean village elders, political leaders and civic organizations, explaining that he would be overseeing the drafting of a constitution for a soon-to-be-federated Eritrea, complete with democratically-elected Assembly. The announcement hardly triggered dancing in the streets. 'The Commissioner gained the impression that the population was mainly pessimistic,' he noted. 'Part of the population had no real confidence in the idea of federation or the possibility of applying it.'[25] As he prepared to set about turning UN Resolution 390 A (V) from abstract concept to concrete reality, Matienzo can hardly have guessed just how testing the next 18 months would prove. He was about to lock horns with a lawyer who – notwithstanding the fact that this was not his country, not his homeland, not his fight – considered it a point of principle never to surrender an inch when negotiating on behalf of his employers, the Ethiopian Crown.

John Spencer, by his own reckoning, was to be more intimately involved in the drafting of the Eritrean constitution and technicalities of Britain's handover of power than any other Ethiopian government employee.[26] Having agreed the broad lines of what he wanted with Spencer, Aklilou was content to oversee developments from Addis. Neither man entertained any doubts as to where their boss stood on the matter.

In public, Haile Selassie, while hardly faking enthusiasm for the federal compromise, had at least agreed to accept the inevitable. 'The formula as adopted by the General Assembly does not entirely satisfy the wishes of the vast majority of the Eritreans who seek union without condition, nor does it satisfy all the legitimate claims of Ethiopia,' he told his nation.[27] But it had become obvious that the formula was the only one that could obtain a two-thirds majority at the UN – better this, than an even longer wait for justice. 'The solution has been adopted and the principle accepted,' he affirmed. The truth was slightly different. 'The Ethiopians did not want federation of any kind, they never believed in any of it,' Spencer told me at his Long Island retirement home. 'The Emperor really pushed our hands, he was all for taking Eritrea immediately. He said, "I insist on the full return of Eritrea to Ethiopia." I told him, "No, you have to ease into it, you can't grab it all at once. Even if you want nothing to do with the Federation, you will have to slide into it gradually, bit by bit." '[28]

So when Spencer flew to Asmara to take his place alongside Matienzo, a team of UN legal advisers and representatives from the British Military Administration, he was hardly negotiating in good faith. The Emperor, he knew, regarded the documents he was labouring over as so much meaningless paperwork, to be trashed as soon as the world looked the other way. 'Was it hypocritical of me? Yes, I'd agree with that. But it was justified. I had said to the Ethiopians, "I'll help you get Eritrea" and that is what I did. They were desperate for access to the sea.'[29]

Spencer's aim was to strengthen the Federation to the point where Eritrea was so tightly bound to Ethiopia, total incorporation was only a blink away. Eritrea would be another province of Ethiopia in all but name. Matienzo and his advisers, in contrast, were committed to a Federation which was more than just a form of words. The relationship should be loose enough to allow independence, should the public ever decide to embrace it. Concession by concession, paragraph by paragraph, Spencer negotiated a better deal for the Emperor, 'easing into it' as only a wily lawyer knows how.

The Federation's very foundations were built on shifting sands. As legal experts pointed out at the time, there was something innately problematic about the notion of federating the Western-style democracy Britain had introduced in Eritrea, in which the rights to an independent press, trade union membership and freedom of religion were guaranteed, with an ancient empire whose criminal justice system dated back to the 13th century and in which all real power lay in one man's hands. Burma's delegate on the UN Commission had recommended a system in which the governments of Eritrea and Ethiopia tackled domestic matters and a separate federal body handled defence, foreign affairs and inter-state trade affecting both territories, a clear carve-up of jurisdictions. Yet by the time Matienzo came to write his final draft, the three-body notion had been rejected in the face of Ethiopian hostility. Once the Emperor had ratified the UN-approved Federal Act, 'the organs of the Ethiopian Government dealing with Federal affairs would constitute the Federal Government', the report said.[30] There was no word on how the promised 'organs' would be created or how 'federal' issues would be differentiated from 'internal' matters. In the end, the only new 'organ' established would be an Imperial Federal Council, composed of Eritrean and Ethiopian delegates, supposed to meet at least once a year to thrash out any problems. It was a murky, ambiguous formula, hard to grasp, easy to ignore.

Spencer's next challenge was winning a key post in Eritrea's government for a representative of the Emperor. As he admits in his memoirs, the UN Resolution 390 V made absolutely no provision for such a step. Yet the Ethiopian delegation, with typical brass face, informed a surprised Matienzo that this had always been 'a generally accepted principle' during discussions at the UN. Eritrea's future Assembly, due to be split down the middle between Moslem lowlanders and Christian highlanders, would struggle to pass legislation, Spencer argued. Granting a Crown representative a role would stabilize the volatile mix. Matienzo put up a stiff fight, knowing the Moslem community would regard this as letting Haile Selassie in by the back door. But Spencer kept up the pressure, and events in South America worked in his favour. As Matienzo pored over the paperwork, he learned there had been a change of government in Bolivia and he had been declared *persona non grata*. With his homeland barred to him, his very status as negotiator challenged by the Ethiopians, Matienzo can be forgiven for being distracted.

By the time Matienzo came to draw up the final version of the constitution, the Ethiopians had won huge concessions. The Crown representative would have the power to formally invest the Assembly's chief executive, send any legislation thought to encroach on Federal affairs back to the Assembly for a second vote and promulgate laws, all while enjoying 'place of precedence' at official ceremonies. Eritrea's future 'autonomy' was beginning to look distinctly compromised.

Spencer was to notch up a number of other triumphs, but Matienzo was not completely made of jelly. He stood firm on the issue that mattered most to Spencer and Aklilou, raised in what must surely qualify as one of the most obvious leading questions ever put to the UN. Once the Federal Act and Eritrean Constitution had come into force, they wanted to know, under what circumstances could the arrangements be amended or violated? Matienzo called in an international panel of legal consultants to settle what was clearly a crucial question. The panel's findings, from Haile Selassie's perspective, were disastrous. Once the legislation had been introduced, it ruled, Eritrea's future could be regarded as settled, 'but it does not follow that the United Nations will no longer have any right to deal with the question of Eritrea'. The UN resolution on which Eritrea's Federation was based would 'retain its full force', Matienzo's final report stipulated. 'If it were necessary either to amend or to interpret the Federal Act, only the General Assembly, as the author of that instrument, would be competent to take a decision. Similarly, if the Federal Act were violated, the General Assembly could be seized* of the matter,' he added.[31]

In his memoirs, Spencer simply blanks out the finding, choosing, just as his employer would later do, not to see the evidence of his own eyes.[32] But Matienzo's conclusion – glossed over by Ethiopian historians but spelled out repeatedly in the Commissioner's reports and speeches[33] – was crystal clear. Ethiopia would not be free to tinker, water down or abolish

---

* to 'seize', in this context, means 'to put in legal possession of'

the Federation, no matter how unhappy it became with the arrangement. Since the UN had guaranteed the Federation, the Federation could only be ended with UN approval. Aklilou raged at the 'perpetual servitude' to which his country was being sentenced, just another form, as he saw it, of neo-colonialism. For once, his words fell on deaf ears. It didn't take a genius, listening to the 'theoretical' scenarios explored by the Ethiopian delegation, to guess Addis' long-term intentions. If he surrendered a future UN role in Eritrea's future, Matienzo must have realized, he would effectively be abandoning a state whose autonomy he was pledged to enshrine to its hungry, impatient neighbour.

As Spencer noted in his memoirs, the Ethiopian team had succeeded far beyond its own expectations in undermining the work of those aiming for a weak federation. 'That the final result was one more closely knit than they had intended was the source of some surprise – even to me.'[34] One of Matienzo's last speeches to the Eritrean assembly carries a palpable sense of foreboding: 'I feel it my duty to sound a word of warning,' said the diplomat, about to start a life in exile in Argentina. 'The strength of a constitution lies in the strength of a people's desire to respect it, and words mean nothing without the spirit and intention behind them.' 'I did my best,' he seems to be saying to critics of the future. 'If the goodwill isn't there, there's nothing more I can do.'

The new constitution was approved by the new Eritrean Assembly on July 10, 1952, and on September 15, British administration of Eritrea came to a formal end. The Emperor cut a symbolic ribbon and crossed the Mareb river as ecstatic highlanders, brilliant in their white *shemmahs*, chanted 'Mother Ethiopia! Mother Ethiopia!' The party staged to celebrate the event was attended by the man the Emperor, in a classic piece of nepotism, had named to the key post of Crown representative –

Andargachew Messai, his own son-in-law. 'The Emperor did, I think, abuse his position,' Spencer told me, with a shrug of distaste. 'But then, you're dealing with a Middle Eastern mentality.'[35] He registered another small, ominous harbinger of what lay ahead. Ethiopia's newspapers did not bother explaining the niceties of the federal relationship to their readers. 'The Federation was presented to the population simply as "We've won back Eritrea".'[36] If it isn't the way you want it, behave as though it were, was the tacit message from on high.

The lawyer had done his work superbly. The process of 'easing into' complete union was already well advanced. Spencer, however, would not be around to see the fall-out of his labours, moving back in 1960 to the US, where he taught international law in Boston. 'I figured I'd been in Ethiopia long enough and I had a daughter who needed an American education.'[37] In truth, he was worried about political developments in Ethiopia and weary of the constant clashes with the British, French, Soviet and US embassies in Addis, where his knee-jerk belligerence had left him 'cordially hated'.

Many years after leaving, Spencer would have the novel experience of being denounced at a US press conference by Isaias Afwerki, then a rebel leader, for his role in Eritrea's historic betrayal. He knows what today's Eritrean government thinks of him, but Spencer is not the kind of man to apologize. In an autobiography in which he bitterly berates himself for missing many a trick on Ethiopia's behalf, the one topic on which he never expresses a moment's regret is Eritrean federation. His mind still runs along the worn grooves of post-Second World War thinking, in which strategy is always seen in terms of Fascist ambitions to be thwarted, British imperialism to be quashed, Arab fundamentalism to be stifled and Soviet influence to be fended off. He does not approve of what Haile Selassie did to the Federation after his departure, he says. He would have advised him to act differently. 'But that feeling of Eritrean nationhood was not there at the time.'[38]

He is probably right. But a man who, by his own account, spent most of his working hours confined to conference halls and cabinet rooms, his social life hobnobbing with Ethiopian aristocrats and foreign diplomats, was never in any position to gauge how ordinary Ethiopians, let alone ordinary Eritreans, really felt. By leaving Addis when he did, he was spared the jarring experience of setting his neat documents against the messy reality of Ethiopian occupation, with its bulldozed villages and public hangings. Distance and time help when it comes to escaping the spectre of regret. Spencer finds recourse in the excuse available to all lawyers granted, for one moment in their lives, the chance to change the world: 'They were my clients. I had to look at everything from their point of view, even if I didn't always share it. I was their advocate.'[39]

Ferdinando Martini, architect of modern Eritrea.

Haile Selassie, the Lion of Judah.

John Spencer, the Emperor's trusted American adviser.

The remarkable Pankhurst family. Emmeline (left) embarking on a lecture tour in 1911, is seen off by daughters Christabel (middle) and Sylvia (right).

Sylvia, unmarried mother, with son Richard, in 1930.

Sylvia Pankhurst shows off her Queen of Sheba award.

Ethiopian army crackdown on suspected rebels in the Eritrean town of Agordat in the late 1960s.

GIs from Kagnew Station bare all on 'Moon River' bridge.

Crippled EPLF Fighter with a picture of her dead Fighter husband.

Lessons under the thorn trees: an EPLF cadre gives classes in the Sahel.

Two Alpha males meet: EPLF rebel leader Asaias Afwerki is interviewed in 1988 in the field by *Daily Mirror* reporter Alastair Campbell, later to become Tony Blair's press supremo.

Aftermath of the decisive battle of Afabet, 1988.

The EPLF liberates Asmara on 24 May, 1991.

The insignificant village of Badme, flashpoint of a war that cost more than 70,000 Eritrean and Ethiopian lives.

A restored steam engine runs along Devil's Gate, last stretch of track up the Hamasien plateau.

CHAPTER 8

The Day of Mourning

'To betray, you must first belong.'

Kim Philby

Running through Asmara is what must surely be one of the most frequently baptized boulevards in Africa. Its many christenings track the stages of Eritrea's busy history. In Martini's day, it was known as Corso Vittorio Emanuele, after the king who ruled from across the seas. Under the Fascists, it was renamed Viale Mussolini, but when the British took over and taught the Italians to be ashamed of themselves, it became Corso Italia. Under imperial Ethiopian rule, it turned into Haile Selassie Avenue, only to metamorphose into Revolution Avenue under the Derg. This was the street down which rebel tanks, cheered by hysterical Asmarinos, thundered when the EPLF captured the city, a victory parade followed by another, inevitable rebranding as Liberation Avenue.

Lined with cafeterias, banks and grocers, this is the heart of the city, the natural starting point for the evening *passeggiata* in which Asmarinos effortlessly reclaim the thoroughfare from which they were once barred. The Italians planted the boulevard with Canary Palms, which now tower high above pedestrians' heads, giving central Asmara the slightly louche flavour of a

Riviera beach resort, an exotic suggestion of desert vistas and Arab sheikhs. Hidden under the brick and tarmac runs Mai Bella, the river on whose banks a pregnant Queen of Sheba, returning from her fateful tryst with Solomon, is said to have stopped and called for water. This brush with royalty has not saved the river from a sordid fate. It is now a sewer, collecting Asmara's stinking waste and channelling it to the outskirts of town.

At one end of Liberation Avenue lies the Ministry of Education, a massive, prune-coloured Modernist block. It was once the headquarters of the Fascist Party and inside the building a stone staircase bearing a wrought-iron motif of flaming torches rises grandly to the first floor. Open a door and you enter a classical lecture room, with tiered wooden seats and a vaulted ceiling. It is a place full of ghosts, whose whisperings keep lost pigeons fluttering anxiously in the eaves. This was once the meeting place for Eritrea's Assembly – the Baito – and it was here that one of the most humiliating chapters in the country's history took place. Humiliating, because if so many of Eritrea's injuries were inflicted by outsiders, this was an episode in which Eritreans themselves were to be forever implicated, the moment when betrayal took on a sour domestic flavour. The collaborationist sins of the fathers would only be purged by the sacrifices of a new generation of Eritreans.

It did not take Haile Selassie long to demonstrate his lack of enthusiasm for federation. Within days of the Emperor's ratification of the new constitution, Moslem residents were complaining that the Baito had allowed the country's flag – blue as the sky, blue in tribute to the UN which had given birth to democratic Eritrea – to disappear from government offices, now sporting Ethiopian colours. The Imperial Federal Council, the promised 'organ' meant to smooth out Federal problems, died a quick death, smothered in its cradle. Haile Selassie was so far above ordinary mortals, it emerged, that even the Council's Ethiopian delegates did not have the right to submit proposals unless asked to by the Emperor, and the Emperor, as it happened, did not feel like discussing Federal matters. The Council's Eritrean delegates twiddled their fingers for four months in Addis before returning, ignored and defeated, to Asmara. The Eritrean executive got the same cold-shoulder treatment from the Ethiopians, who 'never answered letters, never gave answers to specific queries and in fact ignored the Eritrean government', the British consul told London.[1] As the Baito chairman, Sheikh Ali Redai, aptly put it: 'A hyena had been put with a goat and the result was obvious.'[2]

Crucially, Eritrea's impotence also took economic form. On advice from the withdrawing British administration, the Eritrean government had handed Ethiopia control of Martini's railway, the ropeway, telecommunications and the extensive Italian state properties. The Ethiopians showed much the same cavalier approach to Eritrea's remaining infrastructural heritage as the British had done, moving entire industries south. The ropeway was sold by the Emperor's son-in-law and dismantled, its rusting gantries the only evidence of former greatness. But Ethiopia's most undermining move was to take over the collection of customs duties at Eritrea's ports and then fail to remit Asmara's share, which accounted for up to 60 per cent of Eritrean revenue.[3] Leached of its main source of income, having foolishly surrendered control of its assets, the Eritrean government was soon struggling to pay its civil servants. Haile Selassie was killing the Federation by simply pretending it did not exist.

Squabbling and divided, the Baito often failed to meet at

all, thanks to its increasingly dictatorial chief executive, Tedla Bairu. A committed Unionist, the irascible Bairu seemed bent on destroying the Federation from within, vetoing deputies who did not share his political views, refusing to call cabinet meetings and regularly suspending what he dismissed as an 'assembly of idiots'[4] for months at a time. When deputies dared to voice unhappiness over Ethiopian interference in Eritrean affairs, the Emperor's son-in-law put them firmly in their place. 'There are no internal or external affairs as far as the office of his Imperial Majesty's representative is concerned, and there will be none in the future,' said the Crown representative. 'The affairs of Eritrea concern Ethiopia as a whole.'[5]

For those who did not regard a future under Ethiopian rule as paradise on earth, the atmosphere had turned stifling. Debate itself was becoming increasingly impossible: in 1954, Eritrea's only independent newspaper closed, sued into extinction by the Crown representative. Harassed and threatened – a leading activist, Woldeab Woldemariam, was said to have survived seven attempts on his life – pro-independence campaigners fled the country. The public responded to this steady whittling away of freedoms with petitions, demonstrations and general strikes. But Ethiopia now had 3,000 troops stationed in the territory, and they did not hesitate to use live ammunition on the crowds. 'Under the Italians you could eat but you could not speak,' Eritreans muttered amongst themselves. 'Under the British you could speak but not eat. Under the Ethiopians, you can neither speak nor eat.'

Psychologically, politically and economically, the Asmara government was being made to look ridiculous. 'The adage: "Eritrea is dying on the vine" would still seem, at least to the uninitiate, a valid description of the local situation,' reported Matthew Looram, the US consul to Asmara, in a 1959 telegram back to Washington. 'The Government, which in effect acts as a Quisling instrument for the Emperor, is neither trusted nor respected by the people.' Half admiring, half appalled, he added: 'Machiavelli could well have taken a leaf out of the Ethiopians' book, for it seems to me that they have used extremely astute tactics to date in their gradual takeover of Eritrea.'[6] As the 1950s came to a close, the tribal chiefs and aristocrats who sat on the increasingly subservient Baito agreed to formally abandon the Eritrean flag, bowed to a redefinition of the Eritrean government's role as an 'Eritrean *administration* under Haile Selassie, Emperor of Ethiopia' and, in a particularly unpopular ruling, agreed to replace the Eritrean official languages of Tigrinya and Arabic with Ethiopia's Amharic tongue.

In pushing for union, Haile Selassie had enthusiastically applied the stick, while forgetting to dangle a carrot. The seeds of armed revolt were being sown. With their culture denied them and peaceful protest exposed as not only ineffective but dangerous, a group of exiled Moslems met in Cairo to announce the formation of the Eritrean Liberation Front (ELF), committed to winning independence by military means. The first shots in what was to be Africa's longest guerrilla war were fired in 1961 by a well-known former *shifta* chief, who led a band of ELF in a raid on a police unit in the western lowlands. Initially, the movement numbered only a few dozen men, armed with Syrian Kalashnikovs, who had learnt their fighting skills in the Sudanese army. But from such small beginnings, national uprisings grow.

It was against this depressing background that Asfaha Wolde Michael, Bairu's wily successor as chief executive, summoned the Baito members to the vaulted hall off Asmara's central boulevard on November 14, 1962, on what would in future be dubbed 'the day of mourning'.

They mustered in a mood of uneasy, ominous anticipation. The previous day, Ethiopian troops had staged a show of force in Asmara's city centre, chanting 'a bullet for anyone who refuses'. While many of the deputies naively assumed they were being called to debate the budget, the better-informed, warned of what was coming, had tried to skip the event. Some had announced they were on vacation, a handful had actually admitted themselves to hospital, feigning sickness, in a bid to deny Asfaha a quorum. But policemen hauled them from sickbeds and homes, holding them overnight to prevent them leaving town.

Registering that Ethiopian troops had been posted around the building in a way that seemed more intimidating than reassuring, the flustered deputies clattered up the grand staircase and filed into the hall. The doors were closed behind them: no escape now. Without more ado, Asfaha rose to his feet and read out a short statement. Declaring that the notion of a 'federation' had been imposed on Eritreans as 'a weapon of disintegration', Asfaha said it was clear that all progress would be stymied as long as two administrations divided Eritrea and Ethiopia. 'Now therefore, we hereby unanimously resolve that the Federation with all its significance and implications, be definitely abolished from this moment, that from now on we live in a complete Union with our motherland Ethiopia.' After a split-second of stunned silence, there was applause, but Asfaha did not push his chances by risking a vote. Blurry photographs of the event, creased with the years, show the deputies, dressed in white flowing robes and formal business suits, standing stiffly to attention. Their faces are impossible to read.

The Assembly was swiftly declared adjourned and the dazed Baito members were invited to the viceroy's palace to toast the day's work, leaving the hall to its pigeons and its ghosts. Since they had just made themselves redundant, it was the last time they would ever meet together in one place. It had taken them just half an hour to abrogate an internationally-guaranteed compromise that had been to nobody's liking, abolish the constitution they had sworn to protect and lay the groundwork for war.

The foreign press largely ignored the event. In London, *The Times*, which like other newspapers had dedicated reams of print to the torturous discussions leading up to the Federation, devoted just four paragraphs – what is known in the news trade as a 'brief' – to its sudden abrogation. Filed by 'our own correspondent in Addis Ababa', a euphemism for an item taken straight from the wire agencies, the story faithfully presented the version of events Haile Selassie wanted the world to believe, repeated to this day by scholars who should know better: Eritrea's Baito had 'voted unanimously' to abolish the Federation.[7] The account telegrammed to Washington by Looram's successor as US consul was more accurate. 'The "vote by acclamation" was a shoddy comedy, barely disguising the absence of support even on the part of the government-picked Eritrean Assembly,' he reported.[8]

In Addis, Haile Selassie lost no time in ensuring that a highly dubious piece of lawmaking became set in stone. Both houses of the Ethiopian parliament were convened to pass, with indecent haste, a motion welcoming the Baito's resolution. The 10-year-old Federation was declared terminated, Matienzo's careful work redundant. Airbrushed out of history, Eritrea – reduced to Ethiopia's 14th province – vanished from Africa's atlases. November 16th was declared a national holiday in recognition of the fact that the Emperor, after so many years of striving, had achieved his heart's desire. He had made good Menelik II's blunder. Landlocked Ethiopia had acquired a coastline.

* * *

Why had the Baito's deputies caved in so meekly? Was it through cowardice, self-interest or did they simply commit the grossest of historical mistakes? Only the individuals themselves knew the answer. But my inquiries met with flummoxed expressions. 'You're looking for the Baito members? Many of them went abroad, you know, they weren't very popular here. As for the rest, aren't they dead? Some of them were already quite old when they were deputies, and that's fifty years ago.'

It took weeks, but in the end I tracked down the last of the ageing protagonists. Time had whittled away their numbers. Less than a handful remained and most were in no rush to discuss a role they had hoped forgotten. One octogenarian, the well-heeled owner of an Asmara petrol station, backed off from me in his office, shaking his head. 'It was a very painful period. I'd really rather not talk about it.' Another, sitting small and prickly behind his public notary's desk, radiated defensiveness. 'It's too easy to judge things with today's eyes. It may look like a huge mistake now, but at the time it was done in good faith,' insisted 76-year-old Araya Hagos, who served as Asfaha's secretary. 'People thought union was the only way to escape a repressive colonial regime.' As he talked, he became angry. 'The Emperor behaved like a delinquent, in my view. He was a wicked, wicked man.' He would not say more.

So when I was eventually pointed in the direction of a turquoise villa a stone's throw from the sandals monument, it felt like a breakthrough. Originally built for one of Mussolini's sons, it belonged to Gebreyohannes Tesfamariam, former Minister of Economic Affairs in Eritrea's federal government. Aged 90 and deaf in one ear, Gebreyohannes needed his son on hand to serve as translator and loudspeaker. But as he held forth, gnarled hands clutching a polished cane so tightly it seemed the only thing anchoring his frail body to the floor, it became clear there was nothing wrong with his memory.

Behind him in the gloom of the 1930s lounge, hung a portrait of the middle-aged Gebreyohannes in full ministerial regalia of stiff white shirt and black tails. He gazed at the photographer with sad, liquid eyes.

He'd entered politics as a student, a devoted advocate of the Unionist cause, and swiftly climbed the political rungs, eventually achieving the Ethiopian title '*Dejasmatch*' ('Honoured'), one notch down from '*Ras*' ('Duke'). But he began entertaining doubts about the merits of the Unionist cause early on, Gebreyohannes wanted me to know. As Eritrea's post-war economic crisis deepened, Unionists in the Baito had argued ever more forcefully in favour of annexation. As long as it remained semi-independent, Eritrea would not benefit fully from Ethiopian investment, planning and jobs, they said. The logic worked with deputies who never strayed south, who had built up fantastical images of life across the border. Used to complaining about the Italians, they had failed to register how thoroughly European colonialism had transformed their country, pulling it far ahead of Ethiopia. As a minister, however, Gebreyohannes travelled far more than his colleagues, and the trips to Ethiopia opened his eyes. 'I could see the poverty, the general backwardness and I knew about the corruption and poor administration. I came to realize that uniting with Ethiopia would not bring tangible benefits, because things in Eritrea were a lot better than in Ethiopia.'

He passed on his doubts to Unionist deputies. But Gebreyohannes' voice was drowned out as a formidable quartet – Asfaha, Baito vice-chairman Dimetros Gebre Mariam, police commissioner Tedla Ogbit, and the new Crown representative, General Abiyi Abebe – launched a concerted campaign to persuade the Baito to agree to dissolution. 'There were some threats. Mostly, they tried to persuade the deputies it would be in Eritrea's best interests if they united with Ethiopia.' The argument of national interest lent a spurious gloss to a decision in which the all-too-human yearning for personal advancement clearly played its part. The Baito deputies, said Gebreyohannes, were promised land, property and key posts in the Ethiopian administration if they agreed to smooth the way. If they played ball, they would never want for money again, for the Emperor had agreed to pay their salaries for the rest of their lives.

One by one, amenable deputies were called to the palace, where they put their signatures to a document agreeing to the Federation's abrogation. 'The whole process took about a month.' The key November 14 meeting, at which dissolution was formally announced, was only called when 51 of the Baito's 68 members had signed. By tackling the deputies individually, Asfaha had softened up his audience, playing on each man's weaknesses while avoiding the dangers of open debate. The document, with its 51 signatures, allowed the chief executive to argue that, despite the failure to hold a vote, the motion had the backing of 75 per cent of Baito members – a perfect three-quarters majority.

Once the deed was done, it did not take long for the deputies to regret their actions, said Gebreyohannes. If they received their 30 pieces of silver – 'some salaries are still being paid to this day' – that was the only aspect of the bargain Haile Selassie honoured. 'The Ethiopians had made so many promises about building schools, dams and clinics. None of that materialized. They promised land and positions of authority, but with a very few exceptions, those promises were not kept. At that stage, I realized that these people were liars and cheats, and we had made a pact with a power that would not keep its word.' Unionist hopes died as factories were moved south, Ethiopians were sent to fill government posts in Asmara and key administrative jobs were transferred to Addis.

Gebreyohannes went on to hold two ministerial portfolios

in the Addis government. Looking back, he said, he felt he should have resigned, although it would have been no more than a gesture. 'Politically it would have made no difference. Eritrea was already in a trap and it could not extricate itself. Nothing any individual could have done would have made any difference. But resigning would have given me some personal gratification.' How did he judge his former colleagues, the Baito deputies? 'I feel pity for their human weakness. They were naive. It wasn't obvious to them what the result of their decision that fateful day would be. If the deputies had known what was going to happen, they would have acted differently. Many people make mistakes in their lives.'

A leading Unionist who had opposed annexation, a minister with no power to influence events, a man who felt sorry but, by his own account, no sense of personal responsibility – my fragile host had clearly not wasted his time in politics when it came to learning how to sidestep blame. He had resorted to the arguments used since time immemorial by those who belatedly register their small, self-interested actions have allowed a great, overarching injustice to occur. 'It would have happened anyway', 'We were just cogs in a giant machine', 'If I hadn't gone along with it, I would have become another victim.' The standard excuses of those who feel destiny, in forcing them to choose between the roles of hero or traitor, has placed too heavy a burden on the shoulders of ordinary men. Hindsight, unforgiving as a prison spotlight, had pinned him and his colleagues against a wall and exposed their weaknesses for all eternity.

There was another reason, I came to realize, why the Baito members refused to acknowledge a sense of guilt. During the grim years of the Armed Struggle, every Eritrean family paid a price, whether measured in slaughtered sons and daughters, years in detention or blighted prospects. In later life,

Gebreyohannes himself would become a passionate advocate for direct talks with the EPLF, a stance that won him a seven-year jail sentence from an Addis government that did not believe in negotiating with 'bandits'. 'The old guys feel they have nothing to apologize for, because they suffered so much afterwards,' explained a historian friend. 'They feel they paid their dues.' Had Gebreyohannes seen his time in prison as a form of atonement, an act of expiation washing clean a sullied conscience? I would have liked to ask him, but when I next returned to Asmara, I learnt that the old man had died.

It was an approach that held little water with the last participant in the 'day of mourning' I tracked down. In his legal chambers on Liberation Avenue, Dr Yohanes Berhane, a mere youngster at 73, seemed as quietly furious today as on the morning he was called as a young judge to bear witness to the historic events in the assembly hall. Wearing a silk-embroidered shirt and neatly-knotted tie, Dr Berhane belonged to a community of wizened septuagenarians and octogenarians who got up in the morning, put on their pin-striped suits, picked up their Borsalino hats and walked to wood-pannelled offices where there was nothing for them to do. Younger partners now ran their businesses, but they were fighting the good fight, determined to keep up appearances.

Dr Berhane had the clipped delivery of the lawman who sees no need to waste words, but emotion ran just below the surface. Despite his supposed neutrality, Dr Yohanes had been disgusted by what he had seen. Afraid to voice his feelings in front of Ethiopian troops, he had stalked off, refusing to join the others at the palace for champagne. 'It was an abuse of power, you know, they were not supposed to betray us. They had sworn to uphold the Federation.' From then on, he was to snub the Baito members – 'they made themselves a laughing stock' – not an easy feat to pull off in a city Asmara's size. As for acts of expiation, he had no time for them. 'There may be parliamentarians who regretted it afterwards, I wouldn't know. I did not talk to them. I had a grudge against them.'

Through fear and greed, naivety and laziness, the Baito deputies had betrayed their own. Their children would not lightly forgive them, and the rebel movement that sprang up would contain within it a strong element of youthful disgust for a generation regarded as corrupted. But Eritreans who hated the idea of union with Ethiopia clung to a last hope of rescue. They had not forgotten the stipulation Matienzo had made a decade earlier, the formal undertaking Spencer and Aklilou had sought in vain to have dropped from the UN Commissioner's final report. There lay Eritrea's guarantees, spelt out in black and white. The Federation could not be altered or abolished without the UN General Assembly first being 'seized' of the matter. Since the UN had brought about the Federation, only the UN could destroy it.

There is a bitter little story that does the rounds in Eritrea. Like the 'I didn't do it for you, nigger' anecdote it goes – whether true or invented – straight to the heart of the way Eritreans came to regard the outside world. When the embryonic rebel movement was engaged in its first serious firefight with Ethiopian forces on a hill outside Asmara, an old man approached the young men crouched behind their guns. 'Just keep shooting,' he told them. 'If we can only keep this up for 48 hours, the UN will come in and sort everything out.'

Unbeknown to ordinary Eritreans, the UN and its key member nations had already signalled that they took the Federation no more seriously than Haile Selassie. Writing in 1953 to the British ambassador, a Foreign Office employee was toe-curlingly frank about London's faith in the arrangement. 'I think I can say that we never really in our hearts expected the exact United Nations solution to last in the long run,' he said. 'The important thing was to have a solution with some chance of success which would release us from the task of administering indefinitely a territory whose inhabitants did not want us to rule them indefinitely. Such a solution having been reached, our concern was that there should not be an immediate breakdown for which we could be blamed.'[9]

The following year, an exchange of letters between the British registrar of the UN Tribunal in Eritrea and Andrew Cordier, assistant to the UN Secretary-General, was even more revealing. The registrar, Albert Reid, comes across as a self-important busybody, puffed up by his role as head of the UN's only remaining body in Asmara, but what he had to report was important. There was a growing rivalry, he warned, between Federal and Eritrean law courts over their respective fields of competence. The passage of time was throwing up more and more ambiguities in the constitution over where Eritrea's autonomy ended and the Emperor's sovereignty began. 'I feel I should draw your attention to the growing belief . . . that eventually the Eritrean question will again have to come before the General Assembly,' he said. Cordier's irritated response suggests that if he ever bothered to read the documents forwarded to him through the years by Schmidt and Matienzo, he had certainly not retained any detail. 'You should scrupulously avoid creating any impression whatsoever that the United Nations has any interest in the political situation within the Federation,' he warned. 'There now exists no basis on which the United Nations can show any interest in the political problems of Eritrea and the Union. Although the United Nations played the decisive role in the drafting of the Eritrean Constitution . . . that job has been completed to the satisfaction of the General Assembly, and that item has been removed from the agenda.'[10]

Legally, Cordier was in the wrong, but when you are the boss, that doesn't really matter. Only two years after Matienzo left Asmara, congratulating himself on a job well done, there was the sound of an organization washing its hands.

Eritrea's anti-Unionists never got to read these frank exchanges. Blissfully unaware of just how keen the UN was to forget Eritrea, demonstrating a touching belief in the sanctity of international commitments, they sent increasingly frantic messages to the UN's New York offices as the Baito's powers were curbed, asking for Matienzo's undertaking to be enforced. Signed by unhappy Moslem district chiefs, disappointed Christian notables, members of the Baito and high-profile activists who had fled the country, the appeals flew thick and fast across the Atlantic.

Merely filing these petitions exposed Eritreans to enormous risk. After exiled campaigner Mohamed Omar Kadi went to New York in October 1957 to present a 72-page denunciation of Ethiopian policy, he made the mistake of returning home. He was promptly sentenced to 10 years in jail for bringing the Ethiopian government into disrepute. Four-year sentences were handed out to two Moslem League leaders who sent a telegram to the UN the following year; other petitioners were demoted from their high-level government jobs.[11] Determined to crush dissent in its infancy, Addis went so far as to ask the UN to supply it with the identities of those signing the petitions, the better, presumably, to effect a crackdown. The request was refused by the UN's legal adviser, on the grounds that Ethiopia had clearly violated the spirit of the Federal agreement and it was 'not clear', to use his delicate phrase, what it intended doing with the information.[12]

The Eritreans risked their necks. They complained in the 1950s, when the Federation first came under assault, they filed petitions in the 1960s, when the Federation was abolished and they protested in the 1970s, when the civil war was well under way and the complainants had joined the rebel movements. By the 1980s, they had become more sophisticated, and an EPLF representative buzzed around the official heads of mission at UN headquarters in New York, handing out memos and leaflets, occasionally being arrested by UN police after protests by the Ethiopian delegation.[13] It was all for nothing. Occasionally, visiting Eritrean delegations received sympathetic hearings from Middle East representatives at the UN and won a little coverage in the US media. But since Eritrea had officially ceased to exist, however illegally, its case could only be discussed in the General Assembly if either Ethiopia or some powerful sponsor nation wanted the topic broached.

It never happened. In the giant red tomes of the UN Yearbook, which list the subjects placed on the General Assembly's agenda for debate, 'Eritrea' last appears in the index in 1957 and then vanishes, not reappearing again for 35 long, blood-spattered years. The country and its citizens had, as far as the UN was concerned, been expunged from the record. 'Who, after all, is the UN?' Chet Crocker, former US Assistant Secretary of State for Africa, rhetorically asks today. 'The UN is not the General Assembly. It is the Permanent Five and we are the Permanent Five, we, the Brits, the French, the Soviets and the Chinese. And we weren't about to start unravelling an African state.'[14] The powerful nations, for reasons that will soon become apparent, did not want the subject raised.

The Eritreans waited and gradually, reluctantly, they realized that there would be no UN cavalry riding over the crest of the hill. 'The public reaction was one of shock, pure shock,' remembers Dr Berhane, with a sardonic twist of the lips. 'It was as though Eritrea had suddenly become an orphan. The Federal Act obliged the UN to intervene and people waited for that. But the UN never responded in any way.'

By its silence, the UN added Eritrea to what was to become a long list of African countries in which it botched operations originally undertaken in good faith. Like Congo in 1961, Western Sahara in the 1980s, Somalia and Rwanda in the 1990s, Eritrea can argue that its future trajectory would have been radically different had the UN, at a crucial juncture, shown some backbone. Where these countries differ from Eritrea, however, is that while the UN has never been allowed to forget how its troops failed to prevent the assassination of Patrice Lumumba, Congo's elected prime minister, or flew out of Rwanda just as the genocide of nearly a million Tutsis began, only the victims today remember the UN's feebleness in Eritrea.

The UN's cynicism was to leave an indelible mark on the psychology of a people. No UN document can ever have been reprinted more often or more avidly perused than Matienzo's final report, on sale in the clubs and offices the Eritrean diaspora opened around the world. Eyes glazing with boredom, outsiders would marvel at the unflagging interest Eritrean scholars showed in the minutiae of UN charters, articles and clauses. But there was method to this seeming madness. If international law had been violated, then the armed campaign launched by first the ELF and then the EPLF was not, as Addis maintained across the decades, a 'secession rebellion' and the guerrillas were more than thuggish *shiftas*. Eritreans were merely fighting for respect of the law, because no one else would.

When Isaias Afwerki addressed the 48th session of the UN in September 1993, four months after Eritrea had won independence, the new president dispensed with diplomatic circumlocution and used the forthright language of a rebel movement sure in the knowledge of its own moral rectitude. 'I cannot help but remember the appeals that we sent year in and out to this Assembly and the member countries of the

United Nations, describing the plight of our people,' he told startled delegates. 'The UN refused to raise its voice in defence of a people whose future it had unjustly decided and whom it had pledged to protect.' As one high-ranking UN official told me: 'These weren't so much scars, as great bleeding, seeping, open wounds.'

The UN's amnesia not only fuelled a bitter sense of grievance, it marked the start of what was to become a mulish Eritrean insistence on self-reliance. If the international community could so casually brush off its responsibilities, if written agreements counted for so very little, the angry students who were to lead the rebellion concluded, then Eritrea would simply have to turn inwards, tap her own resources, and win justice on her own. They would never depend on anyone else again.

Halfway through the research for this book, I flew to New York. I felt I understood the realpolitik that lay behind the UN's decision not to act, but I was still curious. How had the UN justified such pragmatism to itself? How does an institution go about smothering an issue when it knows it is legally in the wrong? What had happened to all the telegrams sent by despairing Eritreans, whose numbers had faithfully, if pointlessly, been recorded by the independence campaigners? Most large organizations are obsessed with procedure, compelled by their own internal rules to crank out anodyne replies to correspondence, however inane or nonsensical. Had some sorry UN official been assigned the task of telling the Eritreans, ever so politely, to get lost? Had the steady stream of Eritrean petitions triggered, at least, a modicum of internal hand-wringing?

The UN archives are not stored in the waterfront skyscraper familiar from so many television shots. They are kept in a building hidden around the corner, opposite Uganda House and the embassy of the Republic of Korea, where a Chinese laundry does a hissing trade and West African stallholders tout cheap luggage. Inside, where a handful of archivists work silently in their offices, there is a hush. The grey-blue carpets soak up the noise, the bullet-proof security doors keep the street bustle at arm's length. It was clear, from the few tables set aside for readers, that this was a place unaccustomed to receiving members of the public. The week I visited, I was one of only two outsiders to drop in.

I had e-mailed ahead, specifying what I wanted to see – files compiled in the wake of Ethiopia's 1962 annexation – but nothing was ready. As the morning progressed the archivist, a middle-aged woman from New Jersey, began looking more and more unhappy behind her computer screen. 'This has never, ever happened before,' she moaned, stabbing at her keyboard. She was sweating, she had a migraine. 'I have never had these kind of problems calling up a file.' The boxes I had asked for, she eventually explained, had been re-catalogued, and the computer was refusing to recognize the numbers she was typing in.

She could provide copies of Matienzo's speeches in abundance, drafts of his reports and letters to his superiors, but no Eritrean appeals. On the second day, she called a balding male colleague over to help, and he too worked silently away. A few more files surfaced, but never the ones I'd asked for. I had failed, and I found my failure grimly fitting. Somewhere in the UN's vaults lay boxes of correspondence and memos, full of tantalizing insights into the thinking of the day on Eritrea. No one would ever read them now. The system had swallowed them up. They were as thoroughly lost to posterity as if they had been fed through a shredder. Technological modernization had, quite by accident, completed the task of deliberate forgetting launched by the UN in 1952.

I left clutching copies of the sole snatch of relevant

correspondence unearthed during three days of searching, a tiny chink of light in what would now, it seemed, be a permanent darkness. On August 24, 1963, the UN's acting director in Cairo wrote a poignant little letter back to headquarters. He had just received a delegation from the 'Political Committee of the Eritrean People's Legal Representatives Abroad', he told New York, which had handed him a petition for the Secretary-General. 'They complained that although they sent six or seven petitions before, they have never received an acknowledgement, and they are quite upset about it,' AK Hamdy wrote. 'Can you please see to it that an acknowledgement is sent.' Hamdy's request, and the petition, had been forwarded to the UN's legal counsel, a Mr Stavropoulos, with a covering note. Read 40 years later, with the knowledge of the tens of thousands of lives that would be lost in a war in which the UN played a contributory role, its casual indifference chills the heart. 'Would you be kind enough to look these over and let me have your views,' the UN official passing on the paperwork has scrawled in thick black ink. 'My feeling is we should not acknowledge. Do you agree?' And he added a further comment, laden with the exasperation of a teacher sick of a difficult pupil's incessant whining. 'I think that we are really through with the question of Eritrea.'[15]

CHAPTER 9

The Gold Cadillac Site

'Whatever happens we have got
The Maxim Gun, and they have not.'
Hilaire Belloc

Of an evening, Asmarinos like to drive part of the way down
the winding Massawa road to a spot called Durfo. There,
bonnets pointing recklessly towards the sheer drop, handbrakes
cranked, they have sweet tea delivered to their windows and
watch the cloud cover roll across the peaks like a river of white.
A few hundred feet below, villagers suddenly stumble in
the swirling mist, groping for landmarks, voices muffled under
a muggy blanket of vapour. But up here, under a heaven the
colour of forget-me-nots, vision and hearing remain pre-
ternaturally clear. The shouts of children wrestling in a compound, a bridle clinking on a straining mule, an old man
expectorating: unexpected and strangely intimate, the sounds
arrive from across the valley.

Just as South Africa's fate was determined by the realization
that the pretty pebbles children played with on the banks of the
Orange river were diamonds, and Congo's destiny was shaped
by the understanding of what could be done with the sap
dripping from its wild rubber vines, Eritrea's course was to be

197

shaped by the whims of topography and climate. 'Geography has been our misfortune,' a former EPLF commander once ruefully told me. He was referring to Eritrea's location on the Red Sea, but the country's true curse is invisible to the naked eye. For the Eritrean highlands are blessed with a scientific peculiarity – unnoticed by generations of farmers, missed by colonial explorers – of huge value to anyone in the know. It is an idiosyncrasy that goes a long way to explaining why America and its allies at the UN were so ready to turn a deaf ear to Eritrean pleas, so happy to accept Ethiopia's amoeba-like absorption of its tiny neighbour.

Twiddle the tuning knob on a short-wave radio at night and you begin to grasp the nature of the secret ingredient. A sudden Babel fills the room: Israeli voices shouting in Hebrew, dreary Russian monologues, plinky-plonky Eurotrash hits from Germany. Declarations by Libya's revolutionary committees battle to be heard above swooning Egyptian love songs. Gusts of folksy music from Eastern Europe compete with the mellifluous voices of a French phone-in, only to be drowned out in their turn by prayers from Saudi Arabia. The radio shrieks, squawks and hisses like a thing possessed. Local AM radio stations have been picked up as far away as Brazil, Finland and Australia.[1] Transmissions from China, broadcasts from South Africa: they ring out loud and clear. The Hamasien plateau just happens to be one of the best places in the world – some have ranked it *the* best place on earth – from which to receive and transmit radio signals.

Many of the reasons for the plateau's extraordinary reception could be grasped by any schoolchild. The higher the altitude at which a radio receiver or transmitter is set up, the wider the geographical area it covers. Even before an inch of mast is erected, the Eritrean plateau enjoys a vantage point a mile and a half above sea level. Its position, well above cloud cover, means the tropical storms that play havoc with radio waves and monitoring equipment are rare. As for human settlements, whose buzzing power plants and humming factories give off distracting electromagnetic 'noise', once outside Asmara the plains are sparsely populated and quiet. And the fact that the plateau lies just 15 degrees north of the equator means the sun rises and sets at almost the same time all year round, making life simpler for those tuning in to distant frequencies.

Added together, however, such factors still do not quite account for Eritrea's electromagnetic uniqueness. Some experts have speculated that Asmara benefits from another intriguing natural phenomenon – 'ducting' – in which radio signals rise through the earth's atmosphere, bounce horizontally along under the troposphere and return to earth via 'ducts' thousands of miles away from their original source. 'Some said the weather was clearer, the air was purer, there was less surrounding interference, it was the magnetic fields, or Asmara acted as a kind of collector, picking up signals from the ionosphere,' recalls Dave Strand, a US signals analyst who worked in Eritrea in the 1960s. 'Obviously the altitude was important. But I don't believe there is anyone in the world who really knows the answer.'[2]

The Italians had been aware of Asmara's extraordinary broadcast and reception properties, setting up Radio Marina, a station used to communicate with Mussolini's naval fleet. When the British took over Eritrea, they reluctantly ceded the field to their allies across the Atlantic. America's embryonic intelligence industry had mushroomed with the Second World War and Washington was scouring the world for potential reception sites. In April 1943, a US army lieutenant arrived in Eritrea with a six-man crew and testing equipment. The radio traffic samples the team sent back to Washington confirmed what the experts had suspected. The 'levitated white elephant' so derided by the British was, it turned out, the ideal spot to locate a spy station capable of eavesdropping on nearly half the globe. Brushing off British objections, the US army's Signal Intelligence Service started building bomb-proof underground concrete bunkers to hold equipment, while launching an intensive training course in Virginia for the 50 men who would initially be posted to Asmara.

The bet paid off almost immediately. Eritrea proved the perfect place from which to monitor the flow of Axis radio communications between Germany and Japan. In October 1943, when the Nazis were bracing themselves for an anticipated Allied invasion of the European mainland, the Wehrmacht gave the Japanese ambassador to Germany a tour of its defence lines. Baron Hiroshi Oshima's long report, giving priceless details of Germany's military dispositions, was radioed back to Tokyo.[3] On the way, it was picked up by the Americans in Asmara. Decoded, it gave General Eisenhower the information he needed to plan the Normandy landings.

With the end of the war, much of Asmara's equipment was crated up and sent back to the US. But the American military was in no doubt about Eritrea's long-term significance. 'The Joint Chiefs of Staff would state categorically that the benefits now resulting from operation of our telecommunications centre at Asmara ... can be obtained from no other location in the entire Middle East–Mediterranean area,' Admiral William Leahy bluntly told the Secretary of Defence in 1948. 'Therefore, United States rights in Eritrea should not be compromised.'[4] No other military base in black Africa would ever be deemed as vital to American national security.

In 1950, the base jerked back into frantic life with the outbreak of the Korean War. Circuits to Europe, the Middle East and the Philippines were reactivated as the US homed in on North Korean communications. For the next 20 years, the listening post in Eritrea would be steadily expanded and modernized, but never scaled back. Locked into the new, deadly Cold War game, the US needed the reach only sites like Kagnew Station – as the Asmara listening post was eventually named – could offer when it came to listening in on the Communist world.

It was a need that handed Haile Selassie a priceless bargaining tool. Washington, the Emperor was to discover, was ready to jump through a great many hoops to guarantee unhampered use of Kagnew. For more than two decades, Washington's approach to Ethiopia would be essentially that conveyed to an ambassador-to-be during his Pentagon briefing in 1963. Told that Ethiopia's poorly-trained army was trashing most of the American military equipment delivered to it, Edward Korry asked the Pentagon officer how the US was planning to tackle the problem. The reply was cynically revealing. 'He said there wasn't much we could do with the Ethiopians, and it was really Kagnew rent money, and if the Emperor wanted it in "solid gold Cadillacs", that was his term, he could have it that way.'[5]

The Emperor was to play the Kagnew card repeatedly, using US interest in the spy station to achieve several long-held ambitions. Incorporating Eritrea was only part of his master plan for Ethiopia. The time had come for his antiquated empire to make a great leap forward, and it was not something, he knew, it could do on its own.

Well into the 20th century, Ethiopia still went to war in medieval style: *rases* toured their provinces, enlisting fighters with promises of plunder, pulling together travelling armies which dissolved as soon as the campaign was over or when crops needed harvesting. Their motley forces provided Western journalists with picturesque photo spreads for audiences back home. But these were private armies, loyal to individuals rather than any state, and the fighters – barefoot and equipped with ancient rifles – were only as dependable as the *rases* themselves. By the time Evelyn Waugh was covering the 1935 Italian campaign, Ethiopia's military was beginning to seem an absurd anachronism. For Waugh, whose sympathies tended towards Rome, the pageantry had a pantomime quality, a touch of the pathetic. 'They had head-dresses and capes of lion skin, circular shields and extravagantly long, curved swords, decorated with metal and coloured stuff; their saddles and harness were brilliant and elaborate. Examined in detail, of course, the ornaments were of wretched quality, the work of Levantine craftsmen in the Addis bazaar, new, aiming only at maximum ostentation for a minimum price,' he noted.[6]

The trouncing the Ethiopians suffered at Mussolini's hands underlined the lesson. If Ethiopia was to prevent its territory being nibbled away by greedy outsiders and bind its diverse ethnic groups together to form a centralized empire under Amhara rule, it must modernize. It must have a standing army, run by professionals trained in elite military academies and equipped with state-of-the-art weaponry Ethiopia itself could neither manufacture nor afford to buy.

It was only natural that the Emperor should look to the US for that help. Unlike London and Paris, Washington never formally recognized Italy's occupation of Ethiopia, a gesture of solidarity he was not likely to forget. While Britain had pillaged Ethiopia's industrial infrastructure, the Americans generously provided arms and ammunition under their Lend-Lease Act. Given the choice between the tired Old World and a fresh-faced, brash New World, Haile Selassie knew which patron he preferred.

He made his choice clear on February 12, 1945. Robert Howe, the British Minister in Addis, woke in a flurry of alarm at 5.00 am, having just heard a US Air Force DC-3 taking off. It was carrying Haile Selassie to take tea with President FD

Roosevelt on a US cruiser in the Suez Canal, as explicit a gesture of American interest in Ethiopia as it was possible to imagine. The British were left to play catch-up, with Howe chartering a tiny biplane that reached Egypt in a tiresome series of short hops, while Winston Churchill was abruptly rerouted to Cairo to meet an African leader suddenly judged a cause for concern. The undignified scramble achieved little. When a British aide asked the Emperor what points he wished to discuss with the Prime Minister, his reply was curt. 'None.'[7]

As the British withdrew from the Horn, Haile Selassie and Washington plotted what was to be a very pragmatic marriage of convenience. Washington was uneasy with the idea of an independent Eritrea, all too likely, it was thought, to fall prey to a predatory Communist bloc. It wanted a friendly Ethiopian government, a government it could do business with, in firm control of the Hamasien plateau. Even before the UN General Assembly had opted for Eritrean federation as a compromise solution in December 1950, Washington had signalled its interest in reaching a base rights agreement with the Emperor.[8] Addressing the UN Security Council that year, John Foster Dulles, President Eisenhower's Secretary of State, made no attempt to conceal his government's self-interested take on Eritrea's future. 'From the point of view of justice, the opinions of the Eritrean people must receive consideration,' he acknowledged. 'Nevertheless the strategic interest of the United States in the Red Sea basin and considerations of security and world peace make it necessary that the country has to be linked to our ally, Ethiopia.'[9]

It was an extraordinarily frank admission to make. Essentially, Dulles was recognizing the moral dubiousness of imposing a federation no one wanted on the Eritrean population. But it would happen anyway, because it suited US needs.

The Kagnew factor casts a new light on American diplomatic

behaviour during the interminable debate over Eritrea's future that raged during the 1940s and 1950s. Remove Kagnew from the equation and the support the US gave Ethiopia in its campaign for union can be presented as an admirable rejection of colonial lèse-majesté, a high-minded championing of a struggling African nation. Add Kagnew to the mix and the US stance looks rather less noble. Haile Selassie himself was in no doubt as to the pivotal role Washington had played in nipping Eritrean leanings towards independence in the bud. 'If, today, the brother territory Eritrea stands finally united under the Crown and if Ethiopia has regained her shorelines on the Red Sea, it has been due, in no small measure, to the contribution of the United States,' he told Congress in May 1954. No wonder the UN, sensitive to the wishes of its most important member, refused to get involved when the Federation was abrogated eight years later.

Had the spy station been the sole card up the Emperor's sleeve, his leverage might have remained limited. But Haile Selassie knew how to make himself valuable on many fronts. As the years went by, 'Kagnew' would become a convenient mental tag for American policymakers in the know, shorthand for a complex mesh of interests binding the US to Ethiopia.

One of the Emperor's masterstrokes was to volunteer 1,000 Ethiopian troops for the war in Korea. At a time when the Soviet Union was denouncing the UN military operation as a neo-colonial adventure, the announcement that Africa's oldest independent state was joining in on the West's side, black faces fighting for freedom alongside white, presented the US with a glorious propaganda coup. When the US and Ethiopia signed a 25-year rights agreement on the Asmara base in May 1953 – a deal whose terms were negotiated by Aklilou and Spencer – the spy station, tellingly, was christened after the elite Kagnew battalion Haile Selassie dispatched to Korea.

The Korea episode underlined a wider truth. Positioned on one of the world's key waterways, close to the oil-rich nations on which Western prosperity depended and to the bubbling political cauldron of the Middle East, Ethiopia, America's policymakers came to believe, could either bring further chaos to a volatile region or act as a stabilizing anchor. As a Christian ruler hemmed in by Moslem regimes, the Emperor was clearly a natural Western ally. His defiance of Italian Fascism meant he enjoyed enormous kudos in the developing world. Here was a man who could play the role of wise elderly statesman, mediating between the West and the inexperienced politicians emerging to lead Africa's former colonies, a seasoned moderator who could swing delegates' votes in America's favour within international organizations like the UN.

This was not quite as nebulous a role as it sounds. In the post-war years, a vibrant debate raged on the nature of nationhood in Africa, whose territory had been clumsily hacked into states by 19th-century colonial powers indifferent to ethnic realities on the ground. Left-wing thinkers like Ghana's Kwame Nkrumah, regarded in the West as a dangerous radical, called for the creation of a United States of Africa. Haile Selassie, in contrast, was anxious to preserve existing frontiers, however clumsily drawn. His conservative school of thought won the day when the Organization of African Unity (OAU), initially largely funded by the Emperor, was set up in Addis in 1963. 'The Emperor's signal biggest contribution to African history is the creation of the OAU in Addis and the drafting of an OAU charter,' says Chester Crocker, former Assistant Secretary of State for Africa. 'There was to be no messing with inherited boundaries. The Latin doctrine of *uti possedetis* ('boundaries shall stay as they are') was very Ethiopian in its conception. But it also served the conservative interests of a number of other African leaders who were desperately insecure about how they were going to keep the new states together.'[10]

While endorsing a principle that set colonial boundaries in stone, African delegates in Addis chose to turn a blind eye to the most blatant charter violation of all, committed embarrassingly close to home. The border Italy had drawn around Eritrea at the turn of the century had been erased by the very nation championing the principle of frontier inviolability. This act of hypocrisy would not be lost on the EPLF, which would come to regard the OAU with utter contempt during the long years of exile. As far as the Americans were concerned, however, Haile Selassie had delivered once again, shoring up a fragile continent.

Israel, a fledgling state whose survival was a key concern to the US, with its influential Jewish lobby, provided another rationale for American support. Israeli aircraft could only access the African hinterland by flying over the Horn. Given Moslem hostility to the Jewish state, Ethiopia, a country whose foundation myth elided with Israel's own, was the natural spot for El Al aircraft to refuel as they worked their way down the African coast to Johannesburg.

Airspace was not the only issue. Strategists urging Washington in the 1960s and 1970s to increase its military aid to Ethiopia sketched out a doomsday scenario in which, left to its own devices, the Addis government lost control of Eritrea to Moslem insurgents, who promptly joined the Arab camp and cut off Israel's shipping lifeline through the Red Sea. It was a far-fetched sequence of events which sidestepped the obvious fact that Arab states were already, *without* Eritrean participation, perfectly placed to mount such a blockade, but were generally more interested in increasing trade with the West than mounting ideological embargoes.[11] Yet the Israel-as-hostage scenario was still being bandied around to justify support for Addis in the mid-1970s, when it had become clear that a Christian-

206

dominated rebel movement – the EPLF – had seized the initiative from the Moslem ELF in Eritrea, removing a key element in this imaginary equation. Crocker still defends it. 'How could the Israelis, given the sense of total isolation they often feel, be sure that some Arab coalition wouldn't shut off access to the Red Sea?' he argues. 'It might be a worst case scenario, but that's what strategic planners cater for. After all, Nasser did shut down the Suez Canal. "Never again", was the feeling.'[12]

None of these considerations would have carried much weight, of course, had it not been for the little matter of the Soviet Union, the looming threat that formed the backdrop to all strategic thinking at the time. In the 1950s, Secretary of State Dulles had dreamt up the concept of a Northern Tier of anti-Communist states that would serve as a bulwark against Communist expansionism. Why not create a Southern Tier of loyal allies in the Middle East, a secondary line of defence that would keep Communism out of the area and guarantee Western access to the oil-rich Persian Gulf? Haile Selassie adroitly volunteered Ethiopia for the role. 'That type of argument made it possible for the secretaries of state and defence to "find" that the defence of Ethiopia was essential to the defence of the free world,' John Spencer recalled in his memoirs.

Woven together, the various factors combined to form a conviction in Washington – as deep-rooted as it was amorphous – that Ethiopia and its increasingly restless northern province somehow 'mattered' to the US. American thinking on the Horn could be summarized as what one sceptical expert astutely labelled 'the geometer's approach to strategy'.[13] Take a compass, stick a point in the country under consideration and draw progressively larger concentric circles. Slapping your hand to your head, you then exclaim: 'My God, this country's so close to Israel, to the Gulf, to the entire Middle East. It *must* be important!'

If the approach was premised on implausibilities and questionable assumptions, that was hardly unique to Ethiopia. The same logic would be applied around the globe during the Cold War era on similarly specious grounds and with equally damaging results – remember how Vietnam's political future was deemed of vital American interest? For the geometer's approach has a lethal characteristic: it tends to be a self-fulfilling prophecy. By ruling that a backward African nation with few natural resources, far from obvious US zones of interest, was in fact a key ally, Washington helped ensure Moscow reached a parallel conclusion, first in neighbouring Somalia – desperate to match Ethiopia in its military build-up brigade for brigade – and then, when the opportunity presented itself, in Ethiopia itself. A nation's significance, like a diamond's value, lies purely in the eye of the beholder. By deciding the Horn mattered to the West, Washington guaranteed that the region became important to the East, sentencing it to disastrous superpower interference for decades to come.

Haile Selassie had succeeded in establishing the principle of Ethiopia's usefulness, but the price charged for Kagnew was to be the object of heated discussion.

Under the 1953 base rights agreement, in which Washington was granted near-sovereign rights over the various 'tracts', the US agreed to build up Ethiopia's army, providing training and equipment for three divisions of 6,000 men – a deal worth $5m. But this, in the Emperor's eyes, only marked the beginning. He was aiming for an army of 40,000 – a force, American military experts judged, totally out of proportion for a country facing at this stage of its history no significant external challenges. With its underdeveloped economy, they pointed out, Ethiopia would be unable to support an army of that size unaided. It would be better off with a lightly equipped mobile force that could be sent swiftly to crush uprisings in the provinces. 'No encouragement should be given to expand or modernize the Ethiopian forces,' the Pentagon advised.[14]

Their argument made sense, but Haile Selassie steamed on regardless. Africa would rarely throw up another leader so skilled at getting what he wanted from what might be regarded as a position of weakness. The years spent waiting for the throne to come his way, the humiliating exile in Bath, had taught the Emperor the value of patience. 'A Shewan swallows years after he chews,' runs an Amharic proverb. Having seen off both Mussolini and Churchill, Haile Selassie knew that persistence can be the politician's most formidable weapon. Loudly demand an immediate answer, and the response risks being set in stone. Press quietly and relentlessly, and you will eventually get your way, for your opponent will grow weary of the debate.

At every meeting with a US ambassador, during every trip to America, the Emperor's ministers and diplomats would complain that their country was being neglected, its determination to support Washington in the global battle against Communism sorely undervalued. By allowing Kagnew Station to operate, they argued, Ethiopia was putting herself at enormous risk, laying herself open to the threat of punitive action by anti-Western Arab neighbours. The very least Washington could do, surely, was to give her loyal ally the weapons to defend herself. Out of their depth, American negotiators floundered and succumbed. 'By the time our leaders were dealing with Haile Selassie, he'd been in that job for 15 years. Most of our people had been in their jobs for just 18 months,' recalls Crocker.[15]

Crocker, who was later to negotiate with Zairean dictator Mobutu Sese Seko, another supposed bastion in the fight against Communist infiltration, far preferred the Haile Selassie way of conducting business. 'Every conversation with Mobutu was unpleasant because you were reminded of all the things you had not done for him. It was like buying an antique map on the banks of the Seine, a haggling match, really down in the gutter. The Emperor had a more dignified way of doing things. He was very skilled.' But, at heart, the two African leaders' tactics were based on the same simple principle: Squeaky Wheel Syndrome. The potentially unreliable ally always wins more aid than the nation that falls neatly into step behind its chosen superpower. Like Mobutu, Haile Selassie immediately understood that the most effective method of grabbing Washington's attention was to flirt with the enemy.

The blackmail game followed an established routine. Faced with none-too-subtle hints from Ethiopia – Addis was disappointed, Addis was keeping its options open, Addis felt betrayed – a nervous American ambassador would fire off telegrams to Washington urging military spending in Ethiopia to be increased, fighter jets supplied, a navy established. The State Department would back the ambassador, only to find its way blocked by the Pentagon, which would point out that the US had more pressing commitments closer to home. The Emperor would then make a high-profile visit to the Eastern bloc, returning laden with promises of Communist funding. Another batch of anguished messages would fly across the Atlantic – which ambassador wanted, after all, to be remembered as the man who 'lost' Ethiopia? – and the order from a panicked Washington would go out: more military aid for Ethiopia.

The manoeuvre worked superbly in 1959 when, angered by US support for the British notion of a Greater Somalia, Haile Selassie garnered $100m in credits during a visit to the Soviet Union. Appalled, Washington stepped up its military training programme in Ethiopia to cater for the desired 40,000. 'Rent' on Kagnew was now costing Washington $10–12m a year in military aid, in return for which a grateful Haile Selassie agreed to cede an additional 1,500 acres of land. No one would have guessed it from the peeved expressions on the faces of Ethiopian dignitaries arriving in Washington, but Ethiopia, soaking up to 60 per cent of US military aid to Africa, was the superpower's biggest aid recipient on the continent.

The Emperor's expectations stretched well beyond the military. He wanted Washington to take the lead in transforming feudal Ethiopia into a modern state blessed with schools and universities, hospitals and clinics, its own national airline. Given that he was regularly spending between 25 and 30 per cent of his budget on defence – a staggering percentage given that his foreign friends were already footing most of the military bill – he could not fund these projects himself. Washington must provide, and so it did. Not only was the US Ethiopia's major economic partner, buying 70 per cent of its coffee, but it funded agricultural colleges and locust control centres, veterinary projects and commodity improvement schemes. It used its clout to ensure the World Bank and other financial institutions lent Ethiopia hundreds of millions of dollars, sent Peace Corps volunteers to teach in Ethiopian schools, and educated thousands of young Ethiopians in its American universities. The ties that bound the two countries together were being wrapped ever tighter. And the extent to which Washington was ready to go to ensure its favourite's survival was about to demonstrated in concrete form.

In December 1960, two weeks after the Emperor had left Addis on an extended tour of South America, his Imperial Bodyguard took control of the airport, rounded up the royal family and key ministers in the Imperial Palace, and cut communications with the outside world.

The coup attempt was, ironically, the by-product of the very US-funded modernization drive Haile Selassie had launched. Returning from military academies and Western universities – where their courses were often generously paid for by the Emperor himself – a generation of young officers and bureaucrats looked at their own hidebound society, the nepotistic royal court, and judged it all stale and corrupt. Watching Africa's newly-independent states kicking into life, promising their citizens rights and benefits unknown in Ethiopia – supposedly the freest of African nations – these young men felt the status quo had become intolerable. Disappointed and bitter, they blamed their frustration on their former sponsor, the very man who had opened their eyes to Ethiopia's backwardness.

When, on the first day of the coup, the Crown Prince, Asfa Wossen,[16] went on the radio to denounce the fact that life for the average Ethiopian had not changed for three millennia – that Queen of Sheba vision of history once again – and promised radical change, it seemed as if Haile Selassie was already history. US diplomats ordered the embassy's classified files shipped up to Kagnew Station for safekeeping.

But, by the second day, the situation had grown murky. The main body of the army had not joined the Imperial Bodyguard. At a crisis meeting, US officials agreed that the time had come to meet the obligations spelled out in the Mutual Defence pact signed between the two countries. US advisers were sent to stiffen morale at army headquarters, where a handful of loyal Ethiopian generals, summoning reinforcements to the capital, were in danger of collapsing from exhaustion. Soon American pilots were roaring over the Imperial Palace, trying to frighten the rebels inside by breaking the sound barrier. 'Up to that point, it could be said that we were neutral,' commented the US embassy's army attaché, 'but afterwards there could be no question but that we were committed.'[17]

US help went a lot further than advice. Haile Selassie, in Brazil at the time, only heard his throne was in danger because the Americans allowed loyalist generals to transmit a warning from Kagnew Station. Rushing back to Ethiopia, the Emperor stopped off in Liberia, where he used US Air Force facilities to liaise with those preparing a counterattack. The Americans became so embroiled, shuttling between the two sides in a vain attempt to negotiate a ceasefire, that US ambassador Arthur Richards and two of his aides narrowly escaped being killed when the Imperial Palace was eventually stormed. Soon after, the despairing putschists shot their hostages, killing 15 ministers, secretaries of state and deputy ministers to ensure 'Ethiopia should never be the same'. Arriving in Asmara, a jittery Haile Selassie used Kagnew Station's facilities to communicate with his generals and it was only after Richards had personally assured him the situation was under control that he agreed to fly south. When the Lion of Judah landed in Addis, the crowd of dignitaries gathered to greet him included Richards and his military attachés. Given the key role the Americans had played in saving his administration, it was only right they should share this moment of triumph.

The outcome, as far as Washington was concerned, had been a happy one, but the coup attempt had highlighted both how fragile the Emperor's hold on power really was and the dangers of being too closely associated with a fading ruler.

Like so many African autocrats, Haile Selassie always showed more interest in the surface than the substance of change, delighting in prestige projects – the luxury Hilton Hotel, the spanking new university – that drew gasps from foreign visitors while changing nothing for most of Ethiopia's population. American officials fretted over how little of their aid reached the provinces, swept by regular famines, feeding instead a vast bureaucracy of well-heeled civil servants who rarely strayed outside Addis. Still shockingly backward even by African standards, Ethiopia was crying out for land reform and genuine democracy, instead of the simulacrum of parliamentary decision-making the Emperor had introduced to conceal the fact that all decisions were still being taken by one, very small, man.

But to the Americans' dismay, the Emperor did not seem to have got the message. As the coup leaders were hunted down – some committing suicide, others executed and left hanging from gibbets for the crowd's delectation – Haile Selassie announced that there would not be the 'slightest deviation' from his set path. The new government named to fill the gaps left by the executed ministers was recruited from the same batch of aristocratic families that traditionally served the Emperor, while the educated, unhappy bourgeoisie went largely ignored. 'If the Emperor does nothing ... it is not unlikely that a similar recourse to violence will reoccur at some future date, possibly with greater success,' a worried US ambassador predicted.[18] By supporting Haile Selassie so assiduously, the Americans belatedly realized, they risked being identified with all his empire's ugliest features. 'Our assistance programs, especially military assistance, have identified us to a disturbing degree as supporters of an archaic regime,' the US embassy's economics officer warned.[19]

It was business as usual: the putsch attempt presented the Emperor with the perfect excuse to demand – and get – yet more American aid. Three years later, when war broke out on the border with Somalia, he resorted to his usual tactics to win attention, dangling the prospects of defection. 'The United States must either give us the assistance we require or we might have to deal with the Devil himself to save our country,' warned

General Merid Mengesha, the Defence Minister.[20] Terrified of what could fill the vacuum if it pulled out, the US once again upped its spending, also agreeing to send in scores of counterinsurgency experts to help crush the growing Eritrean rebellion.

Both the Emperor and his US friends had missed the fundamental lesson of the 1960 abortive coup. The fact that one of Haile Selassie's first actions was to placate the army by raising wages offered a hint as to where power was shifting. Thanks to Kagnew Station, the Emperor had forged a relationship with Washington that had given him exactly what he wanted. Between 1946 and 1972, Ethiopia received over $180m in US military aid.[21] Sacrificing economic progress on the altar of military expansion – paid for by an obliging ally – he had built the strongest army in black Africa. But in so doing, he had taken his seat on a very hungry tiger, which promised to devour him the moment his grip began to falter.

In a society in which political activity had been banned, a tiny intelligentsia regarded entry into government service as its highest ambition, trade unions were rigidly controlled and the Church allotted the head of state semi-divine status, challenges to the system could only come from one direction. 'By building up the army, the US made Haile Selassie's overthrow a matter of time,' says analyst Marina Ottaway, who was working in Addis Ababa and witnessed the eventual fall-out first-hand. 'In Haile Selassie's Ethiopia the army was the only institution that functioned. It was the only organization that could fill the vacuum the Emperor had deliberately created around him.'[22]

CHAPTER 10

Blow Jobs, Bugging and Beer

'I was 20 years old and damn near indestructible.'
A Kagnew veteran reminisces

On May 3, 1967, George Zasadil, a gangly young GI of Czech extraction from Chicago, stepped off the plane in Asmara, Ethiopia, after a long trip from Boston via Athens and Cairo. The raw 20-year-old, 'a virgin in terms of most experiences, if not sexually', felt nothing but dismay. Having nearly missed both his flights because he'd been so busy sampling the local nightlife, Zasadil, something of a natural rebel, hadn't bothered reading the army handbook introducing new arrivals to Ethiopia. Bound for Africa, he had formed a clear mental image of the lush tropicana that awaited him. Gazing around him now, 'Zazz' – as his army friends would know him in future – brushed away the flies and breathed in the acrid aroma of Asmara: a heady mix of horse manure, eucalyptus, burning charcoal stoves, spices and diesel fumes. He took in the dry, bleak plateau and stark red earth. 'This was desolation. I'd never seen earth that looked that colour. I was used to soil. My first reaction was "What have I done to deserve this?" "Why has my government done this to me?" and "Boy, have I made a mistake." It was pure culture shock.'[1]

It wasn't as though Zazz had any interest in Africa per se, or the US army's operations in Ethiopia. A college drop-out who'd been sent his draft papers the year before, he had confidently assumed he would fail his physical, thanks to major knee surgery. Disconcertingly, he had passed, and suddenly Vietnam loomed on his horizon. 'I didn't want to go there, I didn't want to die. I thought the odds were I wouldn't be around much longer if I went to Vietnam.' So when a recruiter suggested sitting tests for the top-secret, elite Army Security Agency (ASA) as a way of bypassing 'Nam, Zazz was all ears. The only downside, it seemed, was that the ASA signed men up for four years' duty, instead of the usual two. Still, what were two years of a life, when weighed against the threat of premature eternity?

He didn't discover he had been suckered – the ASA *did* send men to Saigon and there were other, three-year alternatives to a Vietnam posting – until it was too late. Perhaps the fact that so many of his colleagues had followed exactly the same path as Zazz, Vietnam-dodgers talked into an unnecessary four-year stint, went some way to explaining the extraordinary atmosphere that reigned in Kagnew Station in the late 1960s. The 'yellow-clawed chicken fuckers', as the ASA men were irreverently known to the rest of the military – a reference to the shoulder patches they wore showing a snowy eagle clutching bolts of lightning in its talons – had been bamboozled. In a minor way, admittedly, but one that could definitely be worked into a feeling of grievance over a few Melotti beers at the Bar Fiore. Throughout their special training back in the US, they had been told they were the 'top 10 per cent', valued for their superior intelligence, problem-solving capacities and communication skills. Well, the army could go hang, if it expected neat beds, sharp haircuts and sharper salutes in Kagnew. If they were so special, then army rules need barely apply. 'Kagnew,' chuckles Zazz, in his gravelly, lived-in voice, 'resembled a cross between Animal House and MASH.'[2]

One of a consignment of 'Norman New Guys' on whom the hoary station veterans could amuse themselves with non-sensical errands and macho initiation ceremonies, Zazz was joining a Kagnew in its prime. By the 1960s, $69.5m in invest-ment had already gone into Kagnew – pronounced 'Kaynew' by the locals – even if most members of the US Congress remained blissfully ignorant of the base's very existence.[3] It held 4,200 men, not counting family dependants. If the main residential base, with its clubs, sports facilities and trademark clock tower, clustered in the centre of Asmara, the entire station sprawled far across the Eritrean highlands. Nineteen operational sites, or 'tracts', held 185 buildings, nearly 700 antennae and embraced 3,400 acres of land. Kagnew's holy of holies, simply known as 'Stonehouse', contained a pair of giant satellite dishes whose constituent parts had been trucked up the road from Massawa, laboriously winched over intervening railway bridges, before being reassembled on the plateau. The biggest dish, 150 ft wide, weighing 6,000 tons and worth $600,000, was estimated at the time to be the largest movable object ever built. It was visible from 30 miles away.

So many facilities, so many men, to do precisely what? Read the old official brochures and you will learn that 'Kagnew', which means 'to bring order out of chaos', was the name of an Ethiopian commander's horse which galloped riderless towards the Italian enemy during the battle of Adua. They will also tell you that the base adopted the head of a horned antelope as its insignia in memory of the deer that once grazed the Eritrean highlands, hunted into near-oblivion by the GIs. Rather less detail is offered as to Kagnew's function, which, we are told, was to relay communications from US diplomatic missions around Africa and pass messages from Europe and the

American mainland on to the Far East and Indian Ocean – a function even the most uncurious of observers might suspect didn't require over 4,000 men to perform. As for Stonehouse, unveiled in 1964, it made 'an important contribution to man's expanding knowledge of the mysteries of outer space',[4] a superbly unenlightening phrase. Time has not loosened offi-cialdom's tongue. Nearly three decades after Kagnew's closure, a paltry six pages of information on the project have been released under the US Freedom of Information Act. In them, the censor's black felt-tip pen scores thickly through every sen-tence of interest. An inquiry to the press office of the National Security Agency, the ASA's parent body – described today as 'America's most secret agency' – yields a swift, polite brush-off: 'Unfortunately, Miss Wrong, we have no information to provide at this time,' I was told by a public affairs officer at Fort Meade, the NSA complex on Washington's outskirts. 'It's been good working with you.'

The laboured explanations don't wash now, and they didn't wash in the 1960s and 1970s either. Locals always sensed there was more to Kagnew than met the eye: Ethiopian students staged protests against what they believed was a secret missile storage site, Eritreans whispered amongst themselves about clandestine gold mining. Both groups were seriously off track. Stonehouse's job, in fact, was to monitor the Soviet Union's expanding missile and deep-space programme. Rotating on its pedestal, the main dish could pick up electronic signals from otherwise inaccessible Soviet regions which had hurtled into space, ricocheted off the surface of the moon and bounced back to earth. By analysing telemetry – the data sent by a travelling missile to its launch centre – intelligence experts could work out a missile's engine type, guidance performance and fuel consumption, establishing a detailed picture of the range and accuracy of the Soviet Union's network. The same principle applied to the probes Moscow was starting to send to the moon, Mars and Venus. So detailed was the information from deep space picked up at Stonehouse, on virtually the same longitude as the Soviet ground station in Crimea, Washington could gauge the weight of the Mars launch vehicles, for example, or pinpoint whether the Soviet lunar spacecraft was using retrorockets.

But Stonehouse was only a part of Kagnew and the base's interests extended far beyond the Eastern bloc. Men sent here had been trained at Fort Devens in Massachusetts, where they had learnt the skills of the technological spy: how to read morse, intercept radio signals and decode encrypted messages. All those talents were put to good use, for Kagnew was eavesdropping on communications from the Mediterranean, Middle East, Near East, Sub-Saharan and Southern Africa. Notional ally or enemy, Kagnew monitored – or 'strapped on' – regardless. 'Not everyone we were listening to was a bad guy,' acknowledges the improbably-named Tom Indelicato, a former morse code interceptor. 'In the day of the Soviet Bear no one trusted anybody, not even the Brits.' Donning their headphones, the men would sit at their stations turning knobs, slowly trawling the atmosphere for morse exchanges, telephone conversations or radio signals going out on frequencies which had not officially been assigned by the UN's International Telecommunication Union in Geneva and were therefore considered suspect. When they thought they had found one such 'illicit communication' they would 'dial down' on the frequency to pick up the detail. If the communication proved to be in a foreign language, a linguist would be called over to decide whether it merited further investigation. If it were in code, hours could be spent transcribing gibberish: the meaningless groups of letters that went to make up an encrypted signal. Then a crypt analyst would set about trying to unscramble the system. 'The trick was to find the flaws or "spikes", patterns we could identify which would allow us to break the code,' says Zazz. 'We were computer analysts, basically.' It would then be up to headquarters back in the US to decide which traffic merited closer attention. The monitoring took place around the clock, and when the men struggled to stay awake during the long night shifts, known as 'tricks', they would pop some methedrine, bought at the pharmacies in town.

Much of Kagnew's work required the kind of abstract, problem-solving abilities possessed by the gifted chess player or obsessive crossword filler. But language skills, whether innate or acquired during intensive courses at the Defence Language Institute in Monterey, were also highly prized. There were 6–8 French linguists, whose job was to listen in on Francophone Africa. In the 1960s, as newly-elected Prime Minister Patrice Lumumba flirted with Moscow, the former Belgian Congo was a particular concern and 10 Kagnew servicemen were specially assigned to track the communications and movements of the white mercenaries hired to fight in Katanga. There were experts in Italian, German and Russian, Urdu and Swahili: 20 languages represented in all. The Middle East merited 20 Arab linguists, who were placed on standby when it looked like they might be parachuted in to the desert to track the Arab–Israeli war.

Details were rarely discussed, for the men had been indoctrinated early in the principle of 'need to know'. If you didn't need to know, you didn't ask and you certainly didn't speculate about the larger political and geostrategic picture into which your work fitted. 'People would go on loan to Stonehouse from our section, work, come back and never talk about it. We didn't know what went on there and we didn't ask,' says Zazz. However much of a strain the men found it, thoughts and opinions on professional duties were supposed to go the same way as the paperwork: placed in a 'burn bag' and reduced to ashes at the end of a shift.

They were global snoopers, keepers – or rather takers – of the world's dirty secrets. Some confidences weighed more heavily than others. One piece of eavesdropping no one of Zazz's generation would ever forget was the moment on June 7, 1967, when Kagnew picked up the panic-stricken radio communications from the USS *Liberty*, a spy ship sent into the waters off Israel's coast as Tel Aviv launched its six-day war on Egypt, Syria and Jordan. Determined to secure victory before any international power could intervene, the Israelis sent fighters and torpedo boats to disable the *Liberty*'s communications equipment, killing 34 servicemen in the process. Kagnew men listened aghast to the explosions, the cries, the unanswered appeals for help from their colleagues as the *Liberty* was strafed, napalmed and torpedoed by America's closest ally. 'We were in contact with the guys on the *Liberty* and had it on our speakers. It was the first time I'd ever wept in front of another man,' recalls Zazz.

With work-related talk ruled out on the base, perhaps it was inevitable that trivia should take over. Aware of the need to keep relations sweet with its Ethiopian hosts, the army did its utmost to keep personnel amused on post, developing a near-obsession with providing an ever more sophisticated range of leisure facilities and recreational diversions. 'Kagnew,' as one EPLF cadre later remarked, 'was like a medieval town – everything there but behind closed walls.' At Tract E, the main base and barracks near the old Italian cemetery, the ASA had done its best to reproduce small-town America. White small-town America, that is, because if black GIs featured disproportionately amongst the troops dying in Vietnam, they were rarities in safe Asmara, a fact attributed to the racial sensitivity displayed by an Emperor who did not appreciate being too closely associated with 'subjugated' blacks. The army's Kagnew leaflet was clearly designed to lure candidates wary of what was classified as a 'hardship' posting. It read rather like a brochure advertising holidays in the sun. 'Basically, you will find Kagnew Station similar to any small community back in the States,' new arrivals were, rather improbably, promised. On top of a school for dependants, dry-cleaning plant and chapel, Kagnew, it boasted, had a 346-seat motion picture hall with the latest sound equipment, a 10-lane bowling alley, gym, swimming pool, court facilities and a library, its own post office, veterinary service and craft shop. For those from less privileged back-grounds here, suddenly, was a wealth of opportunity. You could play in a band, disc-jockey for the Kagnew radio station, shoot skeet, edit a newspaper or fool around reading the news for the in-house television channel. If all that seemed too energetic, you could go to the club, play the slot machines and wallow in cheap beer. Either way, the message ran, there was really no need to venture beyond the clock tower and main gate separating Kagnew's walled-in world of bungalows, clipped lawns and tarmac driveways from the great outside. 'The army really encouraged you to drink and gamble,' says Dave Strand, who worked as an analyst. 'They wanted you to stay on post, get drunk, put your money in the slot machine and not go downtown and get into trouble.' Some complied, sitting out 18-month tours without venturing off base. But for the most part, he recalls, 'people drank and gambled, but they STILL went downtown and got into trouble'.

This was the 1960s, after all, decade of free love, the Rolling Stones and LSD, the time of Jack Kerouac, Jimi Hendrix and Hunter Thompson. It was the pre-AIDS era, when promiscuity might bring moral opprobrium but did not carry a possible death sentence. These were bright boys with attitude, cash and lots of time on their hands and even up on 'the mountain', as it was known, word filtered through of sex and drugs and rock and roll. Many had been to college and they brought the wild fraternity party spirit over with them from the States. With an overwhelmingly male population, just out of its teens, you could almost taste the tang of testosterone in the air.

There were girls for the taking and no one would lift an eyebrow if a serviceman dated a teenage 'café latte', fair-skinned evidence of misbehaviour by a previous, Italian generation of visitors. 'Sweater girls', as the women from the Italian-built textile factory were known, would be bussed in to dance with Kagnew personnel when bands played at the clubs on the base. Otherwise the men went hunting. After dipping their hands into the box of condoms kept at sign-out, they would be off to the Fiore, Blue Nile or The Green Doors to pick up prostitutes, order rounds of Melottis and knock back shots of liquorice-flavoured *zibib* – something of an acquired taste. For many, the Kagnew Station tour would be spent in a blurred alcoholic daze, in which a few exploits: racing through town standing upright in a stolen garry cart, throwing up over a bar-girl's balcony, catching bed bugs, releasing a pig in Asmara's mosque, or waking up in someone's rented apartment to find naked girls strewn across the sofas – stood out with particular clarity. Alcohol wasn't the only way of losing your mind. A kilo of grass cost just $20. Mike Metras, a Swahili-speaker who was twice sent to Asmara, comes across as one of the more sober servicemen. But even he remembers, with a fond laugh, helping four colleagues one evening to roll 600 joints for a Valentine's Day party. 'I sat back and watched the effect.'

Like all communities that pride themselves on their wild iconoclasm, Kagnew had its own brand of rigid conformism. Daily life was dictated by two sets of rules: the army's, to be taken with a hefty dose of salt; and the servicemen's, far more rigid in its way. If you wanted to enjoy your time at Kagnew, it paid to be conversant with both. So you could forget about joining the Top 5 Club, where they played country and western, if you weren't between grades E-5 and E-9. The rock and roll of the Oasis Club was the place for low-lifes like you. Things could get nasty if you ignored that basic fact. And if you went to the Oasis, you should never make the mistake of sitting just anywhere, or moving around as the fancy took you. Seats at tables had to be *earned* – strictly taboo until a personal invitation was issued – and once you had forged those friendships, won your spot, you stuck with it.

But above all, you had to learn the lingo, as peculiar to Kagnew as a taste for I W Harper's bourbon, the on-base tipple of choice. You would never hear it used anywhere else in your life, and what it lacked in subtlety it made up in directness. New arrivals were 'Normans', 'nugs' or 'newks'; military dependants were 'brown baggers'; 'ditty boppers' the servicemen responsible for morse intercept. Men who had signed up for more than one term were 'lifers'. Anyone who got an unhealthy kick out of applying military regulations was a 'puke'. 'Not all lifers were pukes,' according to Zazz, 'but all pukes were lifers.' At the end of a 'trick', or shift, the aim was to end up 'downtown', where you could spend your 'gon zip' (money) in the 'bosch' (market) or pick up a 'bar whore' and 'get scrufty' (get laid – according to Zazz – 'whether you had to pay for it or not'). Otherwise, you could join the list of 'Mama's sons': servicemen who'd been initiated in the delights of fellatio by Mama Kathy, a local madame. Weekends could be spent at Massawa in a hotel known appreciatively as 'the Four Floors of Whores', or in Keren, where the army ran another hostel. If someone gave you a hard time, you'd give them the 'on my root' or 'on my bag' sign, an obscene Kagnew invention which consisted of raising both arms in a V shape, then swinging forearms and hands down to point graphically crotchwards. 'It was a smooth, very coordinated movement,' recalls Zazz. 'People did it to each other if they met in the street. When you left, it was traditional to line up on the steps of the plane and do it in synchrony, and the guys who'd come to see you off would do it back.'

Within a few days of arrival, Zazz had worked out which crowd he wanted to hang out with. It was the group that proudly labelled itself the 'Gross Guys' and its survival philosophy was neatly encapsulated by one veteran as follows: 'Stay drunk, fuck all the whores in Asmara, create havoc wherever you are, stay drunk, fuck all the whores again, get drunk again, stay clean for your last month, get drunk and leave drunk.'[5] The Gross Guys' leader in debauchery – undeclared but recognized by all – was Lawrence D McKay, baptized 'Spook' by the favourite prostitute who deemed him '*sebuke*', Tigrinya for 'good'. Unlike his followers, Spook was a lifer, a man who had discovered that his particular talents were better tolerated inside the army than they ever would be in civilian life. 'He was a conman, a loan shark, a gambler and a card sharp,' recalls Zazz with indulgent affection. 'He drank about a bottle of bourbon a day. The ASA had corrupted him, because that was exactly what he wanted.' His followers hailed Kagnew's version of Sergeant Bilko as 'the King of Gross', or 'His Grossness'. But Spook was more modest. 'I ain't perverted,' he used to say, 'I just know what I like.'

What he liked was Rosie Big Tits, known as RBT for short, a bar whore who would unashamedly service any man, or men, willing to shell out. He liked fleecing servicemen who fancied themselves hot-shit poker players, logging his winnings in a black book for collection on payday. And he liked daring his friends to see how far they would go in setting new standards of unacceptable behaviour. One time at the Oasis Club, a serviceman tried to impress Spook by pulling out his penis and filling a shot glass with urine. 'There, is that gross enough for you?' he asked, slamming it down on the table. He wasn't even in the same league. Spook picked up the shot glass and downed it. Spook entered the realm of legend with the 'Triple Crown', a personal best most men felt no desire to equal. Stage one involved performing oral sex on RBT, stage two involved cunnilingus on a fellow bar whore straight after she'd had sex with someone else. Stage three, which prompted Spook's competitor to abandon the contest, involved walking up to a garry-cart horse, lifting its tail, and running his tongue around the rim of its anus. Zazz was to be one of the last new entrants to the Gross Guys and he won his spurs in the heat of a Massawa bar, where the flies were so fat and lazy you could catch them in your hand, shake them like dice and slap them down on the table. 'There was a bar whore in her 40s sitting at our table, old, skinny and ugly, and I'm just 21 years old. I crawled between her legs and went down on her right there in the bar while it was in operation.'

At some stage in their training, no doubt, the men had been lectured on the importance of cultural sensitivity in a strait-laced local culture, the desirability of winning hearts and minds. No matter. For many, the greatest fun was to be had 'twitching' the Eritreans who worked as cleaners, waiters and odd-job men at the base. The GIs had quickly noted how fastidious the locals were about bodily functions – nothing was more revolting to them than an audible fart – and the aim was to stage a jape so disgusting it would send the 'Ethie' concerned into physical spasm as he tried to ward off evil spirits. A simple raspberry, blown with split-second accuracy, would have a startled Ethie drop what he was carrying. Hours of amusement were to be had driving around Asmara, purring up behind Eritrean cyclists and blowing mouth farts so loud the riders fell off or careered into walls. No one ever discovered the identity of the 'phantom shitter', who left his deposits on officers' desks and inside shoes waiting to be shined. But Zazz gloried in the day a fellow Gross Guy hit on the wheeze of smearing peanut butter inside his underwear and then, in front of the white-faced houseboy, dipping his finger in the lumpy brown mess and licking it off with moans of appreciation. A slightly more elaborate version of this trick involved sucking red tomato ketchup with gusto from a sanitary napkin.

What is it about Anglo-Saxon males and their bottoms? A Frenchman or a Spaniard does not seem to feel the same compulsion to bare his bum, a gesture psychoanalysts would no doubt interpret as blending primeval defiance with homo-erotic bonding. Like generations of fraternity fellows before and after them, the Kagnew men delighted in exposing their buttocks to the world. One night Zazz and his friends climbed several of the 100-foot metal pyramids holding up Kagnew's rhombic antennae so they could lower their pants in unison, setting a new record for high-altitude mooning. One of them fell off and broke his arm. An even more dangerous exploit was performed on a much-frequented river crossing on the main road to Massawa – baptised 'Moon River Bridge' – in the middle of the afternoon. Zazz and colleagues captured the moment for posterity on film: a three-tier, seven-man, simul-taneous moon. You can recognize Zazz in the photograph because he's the one with the cigarette clenched in the hand parting two white cheeks for the lens. 'We were young, we were stupid, we were drunk,' he guffaws, marvelling now that no one was killed staging the prank.

How typical were the Gross Guys of Kagnew Station? When Zazz set up a commemorative website on which their exploits held explicit pride of place, it attracted a death threat from an irate Eritrean and a few complaints from ex-servicemen who insisted the sexual shenanigans and alcoholic excesses described applied to a mere 10 per cent. Zazz scoffs at what he regards as false memory syndrome. Sure, Kagnew men prided themselves on their professionalism and productivity – reporting for duty under the influence was not regarded as cool – but if they worked hard, they played harder. 'Those guys came back and they didn't tell their wives what they'd got up to in their youth. Now they don't like to admit it. I've got nothing to hide. All my children know what I did. In my tenure, I'd put the percentage who had a wild time in Kagnew as being closer to 50 or 60 per cent.'

Would the Kagnew men have behaved like this back home? Of course not, but the whole point was that they weren't back home. As with a Club 18–30 holiday, the frenetic debauchery of Kagnew Station was based on the sheer anonymity of alien surroundings, a city where there was no high school teacher to shock, no neighbour to appal and none of the rules that ordinarily hemmed life in for young American males. 'If you were to take any bunch of kids and tell them: "You know all that stuff you weren't allowed to do back home because you weren't old enough, the drinking and fornication? Well, here it's OK," what's going to be the result?' asks Bob Dymond, a former serviceman. 'That's what they do. Take any army in the world and it's the same.'

Indelicato, who was pressurized into joining ASA by a father scarred by the butchery of the Normandy landings, sees something else behind all the foolishness, floppy haircuts and sloppy salutes: an unvoiced frustration felt by youths who had balked at a discredited Vietnam campaign but were uneasily aware, as they lit their own farts and chased Eritrean houseboys down corridors, that their contemporaries in the Far East were experiencing all the desperate intensity of real war. Kagnew boys were only issued with guns and live ammunition during the rare alarms over possible anti-American rebel activity. The biggest piece of weaponry the base boasted was its salute cannon. 'We weren't soldiers, we were technicians. Essentially, we were REMFs – "Rear-Echelon Motherfuckers" and for some people, there was a stigma attached to that and you had to have a healthy psyche to cope. A lot of the high jinks were displacement activities. In Vietnam, our energy would have gone on other activities. We had no way of spending it apart from drinking and whoring.'

What strikes one, talking to veterans, is not the gallumphing obviousness of their crudity, but how little Eritrea and its subtleties featured in what, for many, were to constitute the most vivid episodes of otherwise humdrum lives. Forging the intense friendships of youth, they barely registered the outside world. Today, former servicemen profess to have developed real fondness for the Eritrean workers who cleaned and catered for them, while disliking the Ethiopian soldiers who, claims one Kagnew man, 'walked around like the lords of all creation'. Yet with the impatience for detail that characterizes the American tourist abroad, they sweepingly referred to all as 'Ethies', seemingly oblivious to the fact that national identity in Eritrea was a touchy subject. The official alternative, equally revealing in its implications, was 'foreign national'. 'I always find it interesting that we would apply this term to people living in the country where we were guests,' remarks Dymond, who was to distinguish himself from his Kagnew friends by marrying an Eritrean-born Italian, an experience that opened up a world of understanding. For Zazz, the moment where he registered that the head waiter at the Oasis Club was a rounded human being encountering prejudice from a system propped up by Kagnew had the force of a religious revelation. 'We got talking and he explained that everyone in his village had pooled their money and paid for him to go to university to qualify as a civil engineer. I said, "So why aren't you working as an engineer?" and he said, "Haile Selassie won't let me. I'm a Moslem and I'm an Eritrean." All of a sudden I wasn't talking to a waiter, I was talking to an educated man. The government was treating people horribly. It was dead wrong.'

Looking back, US diplomats who served in Asmara like to dwell on the good works performed by Kagnew men, the orphanages opened, courses taught at Asmara University, Eritrean hospitals funded. Kagnew undoubtedly had its charitable side, but only the obtuse could fail to register that the attitude of Eritreans and Italian settlers towards the Kagnew men was not always one of gratitude, let alone indulgent amusement. The moral chasm between 1960s and 1970s America and the Eritrea of the day – puritanical, conservative, God-fearing – was too wide to be easily bridged. GIs who went horse-riding in the countryside came to anticipate a stoning by village children. In a strait-laced Eritrean family, a daughter who dared date a serviceman was regarded as bringing shame on the household. And there were plenty of young Eritrean males who enjoyed nothing better than loitering in Asmara's bars in the hope of a scrap with the GIs, arrogant despoilers of their women, supporters of a loathed Ethiopian occupation force. Eritrean historian Alemseged Tesfai remembers trying to rescue an Eritrean friend, a Kagnew worker who picked a fight at the Bar Mocambo. 'An American soldier had grabbed an Eritrean girl who was belly dancing and my friend went across the floor and nutted him, and all the GIs piled in. My friend was beside himself, totally out of control. When the military police arrived, the Italian owner said the Kagnew men had started it, though my friend was to blame, because he knew the Americans would be able to pay for the damage while we could not. There was a lot of bitterness there.'[6]

The Kagnew authorities did not encourage curiosity over

the bigger political picture. They knew the inflammatory impact half-digested information could have on a claustrophobic, navel-gazing community. 'A word or two on the subject of RUMORS,' reads a warning published in the 'Kagnew Gazelle', the base circular. 'These can be very dangerous. They can hurt you, your buddies, your unit and your families if they are not stopped. On a small post like this, rumors start very easily. If you are not sure of something that was said, find out the truth, if it is wrong, get the rumor stopped as quickly as possible.' Arrivals were obliged to sign a 'statement of non-involvement' pledging no contact with dissident elements and no political discussions with 'foreigners'. But for those who cared to look, the grim realities of Ethiopian rule were always on display in those years.

Growing Eritrean anger at the imposition of the Amharic language, the Federation's suppression, and the brutal crushing of dissent had given birth to a lively rural guerrilla movement, which staged hit-and-run raids on police stations and garrisons – symbols of Ethiopian authority and useful sources of weaponry – sabotaged bridges and targeted high-ranking Eritrean 'collaborators'. Initially set up by Moslem lowlanders with strong links to the Arab world, the ELF's ranks were gradually swelled by hundreds of well-educated young Christians who had distanced themselves from Unionism, spoke the language of Marx and Mao, and wanted to fight. The Ethiopian army responded to their activities with predictable heavy-handedness.

You would never get an inkling of it from reading the entries on either the official Kagnew website[7] or Zazz's irreverent alternative, but 1967, the year Zazz started his tour, witnessed three devastating offensives in Eritrea. Drawing on counterinsurgency training provided by the US and Israel, the Ethiopian army attempted to starve out resistance by displacing the rural communities on which the rebels depended for sustenance. Over 300 settlements were burned, tens of thousands of farm animals slaughtered and hundreds of villagers killed. During one incident in the eastern lowlands of Semhar, 30 young men were burned alive inside a house. In the villages of Kuhul and Amadi, the Ethiopians, using what would become a favourite technique, ordered the locals to collect in one place, then called in the air force to bomb the site. These 'pacification campaigns' were horribly reminiscent of American tactics in Vietnam, and their outcome was just as disastrously self-defeating when it came to winning hearts and minds. 'The Second Division is very efficient in killing innocent people,' an Israeli adviser wearily noted in his diary. 'They are alienating the Eritrean and deepening the hatred that already exists. Their commander took his aides to a spot near the Sudanese border and ordered them: "From here to the north – clear the area." Many innocent people were massacred and nothing of substance was achieved.'[8]

At regular intervals, Kagnew men who strayed off the beaten track, venturing out without their assigned escorts, would stumble into 'the shifties' – an ELF patrol – who, perhaps realizing that these noisy, spoilt Westerners were ignorant rather than malevolent, would undertake a crash-course in political awareness. 'They were always very good to us. They would stop people and say: "This is our flag, this is our anthem, this is our policy," and then they would let them go. ELF activity was an issue all the time we were there, but you had to find out about things for yourself, no one would tell you,' remembers Metras. On occasions, Kagnew Station would be closed and security tightened when one of these polite outlaws was publicly executed by the Ethiopians. Even in the cultural bubble behind the clock tower, word occasionally filtered through of atrocities downtown, of how the Ethiopian army had strung bodies of captured Eritrean rebels from the lamp-posts of Keren and left them to rot as a warning to others. The party line from the station authorities was that the rebels were the 'bad guys', but Kagnew's inmates were too bright not to entertain doubts or draw their own historic analogies. 'I always used to think that this was probably what we Americans were like before we threw you Brits out,' says Dymond.

But for most Kagnew men, the political situation was confined to the subconscious, a shadow that only solidified and made full sense in later years. 'We realized the role Kagnew played for our government and we saw some hatred because we were seen as puppets of Haile Selassie,' acknowledges Zazz. 'But we dismissed it. The prevailing mood was that we were in the army, we weren't there to make choices. Vietnam was foremost in our minds, not Eritrea and Ethiopia.' Disassociation came easy – it was, after all, the unstated aim of all the leisure facilities laid on in this home away from home. 'We were living in our own little world,' says Indelicato, 'it was going on all around us, but didn't have anything to do with us.' When His Imperial Majesty (HIM) came to Kagnew Station for his free dental check-ups, the servicemen would make their truculent dislike for the man they dubbed 'Lord of the Flies', in recognition of the vibrant insect life that started just outside Kagnew's sprayed confines, as obvious as circumstances allowed. 'They'd make us turn out for him and the resentment in the ranks was palpable. Everyone was aghast at what the Ethiopian government was doing – but we were in service. In theory we could have stood up and rebelled but, well, we were callow youth,' says Strand. Protest never rose above the strictly puerile. 'Let's sing a hymn for HIM,' a GI would suggest when servicemen marched into base under the eyes of the Ethiopian guards. And the chorus would rise up: 'We like it here, we like it here. Fuckin' A, we like it here; HIM, HIM, FUCK HIM.'

The Kagnew boys eventually followed their various routes out of Africa, some of them discovering on arrival in the West that they had developed drink problems they would need to shed to enter conventional careers. The legendary Spook left the ASA when his boozing finally became an issue and now works in a Colorado Springs motel. His memory fuddled by all the years of bourbon and *zibib*, he retains almost nothing of his Kagnew days. Dave Strand runs a telecommunications business, while Bob Dymond has become a network engineer. Tom Indelicato joined the police and at weekends dons camouflage to train with the National Guard. He readily admits that it is his convoluted way of trying to make up for the REMF role played in Kagnew, which left him with an still unsatisfied need to prove qualities never put to the test in Asmara. 'If you were in the army between 1965 and 1972 and you didn't do Vietnam, then a lot of people will wonder what the hell was wrong with you. That was easy to live with when the Vietnam war wasn't popular, when it was "LBJ's war". But now that it's being romanticized, the attitude is "why weren't you there?"

'Am I rare in being haunted by Kagnew? Maybe I'm rare in being able to admit it. The fact remains that Vietnam was the war that defined our generation. There are 58,000 names engraved on that memorial wall in Washington, and some of those names were friends of mine.'

The birth of the internet has allowed old friendships to resume and every year a group of Kagnew veterans stages an informal dinner reunion. Most prefer not to invite their wives. Poignantly, a light-skinned Eritrean youth made a surprise appearance at a recent such get-together, clutching a picture of his bar whore mother and looking for the GI father he had never met. He left with the long-nursed question unanswered. 'It could have been a number of us,' says Zazz, with a shake of the head. 'His mother went with a lot of the guys.'

Pulled out in 1968, the irrepressible Zazz headed back to college, entered the restaurant business and tried teaching. Now 57, he lives on a sultry stretch of Florida coastline trawled by Hell's Angels and holidaymakers, where the waving palm trees are the only echo of Eritrea. He works behind the bar in an up-market restaurant where, on Saturday nights, middle-aged blondes with big hair and shiny red talons prowl in search of new husbands. Looking surprisingly spry for someone who has submitted his body to such unremitting abuse, he smokes three packets a day, wakes at the 'crack of noon' and on busy nights, stumbles to bed at 6.00 am – an appropriate lifestyle, one can't help feeling, for one of Spook's acolytes. Sometimes, he fantasizes about returning to Asmara. 'There was something about the air there. I can't define it, but it smelled different. I'd like to smell that again.'

Boarding his flight out of Asmara, Zazz, one of the last representatives of the once-great Gross Guys, remembers that he turned, raised his arms to heaven and swept them down to point at his crotch in the 'on your root' gesture that was born and died at Kagnew. Seen from one angle, a mischievous young man was having a final bit of innocent fun, keeping faith with a circle of friends that had already splintered into fragments. Seen from the angle of a jaundiced Eritrean, the departing GI had summarized the underlying theme of America's role in Eritrean affairs – the equivalent of 'suck my dick' – with uncanny accuracy.

CHAPTER 11

Death of the Lion

'A house built on granite and strong foundations, not even the onslaught of pouring rain, gushing torrents, and strong winds will be able to pull down.'

Haile Selassie, writing in his autobiography

Something terrible happened on November 30, 1970 in Besik-Dira[1]. And perhaps the worst thing about the entire episode was that it was really nothing out of the ordinary.

For months, residents of the tiny village, a community 14 km north-east of Keren which included both Christian and Moslem members of the Bilin tribe, had watched with growing anxiety as Ethiopian troops destroyed scores of settlements in the area suspected of ferrying food and messages to ELF guerrillas in the mountains. The Ethiopian plan, it was clear, was to resettle the peasants in conveniently-located hamlets which could be easily patrolled. On November 21, the situation had taken an ominous turn: Major-General Teshome Ergetu, chief of Ethiopian forces in Eritrea and the man spearheading the scorched earth policy, was ambushed and killed by the ELF on the road to Asmara. Revenge was in the air, and Ethiopian planes flew low over the valleys, warning villagers to move to Ona, a designated 'safe' village on Keren's outskirts. Yet the residents of Besik-Dira stayed put. The decision was not taken lightly. The rains had been good that year and it was time to bring the harvest in: moving now might mean starvation later. Looking round, they noted that the Ethiopians tended to target villages which had been abandoned – it made burning them easier. If Besik-Dira handled the situation carefully, residents agreed amongst themselves, they might come through the crisis unscathed. The women buried their jewellery beneath their huts and the fearful village waited, hoping for the best.

When the Ethiopian soldiers finally arrived, they were met by ululating and clapping villagers, who presented them with a gift of cattle. The soldiers were unimpressed. Routinely issued with only a few days' rations, they were already accustomed to 'living off the land', that old military euphemism for plunder. What did two paltry cows matter to men who slaughtered livestock when they grew peckish and searched huts when they needed pocket money? As far as these angry, nervy men were concerned, Besik-Dira was a nest of sympathizers: an outbreak of shooting from the ELF in the hills merely seemed to prove the point. They herded the villagers into a valley, shot the livestock and set fire to the huts. Then, acting on the simple premise that being Moslem meant being pro-guerrilla and being Christian meant supporting Ethiopia, they ordered residents to divide on religious lines.

It must have been at this stage that the villagers realized they had made a mistake most would not live to regret. They took the hardest of decisions. Presented with a choice between individual survival and group solidarity, they opted for the latter. Whether Moslem or Christian, no one in Besik-Dira, they told the soldiers, knew anything about the ELF. The soldiers ordered everyone into the mosque. 'Start clapping,' they said, 'the army commander is on his way and should be welcomed.' As the villagers obeyed, hands shaking with fear, the soldiers positioned themselves at the mosque's six windows and only door. There was a chorus of ratcheting as the men released their safety catches, cocked their weapons and took aim.

The massacre was carried out with systematic thoroughness. On a signal, the soldiers opened fire, paused to see if anyone was left standing, then launched another volley. Every few minutes there would be a terrible silence as the soldiers listened for signs of life and, buried deep inside the pile of bleeding bodies that had once been a community, the wounded bit their lips. Then the shooting resumed. Meskela Berk, a local woman, spotted her eldest daughter lying, alive, below a mound of corpses. 'She was about to die of suffocation. I removed the bodies from her and put them over one another to shield her. I did the same for my second daughter. My third was too young, so I lay over her to protect her.'

When the soldiers finally left Besik-Dira, over 200 of Meskela's neighbours and friends lay dead. Even then, the ordeal was not over. Many of the survivors fled to Ona. The following day, the Ethiopian army encircled this supposedly 'safe' village and opened fire with its heavy artillery. More than 700 people – most of them women, children and the elderly – died in the shelling. A few days later, the soldiers covered Ona with earth and sand, so that it merged seamlessly into the landscape, a village that had never been.

Besik-Dira and Ona were not dreadful aberrations, the work of army units which momentarily lost control. This was policy. Eilet and Gumhot, Kuhul and Amadi, Asmat and Melefso, Om-Hager, Woki-Diba, Agordat: the list of Eritrean villages and towns where massacres were staged and atrocities committed in the late 1960s and early 1970s, while Kagnew's embarrassed authorities looked the other way, runs on and on. The Ethiopian military, in attempting to bring an obstinate

Eritrea to heel, fell into the trap that awaits any army sent to subjugate a land where every rebel looks like a peasant farmer, any street urchin could be an informer and the friendliest of housewives is probably cooking stew for the boys in the hills. 'If you wish to kill the fish, first you must dry the sea,' an Ethiopian brigadier-general who had absorbed something of Mao's teachings told foreign journalists. But how do you drain the sea without creating, in the process, a new generation of hate-filled fighters haunted by the memory of their hanged brothers, raped mothers, and charred homes? How do you crush a secession movement without doing the guerrillas' recruitment work for them? The Americans failed to find an answer in Vietnam, the French fluffed it in Algeria, the question baffles coalition forces in Iraq today. It would always flummox the Ethiopians. The survivors of Besik-Dira, Ona, and every other massacre fled into the mountains, and membership of the Eritrean guerrilla movement soared.

By the early 1970s, the leader who had once looked ahead of his time was beginning himself to seem atrophied and out-of-date. Now in his eighties, Haile Selassie had succeeded in dragging Ethiopia out of the Middle Ages and into the Enlightenment, era of the absolute monarch, but that still left his empire lagging a few centuries behind the developed world.

In their telegrams home, foreign diplomats noted that the Emperor was fond of quoting from Rudyard Kipling's tub-thumbing poem 'If' at the end of his speeches. Its Boy Scout exhortations must have echoed through his mind as, for the first time since the 1960 coup attempt, he began confronting serious challenges to his rule. *'If you can keep your head when all about you are losing theirs and blaming it on you . . .* [army bogged down in Eritrea, Oromos in revolt, Somalia being militarized by the Soviets] . . . *If you can trust yourself when all men doubt you* [drought in the north, thousands starving to death] *but make allowance for their doubting too* . . . [soaring food prices, students demonstrating, teachers on strike] . . . *If you can force your heart and nerve and sinew, to serve your turn long after they are gone* [army officers demanding pay increases, cabinet ministers manoeuvring for greater powers] . . . *Yours is the Earth and everything that's in it, and – which is more – you'll be a man, my son!'*

The radiant charisma Spencer had once remarked upon had evaporated, leaving behind a very small old man in the early stages of Alzheimer's. The first symptom of this demeaning disease that aides registered was the Emperor's disconcerting tendency to change topics in mid-discussion, swerving off down lost, overgrown conversational by-ways where a confused retinue could not follow him. As his brain cells withered and died, the Emperor's speech became less and less articulate. Information sluiced through his mind like water, leaving no residue behind. He was having difficulty recognizing his own ministers and Aklilou, who now held the post of Prime Minister, found himself repeatedly briefing and rebriefing the man whose memory had once been so impressive. When Zaire's President Mobutu was invited to a state dinner, Haile Selassie had to summon a lackey to ask who was sitting alongside him. On his only trip to China, he was afflicted by an overpowering sense of déjà vu, repeatedly claiming to recognize places he was visiting for the first time.[2] 'He became disreputable and somehow undistinguished at the end,' Spencer told me. 'What brought that about, I don't know.'

Given a system in which responsibilities were shared, approaching senility would not have mattered so much – aides could have quietly taken over the running of the country while leaving the Emperor as figurehead to preside, Soviet-style, over official ceremonies. But Haile Selassie had made a point of centralizing power into his own, now incapable hands. Sensing his own weakness, he leant heavily on his eldest daughter. 'In the day, I am the Prime Minister, in the night, it is the Emperor's daughter,' Aklilou complained, 'and she undoes everything done during the day.'

It was not a propitious moment for a power vacuum to open up. By early 1974, famine in Tigray and Welo had claimed the lives of 100,000 peasants. Haile Selassie had always accepted the gulf between the desperate poverty of the countryside and his court's gilded wealth with equanimity, regarding them as contrasting facets of a natural order decreed by God. But an Emperor in full command of his faculties would have known that articulating such fatalism was no longer acceptable in the age of radio and television. Instead, his floundering government labelled European reports of the disaster 'wishful malice' and one minister went so far as to comment: 'If we can save the peasants only by confessing our failure to the world, it is better that they die'[3] – a remark that spoke volumes about the Ethiopian obsession with losing face. Public mutterings over an out-of-touch leadership – more interested in military spending than feeding the hungry – grew louder. Yet the government seemed incapable of action. Spencer saw his former employer for the last time in February 1974, emerging from the audience deeply shaken. 'I had the sensation, still vivid today, that in leaving the private office, I was leaving the cockpit of a 747 after finding both the captain and the co-pilot unconscious. How was the craft to keep flying?'[4] The lawyer flew to London, where he met the Ethiopian ambassador. 'I told him: "Look, in six weeks I'm sure we are going to see the Emperor gone."'[5]

It took a lot less time than that – the day after his prediction, the demoralized Ethiopian army in Asmara mutinied and garrisons around the country began following suit. Aklilou and his cabinet resigned, but they could not stop what had begun. The army, whose young leaders set up a coordinating committee, or 'Derg', to present its demands, seemed impossible to placate. They wanted higher pay, better food, but above all, they were sick of fighting what was beginning to feel like an unwinnable war in Eritrea. Haile Selassie's pet project had turned into a hungry Frankenstein, ready to devour its creator. He had built the army up to deal with Eritrea, now it had turned on him.

Just as he had abandoned Addis to its fate in 1936, the Emperor now sold out his nearest and dearest, agreeing – in the face of furious protests from Aklilou and his incredulous colleagues – to have his cabinet ministers arrested in the hope that he, at least, might survive. The tactic bought him only a little time. On September 12, 1974, the Emperor's 44-year-old reign came to an abrupt end. Army tanks rolled into position around the palace and a delegation of Derg officials informed Haile Selassie, who had donned full uniform for the event, that they were deposing him on the grounds of corruption and neglect. The Emperor put up no resistance. 'We have tried to serve our country in peace and war. If we must serve it now by resigning, we are willing to do so,' he replied. As the old man was walked towards the exit for the last time, he was watched in silence by his terrified retainers, peering around corners and peeping from behind heavy drapery. Once the doors closed behind him, a chorus of wailing broke out. The Derg had sent a Volkswagen Beetle police car to drive Haile Selassie to his place of detention, a choice that underlined the Emperor's precipitous fall from grace. 'What, into this?' he stuttered, as the front seat was tipped forward to allow him to squeeze into the back. It was his last public appearance and came stripped of all dignity: confused, disorientated, he automatically waved at the youths who ran alongside as the tiny car puttered along. But this time the crowd was not singing his praises. It was shouting 'Thief!' and 'Hang the Emperor!'

For Eritrea, the Derg's takeover briefly held out enormous hope. The new administration named General Aman Andom, the country's most popular military leader, as chairman. A Sandhurst graduate with Eritrean blood running in his veins, Aman believed Addis must abandon its heavy-handed military tactics in Eritrea, opt for conciliation and consider reinstating the Federation. Had he lived, the future would undoubtedly have looked very different. But Eritrea was only one of the many issues on which he soon clashed with the Derg's hot-headed officers. On November 23, fired up by an uncompromising speech delivered by Mengistu Haile Mariam, an ambitious young major who believed Eritrea's separatist tendencies should be crushed rather than accommodated, the Derg voted Aman out of office. Later that day, an army tank drew up in front of the general's house and soldiers opened fire. After holding off his attackers for two hours, Aman is said to have dressed in full military regalia, complete with medals and braid, and shot himself under the chin. That same night, the 59 detained ministers, generals and members of the royal family, including Spencer's old boss Aklilou, were led out of their cells and executed under the floodlights, machine-gun fire ripping through their bodies as the movie cameras rolled. What had been cheerfully hailed up till then as 'Ethiopia's bloodless revolution' had just turned nasty. The Derg declared Ethiopia a socialist state and announced its intention of instituting a one-party system and nationalizing land and key industries. The old order – including an entire social class of landed gentry and businessmen – had been swept away.

Haile Selassie himself was held prisoner for nearly a year, while the Derg attempted without success to persuade him to sign over his Swiss bank accounts. Rare film footage taken during this period shows the deposed Emperor staring, baffled, at the camera. His hair, as uncombed as any asylum inmate's, curls in a frenzied corona around his head, his uncomprehending eyes have the milky glaze of a mad, caged eagle. Finally, in August 1975, the Derg announced that he had died of natural causes, supposedly of complications following a prostate operation. Few believed the official version. Exiled family members said the Emperor had been fed cyanide, rumours spread that he had been smothered with a cushion.

It would be another 20 years before the true account of Haile Selassie's death emerged. At the 1996 trial of Derg leaders on charges of genocide and murder, the court was read the minutes in which the ruling military council – which demonstrated, like so many of the world's most murderous administrations, a bizarre bureaucratic fastidiousness in recording every atrocity committed – agreed it was time to eliminate a man who might one day serve as a rallying point for a counter-coup. 'It has been decided that the necessary means should be taken,' read one entry. A personal attendant recalled how the 84-year-old Haile Selassie fell to his knees and prayed when soldiers appeared at his door one evening, crying: 'Is it not true, Ethiopia, that I have strived for you?' The next time the attendant saw Haile Selassie, he was lying dead on his bed, probably strangled while under anaesthetic. 'There was a smell of ether in the air and His Majesty was not lying in his usual position . . . The shawl that he wrapped himself in when he went to sleep was lying in another part of the room. His face was ghastly and there was a bandage around his neck.'

Macbeth-like, king-killers tend to become obsessed with the question of whether their victims will lie quiet in their graves or issue forth as malevolent spirits to haunt them. There was a curious personal vindictiveness about the way Mengistu chose to dispose of his predecessor's remains. Ordering several graves to be dug in order to spread confusion about Haile Selassie's actual resting place, he told a gravedigger – another witness at the trial – to dig a hole below his office window at the Imperial Palace. While Mengistu watched, the Emperor's small coffin was placed there – head down, some say – liquid concrete poured in and a latrine built on the site. Not only could Mengistu, responsible for a regicide which rivalled in boldness the murder of the Czars, verify at a glance that Haile Selassie remained dead and buried. Each day, when he felt the call of nature, he could express his feelings for the Emperor of Ethiopia, King of Kings, Lord of Lords, Conquering Lion of the Tribe of Judah, Elect of God, Light of the World, in the crudest way known to man.

CHAPTER 12

Of Bicycles and Thieves

'How can we fail, when we are so sincere?'

Sign in an Asmara public library

The Derg's ruthless elimination of General Aman Andom served as a massive recruiting drive for the Eritrean rebel movement. Rightly concluding that an aggressively nationalistic regime which trumpeted *Etiopia Tikdem* ('Ethiopia above all') as its slogan would never be interested in a negotiated settlement, tens of thousands of Eritrean students and high-school pupils abandoned their studies and joined the resistance.

These fresh recruits were no longer all heading in the same direction, for the rebel movement had splintered in the early 1970s, torn apart by disagreements over strategy and ideology. The civil war between the ELF – mainly Moslem, based in the Western lowlands, viewing Eritrea as part of the Arab world – and the breakaway EPLF (Eritrean People's Liberation Front) – dominated by Christian highlanders, secular and socialist in outlook – would sap the insurgency's strength and undermine its effectiveness for years to come. But in 1974, with their numbers soaring and Addis in a state of political uproar, the two movements managed to put their differences aside long enough to launch a staggeringly successful joint push

on Ethiopian positions. By January 1975, the two rebel movements were nibbling at Asmara's outskirts. Terrified Ethiopian
army units sat in their rural garrisons, only daring to venture
out in heavily-guarded convoys.

How could anyone have guessed that, far from representing the beginning of the end of Ethiopian rule, this was
just the first round of a conflict destined to drag on for almost
two decades? With the frontline visible from the centre of
town, shooting audible, and wounded Ethiopian soldiers being
ferried through the city, victory seemed imminent. No wonder
so many Eritreans too old or cautious to take up a gun chose
this moment covertly to embrace the cause. The mid-1970s
were the golden age for an underground resistance movement
in which every member of society, from schoolchild to bar
owner and bank manager, could play their part. Ask the classic
question – 'What did you do in the Struggle?' – and you
will discover that across Eritrea, ordinary people were doing
extraordinary things.

The importance of tapping into that public support was first
recognized by the ELF, which began setting up the fedayeen –
mobile units of Fighters who would sneak into Asmara, take
lodgings with sympathizers, work undercover for several
months, then return to the front. Adopting a structure that the
French Resistance or IRA would have recognized, the fedayeen
were organized in small, interlinking cells, each with its own
code name. In theory, each member knew only two other cell
members. Should one member be captured by the Ethiopians
and crack under torture, the theory ran, he would only be able
to betray two other people. Protected from informers and the
indiscreet alike, the network itself would live on.

The fedayeen were divided into three sections. The political
branch was charged with spreading the word: each member

was supposed to recruit five new supporters. The economic wing raised funds. The military wing was responsible for assassinations, sabotage operations and monitoring Ethiopian troop movements, information which was smuggled in hand-written notes across the checkpoints at the city's limits and out to rebel lines.

One of the military wing's most high-profile missions was staged at Sembel, a supposedly impregnable prison just off Asmara's airport road. Sembel's detainees included top political activists – several would become ministers in the post-independence government – and the ELF and EPLF wanted their best brains out. The fedayeen worked on both the prison governor and the Eritrean police force responsible for security. On February 13, 1975, as wild rumours circulated amongst the inmates, the governor summoned the prisoners together. He had already divided the policemen into two shifts: those who wanted to remain in Asmara had been assigned to the day shift, those ready to join the liberation movement would be on duty that night. 'I'm an Eritrean like you, you are my brothers,' the governor told inmates. 'In a few hours' time, we will leave this place together. You may belong to two organizations, ELF and EPLF, but this is a national operation, so we must all go together to an agreed place. Then it is each for himself.'

Under cover of darkness, the fedayeen opened the cells and led the inmates to the guard tower, where each jumped over the perimeter wall, landing on a pile of mattresses stacked on the other side. It must count as one of the least sophisticated prison breaks in history, but it went unnoticed by the Ethiopian guards outside, and that was all that mattered. 'God knows how many people broke their legs and hands,' recalls Tzadu Bahtu. Employed at the Finance Ministry today, he was then doing time for his role in the fedayeen assassination of an

Ethiopian commander. 'If one person had shouted out, the whole operation would have collapsed.' Around 900 prisoners jumped to freedom that night and with them went 70 policemen.

The railway was a favourite target. The Ethiopians controlled the line only as far as the town of Ghinda, a paltry 45 km from Asmara – anything beyond was subject to the whims of the ELF, which relied on the Eritrean drivers to carry messages, relay military information from Asmara and stand meekly aside when they decided to hold up a train. On one occasion, when Haile Selassie was due to visit Massawa, the ELF stopped a train, loaded it with explosives and sent it trundling unmanned down to the coast. This rather symbolic assassination attempt failed, but another operation was more dramatic: stopping a train on Asmara's outskirts, the ELF ordered the passengers off and put the locomotive on automatic pilot. As station chiefs telephoned in panic back to headquarters ('There's no one aboard!'), the rogue engine screamed through its usual stops, thundered across Asmara, smashed into an oncoming loco- motive carrying Melotti beer, carried it piggy-back through another two stations – the Melotti driver applying the brakes to no avail – before finally coming to a halt on the line to Keren, 35 km from its starting point. Retired railwaymen find it impossible to recount the incident without giggling.

This element of sheer cheek characterised many of the feda- yeen's interventions. Cocking a snook at Addis, they demon- strated, with humiliating clarity, the shakiness of Ethiopia's hold on Eritrea. 'The people are with us. Our sympathizers are inside every organization. There is no one you can trust,' ran the relentless message. But there was also a large dollop of prag- matism. Given the world's indifference to Eritrea's thwarted aspirations, the separatists quickly realized the Struggle would have to be largely self-financed. The fedayeen's task was to raise the capital that would allow the rebel movement to stay in business.

One-off stings yielded fantastically rich pickings. The most ingenious of these involved Asmara's main state-owned bank, where the fedayeen kidnapped the Ethiopian employee in charge of money transfers. The ELF sympathizer who took his place arranged a massive transfer of funds from Addis into a smattering of local accounts. Over the next few days, the account holders filed into the bank to write out generous cheques to themselves. 'We were collecting 60,000 birr a day. The whole thing lasted a week and we were using fake IDs and swapping cards around to get out as much money as possible. If we were dealing with someone who was well-dressed, we'd say: "Write out a large cheque", if they looked like a farmer they had to keep it small to avoid arousing suspicion,' recalls Zemhret Yohannes, an active fedayeen organizer at the time. 'There were a lot of people involved and at the end they all had to leave for the front, as their names were known. Mind you, we could probably have continued for another week as the Ethiopians didn't seem suspicious at all about the amount of money leaving the bank.'

But for the most part, the fedayeen looked inwards. Members would quietly investigate the activities of prosperous Eritrean businessmen, drawing up income estimates. Finally, the entrepreneur would be discreetly contacted and a levy in line with his financial circumstances agreed between the two sides. While the middle class was reaching into its pockets, lowly workers in Asmara's factories were just as important. They were perfectly placed to fiddle the books, pilfering raw materials and manufactured goods of use to men at the front. One day a load of spare parts for motors would disappear from a warehouse. Another day 10,000 cardigans would be mislaid from the textile plant, or a delivery of Kongo sandals would go missing from the shoe factory. It was known as 'liberating' supplies, and sometimes the level of planning that went to freeing up these vital goods reached astonishing levels.

If Melles Seyoum, director of the Central Health Laboratory in Asmara, puts his paperwork to one side and walks to the window of his second-floor office, he can see the low outline of a white building across the road.

That view holds a certain personal irony. Melles, a sober, softly-spoken man with a corona of grey-white hair, once worked there, in what used to be the Central Medical Store. The subsidiary of a state-owned pharmaceuticals company in Addis Ababa, it provided all of Eritrea and northern Ethiopia with drugs and medical equipment, supplying in the process Ethiopian army forces stationed in the province.

The couple of hundred yards of dusty tarmac between the two establishments – filled with the soothing cooing of ring-necked doves – represents a massive symbolic distance. It is the journey Melles, in his own mind, travelled from white-collared vassal of an oppressive state to free man. For it was in the building opposite that this trusted employee and respected colleague, a man widely regarded as more 'Ethiopian' than most Ethiopians, turned fifth-columnist, traitor and sneak – or inside-operator, patriot and hero, depending on your point of view.

The decision he made – to exploit his position of trust to burgle his own business on the EPLF's behalf – exposed him to extraordinary danger. Throughout the whole gut-churning episode he had a home-made cyanide pill taped to his groin, ready to die rather than face discovery and interrogation. But while Melles talks about it with a quiet intensity that makes clear those fraught days in 1976 were the most vivid of his life, he regards the decision itself as unremarkable. His experience, in truth, stands out not because of the risks taken but because of the meticulous preparation, a sample of the tenacity that would enable a small rebel movement to bring down an army.

Whatever motivated Melles, it was not a minority's resentment over dashed hopes and blighted prospects. Born in Eritrea, he had moved to Ethiopia at the age of nine and spoke fluent Amharic. Many of his friends were Ethiopian and he had completed his studies in Addis, winning qualifications that placed a successful professional career in Ethiopia within his grasp. But, in his heart, Melles never felt anything other than Eritrean. Having secured a new job in Asmara, the 27-year-old embarked on a life of self-conscious, careful duplicity.

'I moved into the Keren Hotel and there I used to spend my time with Amharas, very rarely going to see my relatives. I did my best to make friends amongst the Ethiopian professionals staying at the hotel, to blend in, to camouflage myself. I'd talk about a girl in Addis I was thinking of marrying. One of them, an Ethiopian bank manager, became very fond of me. He used to say: "You look like a true Ethiopian," thinking that he was complimenting me, when instead he was denying my nationality.'

A plan had been gestating in Melles' mind for years, ever since he had joined an ELF secret cell in the Ethiopian capital and started 'liberating' small drug consignments from Addis hospitals and the German pharmaceuticals company where he worked. It was the reason he had applied for the job as regional manager of Asmara's Central Medical Store. But he had transferred his allegiance in the meantime, joining the EPLF, and he wanted the rival movement to benefit from a haul only someone enjoying his level of access could engineer. So he stalled. 'I was giving the ELF all sorts of reasons why it wouldn't be possible to do anything: security was too tight, the bureaucracy was very complicated – in fact, I didn't want to continue working with them.'

The Ethiopian security forces had established what seemed to be a tamper-proof system. Across the street from the Central Medical Store lay an Ethiopian army garrison, complete with watchtower and roving spotlight. Military and police guards were stationed at the store's high gates. All employees were searched on arrival and departure, and Ethiopian agents had been assigned to work undercover at the site. No consignment could leave the premises without a signed approval from Asmara's Ministry of Health and Eritrea's military head-quarters. When a delivery was dispatched to the airport or a regional hospital, it did so under escort.

Melles knew, better than anyone else, just how enticing the contents of the warehouses were for a guerrilla movement beginning to take heavy casualties in the field. Their stores held more than routine hospital supplies. Earlier that year, he had been involved in a drive to prepare surgical sets for the Ethiopian army's frontline medical units, ahead of a major military offensive planned against the EPLF. The offensive had stalled in its tracks and the surgical sets had been sent back to Asmara's central store and placed under lock and key. There they sat, waiting for the next military emergency.

Working alongside one of the EPLF's underground cells, he spent more than three months preparing the ground. Nothing must be left to chance. As manager, he had access to employees' personal files. 'We knew that there were security people working at the plant and some were armed. I studied the files of every member of the Central Medical Store – I think there were 50 or 60 – working out who was likely to be an Ethiopian agent and who was likely to be sympathetic to the Eritrean cause.'

The files of those deemed possible threats were smuggled

out of the site and their photographs copied. Some staff were reallocated, buttered up by Melles – sympathetic boss par excellence – or sent on carefully-timed vacations. 'I had to do some manipulation. I changed a storeman I didn't have confidence in and put in someone I was less suspicious of, a relatively simple person. I encouraged my driver to take a vacation and sent my secretary off for a few days so I had access to the various documents.'

Keys to offices and storerooms were copied. An old form from a hospital in Keren requesting supplies and bearing all the necessary official signatures was smuggled out to the waiting EPLF. When it came back, the word 'Keren' had been removed and replaced with 'Axum' and the forged form was being brandished by a member of the fedayeen, playing the role of an Axum pharmacist with a rush request on his hands.

'We arranged for him to come on a Saturday at about 11.00, because the store closed for the weekend at 12.00. It meant staff would be in a rush to leave the premises and any news would be unlikely to leak out until Monday. We staged an argument in front of the Ethiopian guards. I was saying, "It's late, how come you've waited until now, it's Saturday and we close soon." He was saying "Please, it's really urgent, a charter plane is waiting at the airport to take the consignment to Axum this afternoon."

While the two men bickered, soldiers bearing standard-issue Ethiopian weaponry drove an empty truck through the gates of the store. They looked like Ethiopian troops, providing the obligatory escort. In fact, they were EPLF Fighters, who had slipped into Asmara in the previous days and donned stolen Ethiopian uniforms.

While his staff prepared supplies for the Axum hospital, Melles led employees to another storeroom where the unused surgical sets had been stockpiled. 'The EPLF had sent me a list of requirements from the field and I'd already worked out exactly which boxes we needed.' When the Ethiopian guards came to check all was in order, they saw porters loading a truck, a waiting military escort, and the standard paperwork. Maybe the fact that the regional manager was proposing to accompany the consignment to the airport in his own car seemed a little unusual. But, hey, this was a rush job, it was long past midday and the manager's driver, as it happened, was on vacation.

Melles' next challenge was to shed the porters before they registered no charter flight actually existed. Stopping the convoy en route, he pretended to call the airport, emerging exasperated. 'I came out and said, "Today's flight has been delayed. It's not leaving now until the evening, so we don't need to unload immediately." Solicitous as ever, the boss drove his porters back to town. Behind him, the truck promptly veered off the airport road and plunged into Asmara's back streets, finally screeching to a halt inside a private compound.

Now might have seemed a good time to exit the scene. But Melles pushed his luck a little bit further. 'There was some important equipment – anaesthetic machines, microscopes, antibiotics and additional drugs – that I wanted.' Knowing this would be his only opportunity, he was determined to wring as much from it as possible. But first he needed to know whether it was safe to return.

This was where the months of preparation really paid off. No one noticed the group of 12-year-old boys, the kind of mischievous teenagers you can see playing ball on any Asmara street corner, loitering at the bus station opposite the Central Medical Store entrance. They had been given a copy of a group photograph of laboratory staff, with suspected undercover agents clearly marked. Since the morning they had been quietly checking the faces of workers going for lunch, logging who had left and who remained. The coast, these teenage informers told

Melles, was clear. The suspected agents were off the premises, and no alarm had been raised.

Melles and the Axum pharmacist staged another fierce quarrel for the benefit of the Ethiopian guards at the gate. 'I was saying, "How come you don't have your own car?" And he was saying, "I couldn't organize one, can't we use yours? This is an emergency." So I said: "We don't normally do this. But, since it's for the nation . . ."

'A month before the mission the Ethiopian sergeant had been fined for being involved in a car accident. He had gone around the staff asking for help to pay the fine and people had been contributing one or two birr. I had given him 60 birr and said "You're an old man, my father." So we had built up a relationship and whatever I told him, he would have believed. I remember that he even told the soldiers in his charge to help us load my car.'

Incredibly, Melles was to return to his workplace one last time, to explain to the Ethiopian guards that since his car had developed a fault during all the toing and froing, he was taking it to a garage and it would not be returned until Monday. One senses success had gone to his head. He was almost courting discovery.

During the night, the fedayeen pushed the car in silence to the outskirts of Asmara – driving a vehicle with government plates always drew curious eyes – then drove it away. While EPLF forces on the edges of the city opened fire to distract attention, the drugs and equipment were smuggled out and Melles crossed the frontline, heading for the guerrilla training camps of the Sahel. His cushioned life as a middle-class professional was over. By the time the Ethiopian authorities discovered the sting, the EPLF had taken delivery of a consignment of antibiotics, microscopes, surgical blades, stethoscopes and surgical equipment worth at least 280,000 birr, or $140,000

– a considerable sum in those days. 'The EPLF was still quite small at that time, so this amount meant a lot.'

Melles was to use many of those stolen supplies setting up the EPLF's laboratories, which served the medical units that went into combat alongside the Fighters. He trained up scores of health workers, teaching them how to take samples, prepare petri dishes and test blood. He proudly demonstrates the fold-away microscopes, made of lightweight plastic, patented by the EPLF. 'Eighty per cent of lab technicians in Eritrea today are former Fighters I taught in the bush, and that fact gives me huge satisfaction,' he says.

If Melles never needed his cyanide pills, he still paid a high price for his fateful decision. Stationed in the Sahel, he was not to see the city, his old friends and family, for another 16 years. 'My father died without me ever seeing him again. With my mother, I was luckier. We met at independence.'

Looking back on his 27-year-old self, the older man knows he would not now be capable of such deeds of sangfroid. Not through fear, but because of the gritty appreciation of statistical probabilities that comes with experience. 'When you are young you only see the positives, you underestimate the risks. At this age I wouldn't do it, not because I wouldn't want to die, but because I'd calculate the odds and think we probably wouldn't succeed. When you're young you calculate the odds in a very different way. You think you always stand a chance.'

He feels little regret about the aspect of the operation that would trouble some the most: the small deceptions practised towards colleagues, apparent friends: his bank manager buddy, the old sergeant who helped him load the car, the secretary whose keys he copied. Looking back, did he ever feel he'd shown a certain ruthlessness?

'You know, a lot of people at the time made what seemed to the enemy like ruthless choices, but from our point of view represented sacrifice. Many of my contemporaries who went to the front left behind wives and children. They were the family's only breadwinners, yet they abandoned them. Naturally, you feel bad. But once you are committed to a certain course, it becomes easy.'

The Ethiopians did not make the mistake of trusting an Eritrean with such a key post again. When the new regional manager was chosen, the job went to an Ethiopian, not an Eritrean. Long after the mission, an acquaintance told Melles his name had come up during a conversation with the bank manager he once befriended at the Keren Hotel. 'They got talking and he told my friend: "After Melles, I will never trust anyone again. What he did shows you can never truly know anyone until they are dead."'

Sitting in his office, hunched over a desk, Asmerom[1] is leafing through a drawer in search of some photos he just *knows* are in there, somewhere. Waiting, I'm finding it hard to concentrate. There's an old girlie calendar hanging behind his head, nearly covered up by more recent accumulations of posters and adverts. But my eyes keep being drawn to one hairy pudenda peeping through. This is very much a man's world. The desk is scattered with spanners, spark plugs and other debris. It's not really an office at all, more an offshoot of his workshop, a place where tea is drunk, paperwork signed with oil-smeared fingers, prices negotiated and machinery examined.

'Here they are,' says Asmerom and I drag my eyes away to see he is brandishing a handful of snaps. The stack contains pictures of him, smart and besuited, at his wedding. But he whips through those, as though they are of no great interest, then stops to linger over the ones that really matter. Here he is, in his late teens, photographed in black-and-white with a group of 20 classmates. It is the early seventies, and they all sport grandiose flares. Collars stretch as far as breastbones, these are funky young dudes. Arms are draped lazily over shoulders in a moment of easy male camaraderie, grins are wide and relaxed. 'There, that's me,' says Asmerom, pointing to the middle row. There is a long pause and he sighs. 'I am the only one of that group that is still alive.'

Other pictures. Asmerom as part of a group of three, taken in a studio. The flares are still in evidence, the hair styles verge on the Afro, but the clothes this time are less modish and more practical – clothes designed for long treks and nights spent sleeping rough – and the expressions are solemn; no smiles now. Another group of three, this time dressed in camouflage and snapped outdoors. Each young man is holding a rifle in one hand. With the other they embrace, but this is not the affectionate embrace of untroubled youth. Their arms are rigid, three hands clenched in a gesture that needs no words. It spells, unmistakably: 'All for one and one for all.' 'To the death.' 'All gone, all gone,' mutters Asmerom, with a shake of the head.

He lingers longest over a snap of a tall, rangy young man in camouflage, clutching a rifle. It is not a good photograph. The young man is in mid-stride and half his body is out of the picture, while the image is so fuzzy you can barely make out his features. But the photo is clearly a favourite – as Asmerom leafs through the pile, copies of that blurry image keep reappearing at regular intervals, outnumbering even the wedding pictures. 'That is Abraham,' explains Asmerom. 'He was a great fighter. He spoke French and Italian. Very bright, very handsome, a really intelligent man.' Were you close? I ask, although the question barely needs posing. 'Yes. He was my best friend. He was a great man.'

Mournfulness clings to Asmerom like strong aftershave. When we meet later for coffee, he chooses possibly the quietest venue in Asmara, a tiny hotel veranda facing the back streets. He avoids cafés or bars where he might bump into people he knows. 'I don't like meeting people. To be honest, I don't feel safe in crowded areas. We aren't the only people who achieved independence in 1991, you know, and people have long memories. It's best to keep your mouth shut.'

He prefers the intimate tête-à-tête, lighting up at the chance to talk to a Briton about London. London, where he lived in Tower Hamlets and worked as a store manager and mini-cab driver; London, home of the Hope & Anchor pub in Brixton, where he learned from his Irish girlfriend the devastating effect of mixing beer and spirits. He made friends there who had no idea where either Eritrea or Ethiopia were on the map, and couldn't care less. London had bestowed upon him the gift of anonymity, allowing him to forget he ever belonged to the most lethal breed of fedayeen, the hit man.

Not everyone is suited to undercover work. It demands more complex qualities than pure soldiering. Physical courage, that quality most human beings spend their entire lives never certain they possess, is not enough. A level of guile, an ability to think on one's feet, is essential. The empathetic capacity to befriend the enemy, to feel in one's bones – for a half-hour or an afternoon – that one actually *is* the enemy, with all the enemy's beliefs and opinions. And it helps if you can hold your liquor and your tongue.

Asmerom possessed those dark skills. Crucially, he had also grown up in Asmara. In a city of less than 500,000, where it was scarcely possible to walk out of the door without bumping into an old acquaintance, only someone with an established network of loyal friends could hope to be a successful fedayeen, swimming like the Maoist fish in waters of silent public support. 'Assassinating in Asmara at that time was very easy because the people were with us,' Asmerom says, in a matter-of-fact sort of way. 'Half of Asmara knew I was fedayeen but they didn't say anything.'

He was recruited by Abraham Tekle, a classmate who had joined the ELF. Just 18, Asmerom was awe-struck to see how Abraham's physique had changed since their last meeting, the lanky student transformed into hardened fighter. He, too, wanted to be reborn. His baptism was a night attack on the High Court, where the Ethiopians had placed a stash of confiscated weaponry on display. Armed only with knuckle-dusters and home-made bombs, Asmerom and his friends knocked the guard unconscious and snatched 28 pistols and a submachine gun.

Asmerom slipped off to the front to do his military training. When he returned, he teamed up with Abraham to begin his new life as an undercover agent, a man constantly on the move. 'I used two fake names and had forged ID cards and documents. If I slept in northern Asmara one night, I'd sleep in eastern Asmara the next. One day I would dress like a business-man in a suit, the next I'd look like a workman in overalls and I'd have dyed my hair, or be wearing a wig. You made sure never to write things down. And you used different entrances and exits each day, because you knew that the Ethiopians were watching every zone. They were very good at counter-espionage.'

He was always armed. Asmerom shows me how he used to keep a loaded pistol on one side of his belt, held snugly in place with a knotted shoelace, and a Russian or Chinese grenade on the other. A sweater, pulled down low, concealed the tell-tale bulges. If facing capture, the fedayeen were expected to turn such weapons on themselves. 'If there was no prospect of escape you shot yourself. A lot of my friends killed themselves that way. You knew you would be questioned and you might give names away. You have to be very bright, very systematic, to resist cross-questioning, and the Ethiopians would kill you in the end anyway. So, for the sake of the others, you killed yourself.'

The fedayeen's high-profile targets were the Ethiopian army commanders. However many bodyguards they took on, however many times they might change cars or alter their routines, these men could never feel safe. One top general was gunned down with his driver and bodyguards outside an eye hospital by fedayeen who disguised themselves as eye-patched patients.[2] Another colonel was shot outside the bar run by his girlfriend, an EPLF member in league with his killers.[3] But for the most part, this was a war between the fedayeen and the spies: the Ethiopian undercover agents and Eritrean collaborators trying to infiltrate and destroy the independence movement.

'Most of our targets were Ethiopians. They were air force or navy or policemen who were pretending to be civilians but were actually working as spies. We had a list.' Many of those who featured on the list had been spotted surreptitiously entering or leaving an Ethiopian detention centre opposite one of Asmara's two main cinemas. 'Little boys noted down the registration plates of drivers pulling up. So we knew who the spies were. Or sometimes they'd have been overheard boasting about their exploits in bars.' No wonder Asmerom prefers deserted meeting places.

'The most important thing was to study everything thoroughly, the place, the person, the bars he liked to frequent, what time he left work. Timing was critical. Lunchtime was good.' Asmerom will not say how many of these operations he was involved in. Every account is prefaced with a vague, all-embracing 'We'. 'We did this,' 'We did that', never 'I'. There is more at play here than the self-effacing modesty of the Fighter. The relatives of several fedayeen he knows met strange, mysterious deaths in subsequent years. One was stabbed to death while swimming at the coast. Vendettas can stretch across the decades.

The fedayeen's tactics were dictated by Asmara's confines. The city was tiny, it could be crossed by foot in an hour, and about a quarter of the population was Ethiopian and therefore not to be trusted. How to stage ambushes and executions, then vanish into thin air, in one of the most parochial capitals in Africa, where everyone knew each other's business? Cars were highly visible, because so few Eritreans could afford to own them, and therefore ill-suited for getaways. The fedayeen fine-tuned a tactic that would have seemed comic, had it not proved so grimly effective: the bicycle assassination squad, or 'bandit's tanks'.

Asmerom remembers the last time the technique was used. It was against a young man who had been working as a pusher, trying – the fedayeen believed – to hook clean-living young Eritreans onto drugs. 'He was a very handsome boy, with lots of money, and he was always hanging around young men and ladies. He was very dangerous.' Two youths were placed on lookout, casually loitering with their bicycles in their hands as they monitored the prospective victim, sitting in his car. The two designated killers strolled up and shot him dead. Seizing the bicycles held at the ready, the fedayeen bicycled furiously off into the distance. 'It was always a good idea to use four people for an operation. Two on watch and two to carry it out.'

In the years Asmerom worked for the fedayeen, 25 assassinations were successfully staged, half of them carried out by bicycle squads. The Ethiopian authorities reacted by first banning bicycles from the main thoroughfares that sliced across the city, a ban which remains in force to this day, and then outlawing bicycles altogether. Asmerom was none too bothered. 'I preferred operating on foot anyway.' But the Eritrean population paid a high price for the fedayeen successes. Adopting similar tactics to the Nazis in occupied Europe, the Ethiopians applied a 10 to 1 ratio when punishing local residents for a fedayeen operation. Students and high-school students considered likely rebel sympathizers were garrotted with piano wire, women snatched away and gang-raped in the army camps.

He lived the underground life for four years. He was nearly caught in 1976, when Asmerom and a woman fedayeen tried to hijack an Ethiopian Airlines plane to fly to Sudan, where they planned to demand a ransom that would go into ELF coffers. The two had smuggled a pair of pistols into secret compartments sown into their bags. Passing through airport formalities, Asmerom watched in horror as an Ethiopian security woman inserted a sharp fingernail into a seam and began pulling the bag apart. 'I seized my accomplice's knee and said: "They are going to discover us."' She froze, paralysed with fear. 'She was only a little thing, very quiet and shy. She just threw her hands up in the air. So I left her.' Pretending to have forgotten a piece of his luggage, Asmerom ran for the exit, jumped into a car parked with its key still dangling from the ignition, and drove to safety. His conspirator paid for her hesitation with a 12-year prison sentence.

It had been a close shave, and there were others. Once, walking on the city outskirts, two trucks of Ethiopian soldiers pulled up suddenly on either side. 'I thought, this is it, I have to die.' But onboard was an Ethiopian spy Asmerom had once shared a beer with. 'He called out, "Don't worry, he's one of ours." It was a moment from God.'

But Asmerom was running out of chances. The Ethiopians were looking for him, his sisters had been jailed and his father beaten so badly about the shoulders he needed a skin graft. In any case, by the late 1970s the era of the underground movement was reaching a close. The war had not followed its anticipated course and amongst Eritreans, the bleak realization that the conflict would be long, testing, and possibly unsuccessful was sinking in. The number of those ready to work as informants for the Ethiopians rose, and the list of fedayeen falling into enemy hands lengthened.

It was time to end it, and Asmerom left for the front, moving on to Kenya and then London. There, he slept quietly in his bed for the first time since his teens. When Darren, his Hope & Anchor drinking partner, asked him if he had ever been a militant, ever seen someone shot, he changed the subject. How could his well-meaning friend, who had never encountered violence outside a football stadium, ever hope to understand? 'British society was so different. Darren would have looked at me in a different way. I'm not a murderer. We had to win our independence or die.'

Was he, in fact, a murderer, or the Eritrean equivalent of a French Resistance hero? Talking to Asmerom was like looking down a kaleidoscope. A slight turn of the wrist and the shards of colour shift, the pattern changes, taking on a new configuration. Here I sit, sharing a beer with a middle-aged family man reminiscing with rueful charm about hangovers and pubs. But for a split-second, imagining what it would have been like to be a passenger on a hijacked Ethiopian Airlines flight, with a pistol-waving Asmerom striding down the aisle, single-minded deliverer of justice, I shudder. 'My God, this man's dangerous, a terrorist.' It's a switch in perspective even Asmerom seems to find problematic. 'Sometimes, thinking about the past, I frighten myself.'

But what haunts him is not guilt, but loneliness. He returned to Eritrea after liberation to save the crumbling family enterprise but found himself ill at ease, no longer attuned to doing business in Africa. Abraham is long gone, killed not by the Ethiopians, ironically, but during the internecine fighting between the Eritrean liberation movements that only ended in 1981, when the EPLF emerged triumphant.

Asmerom scoffs at Eritrea's new generation of soldiers, who he suspects do not possess the wiry capacity for suffering, the potential for desperate acts, of the dead young men whose photos litter his office drawers. 'They go to bars, to discos, they always seem to have money to spend, although no one knows where it comes from. The old guys,' he says approvingly, 'were different. The old guys were really violent.' Like the Kagnew GIs, Asmerom knew what it was to be part of a brotherhood, young men sharing moments of heightened emotion, more besotted with each other than they would ever be with their women. But unlike Zazz and his buddies, Asmerom cannot exchange e-mails with those who shared his underground life or attend backslapping reunions. And you sense he is asking himself the question: was it really worth it? Did his generation make, too lightly, in the careless ways of youth, sacrifices that were greater than they were capable of grasping at the time? Dying for a cause comes ridiculously easy to the young. They offer up their lives with such casual generosity. It is in old age that the gesture becomes astonishing, and even the lesser steps along the way turn grindingly difficult, punishingly costly. 'You get so angry when you're young. You think you have to fight. But later on, you look back and you realize you've gone from country to country, you've had many experiences, yes, but you've never had a career in your life, you have no classmates, no one to go and have a drink with.'

CHAPTER 13

The End of the Affair

'What once seemed a she-lion now looks like a dog.'

Tigrinya proverb

Haile Selassie's ousting did not take US policymakers by surprise, although the bloodshed that followed did. Washington had become so sceptical about the imperial government's future that the embassy in Addis had been instructed to cultivate good relations with any of the Emperor's likely successors.[1] In August 1973, a year before the coup, President Richard Nixon felt uneasy enough to order a prompt review of US military assistance to Ethiopia and recommend the withdrawal of all but 'designated residual functions' from Kagnew by the end of fiscal year 1974.[2] But the Watergate scandal was looming, and Nixon was soon caught up with events closer to home. Its mind elsewhere, his administration convinced itself Washington would be able to enjoy the same cosy, mutually advantageous relationship with Ethiopia's new leadership as it had with the old. The Derg's fondness for denouncing capitalism and railing against the CIA did not go unnoticed, but was dismissed as crowd-pleasing rhetoric. Determined to woo the new power in the land, the US actually chose this moment to hike funding to levels Haile Selassie had only been able to fantasize about, arguing that this was the best way to ensure a pro-Western government emerged. Amazingly, more than one-third of the military aid Ethiopia received from the US over a 25-year period would be given *after*, not before, the Derg's takeover.[3] 'Suspension of these shipments would only strengthen the hands of radical elements among the military and further frustrate the moderates, perhaps leading them to concur in more radical initiatives,' the State Department explained to the incoming president, Gerald Ford, in August 1974.[4]

With the Soviet Union firmly implanted in Somalia, this really was not the time to lose a US foothold in the Horn of Africa, argued policymakers, reluctant to turn their back on decades of carefully-fostered relations. US influence had played its part in the creation of a generation of Western-trained technocrats and military officers in Ethiopia and it seemed only logical to assume that these moderates – men the US would be able to do business with – would now come to the fore, Washington reasoned, making a mistake it was to repeat a few years later in Iran. Yes, it was true that the statements coming from the new regime were somewhat radical in tone, but a soft form of African socialism, one that fell well short of classical Marxism-Leninism, would be acceptable. The Derg had promised Ethiopia's foreign policy would remain unchanged. A little flexibility on ideological issues therefore seemed in order, particularly in view of the battering US global prestige suffered when its forces were sent scrambling out of Saigon. 'After the withdrawal from Vietnam, the message coming from Washington was that the status of the US would suffer if it stopped supporting Addis,' recalls Keith Wauchope, deputy principal officer at Asmara's US consulate between 1975 and 1977. 'Our credibility, we were told, was at stake.'[5]

It was not a view staffers at the consulate shared. Those on the ground were becoming convinced not only that Kagnew was no longer worth the candle, but the entire US–Ethiopia relationship was untenable.

The words 'Kagnew Station' and 'Top Secret' had always worked like a charm when it came to winkling no-questions-asked grants out of Congress. But the standard arguments voiced in Washington to justify policy in the Horn were lagging dangerously behind the science. Advances in satellite technology were making Kagnew's unique geographical and climatic conditions irrelevant with a speed no one had anticipated. In 1960, Eisenhower had approved the first launch of a spy satellite. The details gleaned on Russian radar systems delighted US scientists, and 10 years later Rhyolite, a satellite which orbited in time with the earth, picking up microwave signals, was launched. After that had come a generation of satellites designed to hover over Russia's most inaccessible regions.

Now that the Soviet space and missile programmes were being tracked from the darkness of space, Stonehouse was obsolete. Too unwieldy to be shipped elsewhere, its vast dish antennae were sold for a paltry $6,000 to an Asmara merchant who laboriously dismantled them – no mean feat in itself – and sold the metal as scrap. In early 1972, in recognition of the new status quo, the ASA pulled out of Eritrea and control of Kagnew passed to the navy, which used it to relay messages to the US fleet in the Indian Ocean and American bases in the Middle East. Kagnew staff had less and less to do with each passing year. By 1975, only a handful of Kagnew's 19 sites were still operational. A base that had once housed more than 4,000 men, women and children, linguists and decoders, army officials and contractors, had shrunk to a skeleton crew of 13 naval personnel and 45 civilian contractors. With two-thirds of the buildings deserted, the men roamed the now-oversized base like ghosts.

But if there were purely practical questions to be raised

about Kagnew, whose future was in any case due to come up for formal reconsideration when the 25-year base rights agreement expired in 1978, larger moral and strategic issues could no longer be ignored. The original deal America had struck with Addis – guns for Kagnew – came to seem positively Mephistophelian as the Derg, outnumbered and outmanoeuvred by the rebel movements, struggled to regain the upper hand.

'The Ethiopians kept justifying their need for arms on the basis of the continuing threat from Somalia,' remembers Wauchope, a diplomat with a keener moral conscience than some of his peers. 'But in reality we saw them using the weapons against the separatists in Eritrea, something we didn't think they should be doing. We could see the jets the US had provided taking off from Asmara airport and we could hear the bombs going off and the artillery, mortars and anti-tank weapons we'd sent being fired. We knew they were using those weapons in Eritrea and we reported that back to the embassy.'

As the EPLF and ELF launched what was meant to be their final drive on Asmara in early 1975, the Ethiopian army went into action with a ferocity born of desperation. Looting and shooting, Ethiopian soldiers rampaged through the city, opening fire on passing minibuses, dragging residents from their houses and gunning them down in the street. When the bullets ran out, they used bayonets. During a four-day killing spree, hundreds of civilians – some put the death toll as high as 3,000 – were slaughtered. Asmara had become an oppressive city of horror, hemmed in by army roadblocks, paralysed by curfew.[6]

Wauchope began keeping a tally of the murders carried out by Ethiopian forces, so indiscriminate they even included members of his own staff. 'In the two years I spent there I counted 450 murders, almost exclusively of Eritrean civilians. One of the victims was the brother of an employee, who was hung, another employee was shot on the road for no reason.

271

The Ethiopians would descend on a neighbourhood and the eviscerated bodies would then be left out on the street to intimidate people,' he remembers. Once the capital was secured, the army shifted its focus to the countryside, where it burned 110 villages, bayoneting and shooting hundreds of peasants suspected of collaborating with the rebels.

The unrelenting brutality, the diplomat saw, was transforming Eritrean attitudes to the Americans. In the 1960s, respectable Eritreans might have been shocked by the loutish behaviour of the Kagnew boys, but they nonetheless appreciated the hundreds of jobs created by the base and the schools and leper clinics opened by US personnel. By the mid-1970s, deep bitterness had set in. The Eritrean public was making the connections so many Kagnew men had preferred not to register. When the Ethiopian army blew up the houses of suspected collaborators, they used US rocket-propelled grenades. When they strafed rebel trenches, they did so from American F-5 jets flown by pilots trained in the US. When Ethiopian convoys trundled through towns, they did so in US Mack trucks, and whenever the rebels captured weapons, the prize included spanking new US-made mortars, machine guns and rifles.

If Washington was going to support Mengistu with such gusto, Eritreans decided, then the Americans were no longer welcome. 'People were telling us, "We have been your friends and hosts for many decades. We have allowed you into our homes and country. Now these people are coming after us and they are using your weapons against us, your friends and your hosts,"' recalls Wauchope.

Wauchope and his colleagues registered a shift in tactics by the up-till-now forgiving rebel groups. Two Americans working on one of Kagnew's more remote facilities were kidnapped and taken into the bush. Then the British honorary consul,

Basil Burwood-Taylor, was taken hostage and smuggled out of Asmara on the back of a camel.[7] This kidnapping, staged in Burwood-Taylor's office in broad daylight, brought home a fundamental truth. 'We realized that these people had access wherever they wanted, because the population was supporting them.' Two more Americans were seized, and the rebels threatened to try their prisoners for deaths caused by US arms deliveries. Under pressure from anguished parents of the kidnapped men, US Congressmen woke up to the realization that their country was heavily embroiled in the Horn of Africa and began peppering the then Secretary of State, Henry Kissinger, for explanations.

Then the atmosphere darkened dramatically, as far as Washington was concerned. 'Two American contractors hit a landmine that had been deliberately placed on a road leading to a remote Kagnew site. They were blown through the roof of their car and killed. That changed things very substantially,' says Wauchope. Kagnew decided to draw in its horns, closing down hard-to-police outlying facilities and running both broadcast and reception sites from the same location – something no radio technician would normally recommend. But the Americans could not fence themselves off from reality: a fifth US employee from Kagnew was kidnapped. 'We had feedback from the Eritreans to the effect that they didn't intend to torture or kill or put the hostages on display. They simply wanted to send us a message that they were unhappy with our support of the Addis government.'

It was Wauchope's job to report to the US embassy in Addis on local conditions and report back he dutifully did, but he noticed that his superiors were growing increasingly unhappy with the recommendations made in his outspoken telegrams. Organizations facing the prospect of their own sidelining rarely want to look reality in the face. Large projects, once launched, develop a momentum all their own. The Derg moderates on whom Washington had pinned its hopes were being shot, one by one, as Mengistu saw off a succession of challenges from rivals alarmed by his hard-line tactics. Mengistu, who had put exploratory feelers out to Moscow, was adopting an ever more pronounced anti-Western stance, histrionically smashing a Coca Cola bottle full of blood on the pavement at one rally and promising to spill the blood of 'American imperialism'. Hatred of the regime was spreading: rebels in northern Tigray announced the formation of the Tigray People's Liberation Front (TPLF), dedicated to Mengistu's overthrow. Yet the embassy in Addis balked at reaching a conclusion that would lead to a once-key mission being downgraded, while the defence establishment choked on an admission that would mean budgets were slashed. 'The Department of Defence simply wasn't seeing this as a problem. The message was: "Let's keep it going, let's keep it going,"' remembers Wauchope.

The strains were becoming unbearable, especially once Jimmy Carter, a Democrat who had campaigned for an 'ethical' foreign policy, moved into the White House. 'We are very much concerned by the use of American military equipment in suppressing indigenous movements inside Ethiopia,' the deputy assistant secretary of state for African affairs told a Congressional committee. 'We would be reluctant to abandon Ethiopia to total Soviet domination. On the other hand,' he confessed, 'we do not want to see our weapons used in this fashion.'[8]

One wonders just how long, if left purely to its own devices, Washington would have continued to deny the blindingly obvious: that American military aid was being channelled, as the Cold War raged, to a brutally-repressive Marxist African regime whose values were anathema to ordinary Americans. Western accounts of this period have been kind to the

Americans, with some attributing the eventual break in relations to a supposedly high-minded decision by the US to 'sacrifice' a client that failed to meet Carter's moral standards.[9] If only it were true. As Mengistu revealed his true nature, Washington certainly voiced increasing concerns about Ethiopia's appalling human rights record. It dragged its feet over pre-agreed arms deliveries – a backtracking that drove Addis wild – and served notice that Kagnew would soon close. But it was Ethiopia that finally brought the jarring ideological and moral contradictions to an end, pushing the relationship to collapse and formalizing the break.

By early 1977, Washington was receiving horrifying reports of purges in Addis. In what would later be dubbed the Red Terror, revolutionary squads armed by Mengistu were massacring the very student leaders and educated Ethiopians who had propagated the ideals of the socialist revolution. Mengistu chose this moment to make his move, putting in an arms request so large he must have known it would be rejected. With Somalia looking ever more threatening, it was time for America to show its mettle. 'It may have been a test,' says Wauchope. 'Washington felt very unhappy with it – was the maintenance of a traditional relationship worth supporting a movement of this kind? We came back with a feeble counter-offer, full of caveats and strings attached. Maybe we could provide a third of what they were asking for, we told them. The next thing we know, they've gone to the Soviets, who had no hesitations.'

It was a moment replete with historical irony. For decades the US had poured money into Ethiopia, helping a leader regarded as a bulwark against Communism to pummel an Eritrean rebel movement regarded as a likely route for Soviet infiltration of the Horn. By propping up the Emperor, Washington had ensured the US was indelibly associated with all that was corrupt and outdated in the minds of Ethiopia's new rulers, who now wanted nothing more to do with the West. When it came to Communist penetration, trusted, cosseted Ethiopia, rather than the despised Eritrean liberation movement, had proved the weakest link. The US had contributed not only to the downfall of its champion Haile Selassie but, by building up the organization that destroyed him, had also enthusiastically armed a government bent on spreading Marxist ideology across the region. Measured by even its own self-interested standards, US policy had backfired. Looked at from the viewpoint of ordinary Eritreans, the judgement was far harsher. 'The Americans had no idea what they were doing,' scoffs Dr Aba Isaak, an Eritrean historian. 'They were like the elephant – when it treads on small creatures, it doesn't even notice. It just goes, and goes, and goes along its way.'[10]

Ethiopia had swapped sides in the Cold War and the consequences for Kagnew were immediate. On April 17, the Derg gave the US just four days to close five facilities in the country, including the communications base. The defence agreement signed in 1953 – the most astute piece of diplomacy Haile Selassie ever negotiated – was to be terminated a year ahead of schedule, Addis decreed. The news was delivered on a Saturday, timing designed to make the pullout as practically difficult as possible for Ethiopia's erstwhile friends. Amazing as it may seem in retrospect, the Derg's decision hit US officials in Addis and Asmara like a thunderclap. 'We'd thought things had stabilized, we didn't expect the Ethiopians to make the break that radically,' says Wauchope. 'They had been adept at playing the superpowers off against each other. The Ethiopian government still had links, don't forget, with Yale, the Ford Foundation, TWA. To suddenly opt for one side was quite a jump.'

Wauchope found himself in command of an emergency operation to ensure nothing of any strategic interest – no coding information, no classified circuit boards, no transmitting or receiving equipment – remained intact for the East bloc experts he knew would be invited to pick over the site by the Ethiopians once the Americans left. 'It was an exciting, hair-raising time. We had to scramble like hell, but by Monday we'd worked out how to destroy or dismantle everything of strategic value. I tried to organize the closure with a maximum of dignity, but while I was out loading the transport plane we heard that Ethiopian soldiers had attempted to storm the consulate. The marines were put on alert and the Ethiopians were told this was a violation of the Vienna Convention. Then they demanded a tour of the Kagnew facilities and we had to conceal the backup machines that were still in use. The Ethiopians were telling us we couldn't remove any hardware, so we destroyed the classified circuit boards by slipping them into the water collection point. In the end, they got nothing. When we left the consulate, we placed the instructions to the game Dungeons and Dragons amongst the procedural papers, just to sow confusion, and we left the calling cards of prominent people in government we had no use for lying around, to give them something to think about.'

International pressure on the Addis government secured the Americans an extra two days' grace. On the last day, the Stars and Stripes were ceremoniously lowered at the consulate and staff joined a convoy of vehicles from Kagnew Station which threaded its way through Asmara to the airport. Many of Asmara's residents lined the streets to watch the convoy pass and some, Wauchope remembers, were weeping. The grief was prompted not so much by fondness for foreign guests whose role had been, at best, contentious, but by fear of what the future held. 'All the other Western consulates in Asmara had been ordered to close at the same time. They felt that the last international witnesses to their sufferings were leaving and now there would be massive ethnic cleansing by the Ethiopian army.' At the airport, the Ethiopian military was waiting and a last attempt was made to board the American flight and inspect what was on board. 'We faced them down, told them they were violating international law, and then we were out of there.' If Wauchope, who won an award for the role he played, felt more than a pang at abandoning the Eritreans, there was no doubt in his mind that Kagnew's closure was overdue. A morally untenable partnership had been brought to an appropriate end. 'It was what we had been lobbying the embassy in Addis for. We had lived in that country and we had understood what the Eritreans felt.'

America's need for terrestrial spy stations did not disappear with Kagnew. Washington had had a backup ready since 1974, having signed an agreement with Britain to set up a communications facility on the coral atoll of Diego Garcia, a remote former dependency of Mauritius. Diego Garcia was never going to be able to rival Asmara for reception, but it was ideally placed for the US fleet operating in the Indian Ocean, and monitoring equipment from Kagnew was transferred there. In retrospect the entire Kagnew operation, unknown to most Americans today, barely mentioned in writings on the topic, would seem blessed with a miraculously low public profile. Washington proved far less lucky with Diego Garcia. There, islanders evicted by the British government to make way for the American military waged a vociferous legal campaign, ensuring that, while Diego Garcia remains in operation as a logistics, military and communications base, a key part of America's 'war on terror', no one has ever been left in any doubt as to its role.

Vertical antenna poles – the debris left behind by the ASA – are still scattered around Asmara. Not far from the airport, where young boys tend their goats, a round golf ball of a building rises incongruously from the grassy plains – a giant receiver built by the Americans, now used by the Eritrean military. Kagnew's Tract E, with its solid bungalows, sports facilities and central location, was too good a site to allow to go to waste. It was used as a garrison by the Ethiopian army and, after independence, the EPLF moved its Fighters in. When I visited what is now known as Den Den Camp, baptized after the mountain peak the Ethiopians never managed to wrest from EPLF control, the giant dry-cleaning unit vaunted in the US army brochures was working full blast, processing dirty laundry for UN troops in town. But the clock tower was telling a time of its own making, washing was draped over fences, and the once-neat lawns were littered with rusting containers, temporary homes for soldiers too poor to afford lodgings in town.

Perhaps the Gross Guys would have recognized the atmosphere in the two clubs, where the Melotti beer and *zibib* start flowing early in the morning, a cheap anaesthetic for the frustrated and bored. It is ordered by Eritrean paraplegics of both sexes, wounded in both the new war against Ethiopia and the conflict Kagnew's servicemen were determined to ignore. Wheelchairs crunch along the walkways where young GIs once strolled and the bars in which Spook and his acolytes staged their silly pranks are now run by young ex-Fighters with pinned-back trouser legs, survivors of a series of wasteful wars.

'Come, we should go,' my Eritrean friend told me, as we watched a young amputee in what was once the Oasis Club pick a fight with a barmaid who thought he had already had too much to drink. 'Otherwise you will see something you should not.' Not state secrets, but a spectacle he wanted no foreign visitor to witness: Eritreans behaving badly.

* * *

For the Eritrean rebels, who by late 1977 controlled 95 per cent of Eritrea, nothing could have been more disastrous than the Derg's formal entry into the Communist camp.

The banal rules of the Cold War dictated that the Soviet Union backed Marxist rebel movements fighting right-wing governments supported by the US. Here, thanks to Ethiopia's ideological flip-flop, was a Marxist rebel movement fighting for independence from a Marxist government. For Soviet strategists, it made no sense: national frontiers and ethnic hostilities were surely destined to fade into insignificance once scientific socialism conquered the world. Previously sympathetic to the Eritrean cause, Moscow decided it had no time for a rebellion that refused to fit its ideological paradigm. 'There is no insurgency in Ethiopia,' declared one Soviet observer[11] – obliterating Eritrean history with the same breathtaking high-handedness once demonstrated by the UN's bored bureaucrats. If the US eventually decided to return to the Horn on Somalia's side,[12] no one wanted to touch the Eritrean rebels. Rejected by the East bloc, spurned by the West, they were on their own.

Mengistu made his first trip to Moscow in May 1977, returning in a state of near-euphoria. Unfazed by their long-standing role as Somalia's military supplier, the Soviets had offered Mengistu the weaponry that would allow him to wage 'total war' on Ethiopia's enemies. 'He was absolutely ecstatic,' remembers Ayalew Mandefro, defence minister of the day.[13] 'He told me: "We are going to need large warehouses and a big storage capacity."' In the middle of the year, Somalia launched a concerted grab for Ethiopia's eastern Ogaden and Somali President Siad Barre, exasperated by Moscow's clumsy attempts to back both horses in the same race, moved to expel 1,700 Soviet advisers. The Soviet Union was free to make good its promises to Ethiopia, ferrying in $1–2 billion of armaments, 12,000 Cuban combat troops and 1,500 military advisers during a six-week air and sea lift. The value of Moscow's arms deliveries outstripped in a matter of months what the US had supplied during all its dealings with Ethiopia.

The oversized delivery changed everything in both the Ogaden and Eritrea. The turning point in the north came in Massawa in December, when the EPLF's seemingly unstoppable advance ground to a sudden halt. As Soviet ships moored offshore opened fire and Soviet advisers took the controls of Ethiopia's spanking new artillery, hundreds of Fighters were mowed down or drowned on the flooded salt pans. The EPLF was no longer fighting a panicking African army, it was pitted against a superpower boasting seemingly inexhaustible resources. By July 1978, after a campaign of saturation bombing by the freshly-equipped Ethiopian air force, all the towns the ELF and EPLF had won in southern and central Eritrea had been recaptured. Forced to accept the inevitable, the EPLF pulled out of Keren, its Fighters stripping the town of every object of potential use as they headed up into the only area that now seemed safe: the mountains of the Sahel, where they would spend the next 10 years.

Their leaders called it 'strategic withdrawal', but to those who took part in it, this had the metallic taste of defeat. 'It was our Dunkirk,' acknowledges Zemehret Yohannes, today a leading member of Eritrea's ruling party. 'It was a complete defeat, a defining moment. The biggest army in black Africa, with modern equipment and Soviet help, had pushed us out of the territory we'd been holding. Every expert was saying, "This fight is hopeless, it's a dead movement." In retrospect, it baffles me – how, in those circumstances, we could say: "We can prevail." It seemed a kind of stupidity.'[14]

CHAPTER 14

The Green, Green Grass of Home

'They were madmen, but they had in them that little flame which is not to be snuffed out.'

the painter Renoir, remembering the French commune

Towards the end of the eighth century BC in Ancient Greece, a revolutionary society was born where the limestone fingers of the Peloponnese mainland reach into the blue Aegean. It was founded by the Spartans, who had invaded neighbouring Messenia in search of fertile land. Sparta had won the territory it coveted, but with it came a population of rebellious subjects who outnumbered their new masters and did not take kindly to being used as forced labour to work the fields. To protect themselves against the *helots*, or serfs, the Spartans came up with the concept of the military state.

Not for them the dissolute habits of the Athenians and soft comforts of family life – Spartan men prided themselves on their self-denial and iron discipline. They lived in barracks, their closest comrades were fellow fighters. Sent as children to run barefoot on the chilly mountainsides, they learnt to bear pain without a whimper. Warriors until the age of 60, their greatest ambition was to achieve 'a beautiful death' on the battlefield, defending the Spartan state. Women were no exception: they too were expected to espouse the Spartan virtues of simplicity, moral rigour and extraordinary physical toughness. Feminists before their time, they enjoyed, in the absence of their menfolk, levels of freedom unheard of in the Ancient world. The other Greek city-states scorned the Spartan model, regarding it as totalitarian and brutalizing. But when Ancient Greece was invaded by the Persian army in 480 BC, it was a group of 300 Spartan warriors, embracing their fate in a doomed last stand at Thermopylae, who showed their effete fellow Greeks what it meant to die for a cause.

There was something very Spartan about the society that took shape in the late-1970s in the Rora mountain range that rises from the plains north of Keren and runs north-west to the border with Sudan. The Spartans built their militaristic state on a victory: they had subjugated the Messenians but knew they would not remain forever supine. The EPLF, in contrast, built their society on defeat: with Moscow's entry into the war, they had gone from holding independence in the palm of their hands to confronting total annihilation. Like the Spartans, the fighters of the EPLF adopted a rigidly puritanical lifestyle. The need to create a lean fighting machine meant that conventional family structures were rejected, traditional roles recast. Not only was the Eritrean woman Fighter, accounting for 30 per cent of rebel forces, often as deadly as the male, their children – separated early from their warrior parents and inculcated with the Movement's dour values at the Revolution, or Zero, School – would grow up to outstrip them both.

Their kingdom was what was loosely known as the Sahel and its contours were dictated by the gradient, for the mountains form a more effective barrier against an advancing army than anything man could contrive. This was the part of Eritrea no one really wanted – most of it too steep even for Eritrean farmers, adept at tilling the narrowest highland ledge. The first foothills erupt from the dun-coloured plains around the town of Afabet like whale humps breaking through the surface of a calm sea. Strange stone excrescences form impossibly perky breasts. Then the Rora escarpment starts in earnest, the peaks – giant slabs of brown rock – crowding in upon the dry water-courses. The only fruits that grow here are the fig cactus and fleshy green pods of Sodom's Apple, which give nothing, when you tear them open, but a cobweb of poisonous white sap, said to be strong enough to stop a man's heart. It is a landscape of scrawny thorn trees and spiky pink baobabs that gesticulate nightmarishly across the narrow valleys, like witches crazed with grief. Baboons thrive, gambolling along the ravines in thuggish packs of 40 and 50, red in face and rear. So do foxes, which hunt hares with long, white-tufted ears. But few humans would want to set up base here, unless they had no choice.

Choosing their sites close to the border and the vital access routes to Sudan – a neighbour whose cooperation could never be taken for granted – the EPLF dug in. As the realization set in that the guerrilla movement was in for the long haul, the structures grew ever more complex, ghostly echoes of the institutions and organizations the Movement hoped one day to establish in peace-time Eritrea, a practice run at administering a state.

Up in the north at Orota, they built an underground hospital in whose white-washed rooms most major operations, short of heart surgery, could be staged. Sprawling for 5 km along a valley, equipped with 3,000 beds, it was known as 'the longest hospital in the world'. Learning that survival rates depended on the time it took a wounded Fighter to receive emergency treatment, EPLF doctors took huge pride in the speed with which the injured were tended by frontline medical units and then whipped, tier by tier, to hospitals whose sophistication was in inverse proportion to their distance from the battlefield. At the very most, it was decreed, no Fighter should be more than two hours' distance from a surgeon.

In Orota too, a small pharmaceutical plant turned out essential drugs, bandages, even sanitary pads for the women Fighters. In other workshops, Fighters manufactured bullets and repaired the damaged trucks, tanks and radio sets captured from the Ethiopians. Once the supply of arms from Arab countries dried up – former supporters belatedly realizing they were dealing with a secular Marxist movement, rather than an Islamic revolt – the EPLF was forced to rely on theft, careful recycling and mechanical cannibalization for most of its weaponry. Orota was also the location of the famous Zero School, where the children dubbed 'Red Flowers' were taught. The logistics and military base – the EPLF's de facto citadel – lay further south at Nakfa, where, in 1976, its Fighters occupied a small settlement on the high plateau enjoying panoramic views down to the bleak Naro plains and the distant haze of the Red Sea.

Talking about these installations is always confusing, because the EPLF deliberately muddied the water in an attempt to protect their bases from bombardment. Sensitive sites were often given the same names as villages in the lowlands, or identified by code words, stripped of associations which could give Ethiopian strategists helpful clues. Even today, if you present a former Fighter with a map he will struggle to place them, brows puckered, fingers wandering lost and uncertain across the grid.

The same deliberate confusion surrounded the Fighters themselves. Adopting nicknames is a very Eritrean habit, perhaps because the fund of traditional local names is so limited, it's easy to get confused. But the practice was taken to new extremes at the Front, a form of initiation ceremony that underlined each individual's entry into a testing new existence where no one could afford the luxury of homesickness or regret. You arrived an ordinary civilian, with your own petty, small-minded concerns and were reborn a warrior, or *tegadlai*, dedicated to the Cause. If a Fighter was stocky, he would be known as 'The Body' or 'Sack', if he was thin and ascetic, he became 'The Priest'. A veteran activist was baptized 'The Movement', a macho man reborn as 'Rambo', a rebel who sported a moustache became 'Charlie' (as in 'Chaplin') and, with the thumping inevitability of a group of British squaddies dubbing the giant amongst them 'Titch', a particularly dark-complexioned Fighter would be known as 'Tilian' ('Italian').

A jokey form of shorthand, the nicknames not only made things simpler, they served as a security precaution. In a nation of less than 4 million people, where names usually tell the world whose son you are, it was alarmingly easy to track down a Fighter's village and extended family, vulnerable in Ethiopian-controlled territory. If a Fighter was stripped of his ancestral identity, known only by his nickname even to his rebel comrades, then his relatives were protected from retribution. For the same reason, Fighters were encouraged not to discuss religion or tribal affiliations: such differences, which had done so much to deliver Eritrea into Ethiopia's hands, should in any case be put aside in the fight for a united, independent Eritrea. The result was intense friendships based on the here and now. 'When my best friend was killed, a man I had fought alongside for years, a man I really loved, I buried him without knowing where he came from,' remembers one ex-Fighter.[1] 'Today I would like to go and see his mother and tell her what a hero her son was. But I don't know who she is. He was my brother, but I knew next to nothing about him.'

In a country infatuated with its own history, this is the

episode Eritrea loves the most; it is remembered with the same perverse nostalgia British veterans reserve for the Blitz. A tribute to pig-headed determination and endurance, Nakfa was so integral to the Eritrean experience it seemed only natural, when the post-independence government launched a national currency, that it should be baptized the Nakfa. During the decade the EPLF spent in Nakfa, the Ethiopian army launched eight large-scale offensives in a vain attempt to break the rebel movement's hold on the plateau. A journalist visiting the front once counted 240 bombing sorties by enemy aircraft in a day, with one shell landing every minute. Even today, when a new Nakfa is rising from the ruins, much of the plateau remains an ugly moonscape, each acne scar on its pitted face representing a massive explosion. The original town was reduced to rubble, its simple white mosque the only piece of masonry extending higher than waist level. Some say the pilots who flew the MiGs and Antonovs spared the mosque because it served as a useful compass point, allowing them to establish their bearings above the escarpment before releasing their bombs. Others say they tried to hit it but always failed, a sign, perhaps, that the town held a special place in Allah's affections. A victory for sheer obstinacy, constructed on a military rout, Nakfa was Eritrea's Verdun. It would never fall into Ethiopian hands, and the failure to capture the EPLF's de facto capital would prove the Mengistu regime's ultimate undoing.

The Eritreans survived by moving underground. Hospitals, technical colleges, theatres, guest rooms for visiting VIPs, parking places for the camouflage-painted jeeps, offices where young leaders like Isaias Afwerki, Petros Solomon, Mesfin Hagos and Sebhat Ephrem planned military strategy: they were all painstakingly dug into the rock and carved into the sides of valleys, the narrow entrances then covered by screens of under-growth and foliage. From the air, they were virtually invisible.

The Fighters became creatures of the penumbra, for it was best to go about your business in the hours of dusk when pilots found it hard to focus in the half-light, and during the chill of night. 'When I think of my visits to the front in the 1970s, all I remember is darkness, darkness, darkness,' says Koert Lindyer, a veteran Dutch journalist.

The tentacles of their underground network extended far beyond Nakfa itself. In a giant semi-circle that swept 240 km from Karora on the Sudanese border, down to Nakfa and south to Halhal, north of Keren, the Fighters dug parallel lines of trenches. It was backbreaking work. But, as the Fighters said, whipping themselves on to greater efforts: 'Better to sweat than bleed.' In a conflict in which Ethiopians usually outnumbered Eritreans 10 to 1, and the enemy tapped a seemingly unlimited supply of weapons, the contours of the land were the one obvious advantage the Eritreans enjoyed. If they were forced to surrender one position, there would be another trench line to fall behind, and then another, and another.

If an ex-Fighter shows you how to pick your way through the uncleared minefields, marked with a grinning skull and crossbones, it is possible today to walk the trenches, which trail across Nakfa's slopes like worm casts on a sea-washed rock face. The camouflaged screen roofs are gone now, burnt as firewood by local shepherds. But the thick stone walls, rising higher than a man's head, still stand proud. They are interspersed with pokey underground antechambers where meals were eaten, classes held, sleep snatched and marriages consummated. The slabs of stone fit together with a neatness any Yorkshire dry-stonewaller would admire – the knowledge that a crack could let a bullet through encouraged a certain mathematical precision. At regular intervals there are neat gunsights. If you peer through one of these slits, you will be brought up short by the sudden, seemingly magnified glimpse of what kept the Fighters on their toes. Across the valley, with its own gun-sight trained precisely upon you, sits an Ethiopian position. As I knelt in the dust and imagined what it must be like to stare at your would-be executioner every day, I was reminded of the brown line of dust that marked my T-shirt every time I belted up in a friend's car. Seat belts are always either filthy or broken in Eritrea. Having already braved death in so many more obvious ways, no one fusses over the dangers of flying through a windscreen.

Like the British squaddies at Keren, the Fighters came to know every inch of their natural fortress. Facing Nakfa to the south was Den Den, the mountain the Ethiopians struggled most fiercely to capture, as controlling it would have allowed them to turn their artillery directly onto the plateau. The place where Eritrean and Ethiopian lines lay barely 50 m apart – so close that Eritrean Fighters could hear the news being announced on Ethiopian radio sets – was known as *Testa a Testa* ('head to head'). *Fornello* ('oven') was the area behind Den Den which saw the heaviest bombardment, *volleyball* the sport which involved Eritrean Fighters scrambling to lob Ethiopian grenades back over the parapets before they exploded. To the east rose *Sulphur Mountain*, regularly doused in flaming napalm. The winding track up from the valley was known as *Teamamen* ('We are certain') – Eritrea's equivalent of 'No Pasaran'. No Ethiopian force, the EPLF decided, would ever be allowed to breast its hairpin turns. Upturned in the ravines, the rusting Soviet trucks and tanks show it was a promise they kept.

What is it like to live a subterranean existence, one eye always on the sky for possible danger? You learn to merge with your surroundings as swiftly as a chameleon. Put yourself in the mind of an Ethiopian pilot, and see what he sees, roaring by so fast in a MiG his eyes can hardly fix on objects on the ground.

High above you, scouring these drab expanses of rock and scrub, he is wondering how anyone can survive in this dry landscape. He is looking for something out of the ordinary – a sudden movement that is more than a baboon or a goat, a bright flash of colour, swirling tyre tracks in the sand, some clue that there are humans below. So do nothing to attract his attention. Wash and dry your clothes indoors – a flapping shirt will draw him to you. If you must light a fire, remember that dry wood gives off little smoke, but damp wood is dangerous. If a truck comes to your area, sweep away the tracks immediately. Bury food tins – they can catch the sun and flash up a deadly signal. Before you go out, make sure you turn your watchstrap so the face points towards your body, not outwards, or pull down your sleeves and tuck them in to cover the glass. Forget about rings and jewellery – they glint too. Never wear white, or red, they are visible a long way off. Choose khaki, grey, anything dull. Avoid unnecessary large gatherings, that way, if a bomb hits home, the casualty toll will be lower. And if you're unlucky enough to be caught outside when a plane appears, then do as the rabbit, lizard and snake: freeze in your tracks, hold yourself as still as stone, so still you become part of the landscape, impossible to distinguish from the boulders and bushes, until the plane has gone beyond the horizon and the sinister game of Grandmother's Footsteps can resume. It can drive you close to madness, all that time in the sweltering darkness under the rock, like an antlion in its hole. When the bombing starts and you feel each boom vibrate through the earth's bowels, you want to break out of your living tomb and take a deep breath, because it feels better to see death coming out in the open, than to sit silently and wait. But this way you get to outsmart him, that Ethiopian pilot who wants to kill you. This way you get to live.

There was a seductive simplicity to this existence, for the

knowledge of your own righteousness brings with it a deep sense of peace. For the 30,000–40,000 students and school graduates who became Fighters in the 1970s, the hardest step was deciding to join. But once that route had been chosen, everything acquired the clarity of absolutism. Choosing the hero's path not only snuffed out the existential questions that torture young people, it put paid to the humdrum tribulations of daily life. The Movement expected you to be ready to die for it, but if you happened not to, it took good care of you. Financial worries became a thing of the past, for Fighters, while never paid, were fed, watered and issued with sandals and uniforms. Anything they owned, or were given, was handed in to central stores for general distribution. Commerce was to become such an alien concept that, on walking into bars in Sudan when on leave, Fighters would be nonplussed when they were asked to pay.

Sex was initially taboo – no Kagnew debauchery here – and men caught breaking the celibacy rule were sent to collect salt in the white heat of the Red Sea shoreline, while the handful of women Fighters who made the mistake of falling pregnant were ostracized. But as the campaign stretched on, the Movement accepted the principle of relationships so long as Fighters submitted them for prior approval to their commanders, who decided whether to issue the woman with the pill. Those bent on marriage were obliged to complete a long form showing they had thought through their choice of partner; the Head of Department retained the right to defer a marriage application. Even in this most intimate of areas, nothing was left to chance, and since the Zero School took care of any resulting offspring, Fighters were spared the domestic strains that usually go with parenthood.

Whether you were a Moslem or Christian, middle-class professional or illiterate peasant was deemed irrelevant, so there was no room for snobbery or prejudice. All were equal, and if a relative sent a packet of cigarettes, a bag of sugar or tea, it was automatically shared around. Prior claims of parents, fiancées and children gradually faded, as trench companions came to play the roles of friend and confidant, protector and brother. Life at the Front would be hard in terms of physical suffering, loss and privation. But, for the average Fighter, it would not be plagued by squabbles over promotion, worries about school fees, or the pressure to keep up with the Joneses. 'In a way, we were not fully human,' one former Fighter later ruminated, 'because all the things you associate with being human – setting up a home, bringing up children, holding down a job – we did none of that.'

There was an androgynous sameness about the Fighters which reflected the Front's rejection of sexual stereotypes. With no time or money to waste on the frivolity of hair-straightening, the women, like the men, let their hair grow into wild Afros. They dressed *not* to impress, for in the era of chastity, any hint of flirtatiousness – an undone shirt button, for example – was viewed with disapproval. This egalitarian army, which prided itself on its collective leadership and absence of ranks, adopted the most utilitarian of wardrobes. Each Fighter was issued with a Kalashnikov and a belt with three pockets: one held a Chinese grenade, the second a hand-ful of bullets and the third a folded *netsela*, a thin cotton cloth which served as blanket, shawl, towel, rope and carrier bag. The EPLF could not afford more. Once it lost its Arab backing, the Movement depended on the Eritrean diaspora, exiled pro-fessionals in Italy, Sweden, Canada and Washington, to cover its running costs. From shabby offices in Western capitals, EPLF members coordinated the fund-raising, one of the most efficient and sustained tithing operations ever set up by a rebel movement. This was insurrection on a shoestring, and the

Front quite literally cut its cloth to meet its limited budget.

I used to puzzle over the detail of a famous photo taken during the Nakfa years, showing a group of young *tegadelti* striding towards a mountain summit, the EPLF's green and red insignia billowing behind them. The image has become iconic, gracing the walls of Eritrean embassies, stamped on coins, reproduced on posters. In it, the Fighters' shorts end at about the level of a pair of 1960s hot pants, exposing a generous expanse of rippling brown thigh. 'Why are their shorts so short?' queried a friend. 'I've never seen any other African force with shorts that length.' Even the British army, which complained of the impracticality of its uniform in Eritrea, wore shorts that fell to the knees. The truth, I discovered, was that the Fighters had made a virtue out of necessity. If shorts were cut high on the leg, EPLF tailors could get that many more pairs out of a length of cloth. In the same spirit, a Fighter sent a pair of long trousers by his relatives was expected to sacrifice them to the Cause. 'They'd cut off the legs and give you back a pair of shorts, then use the rest to make more,' recalled a former Red Flower. 'At first it made my heart bleed, to see the nice new trousers my mother had sent me on the backsides of other Fighters, people I didn't even like. When I went home I'd hide indoors, just so that I could feel material on my legs without anyone making fun of me. But then it became a sort of competition, to see who could wear the shortest shorts.' The skimpier the shorts, the braver the Fighter. Ethiopian troops, it was said, paled with fear when they saw the physiques of slain Eritreans, displayed in all their muscular glory. And so the tailors wielded their scissors, and hems inched upwards.

The *tegadlai*'s other main item of clothing also came with a story of ingenuity attached. Black plastic sandals were first introduced to Eritrea by Raffaello Bini, a Florentine who arrived in the region as a photojournalist in Mussolini's invading army.

Falling in love with Eritrea, he stayed to set up a shoe company in Asmara. Registering that ordinary Eritreans were in desperate need of footwear but couldn't afford leather shoes, he imported machines and designed a cheap PVC sandal, nicknamed the 'Kongo' after Hong Kong, traditional source of plastic bric-à-brac. When the Derg regime seized Bini's factories, his workers volunteered for the Front, taking their skills with them. In first Sudan and then Orota, they installed machinery to turn out the Kongo, melting down old car tyres instead of importing PVC. A dirt-cheap sandal that could be rinsed free of grit and repaired over a camp fire suited the EPLF's requirements perfectly. 'We used to play mind games with the Kongo, because we knew the Ethiopians were reading our tracks. The Sudanese, Orota and Asmara versions of the sandal were all slightly different, so we'd cut them in two or melt them over the fire and leave footprints that would make them think a unit from the Sudan was operating in the area, or that there were only farmers around when, in fact, the district was teeming with EPLF,' remembers an ex-Fighter.

The British explorer Thesiger, drawn to austere vistas, believed that the harsher the landscape, the purer the mettle of those who live off it. Few experiences came harder than the Sahel and it encouraged a puritanical earnestness that lay at the other end of the moral spectrum to Kagnew's drunken japes. Drugs, alcohol and gambling were all shunned. There was a spiritual element to this disapproval. 'To pick the word freedom, you have to pray, you have to cleanse yourself, in the Biblical sense, wash your sins away,' said one former unit leader. Another commander tried to explain the intensity of focus that developed amongst the Fighters. 'It's like Pele,' he said. Registering my baffled expression, he expanded: 'Pele's whole life has been football. He cannot do anything else, he can't suddenly find a new career. Football is what he does. Pele will always be football, even when he's dead. We were the same.'

It was both the best and worst of times. Looking back, ex-Fighters remember this as a period of supreme happiness, the unthinking happiness of the very young. But it was also a time of tragedy and heartbreak. When the war swung against them, the comradeship that had developed in the trenches made the pain of bereavement unbearable. A readiness to make the supreme sacrifice tipped easily into a love affair with death, an impatience to get the whole tricky thing that constitutes this, our human existence, over and done with. 'What did you feel when one of your friends was killed?' I once asked an ex-Fighter. 'Did you think that maybe you had made a mistake, maybe you should not have joined the Front?' 'No,' he replied. 'When one of our commanders died you just thought: "I wish I had died alongside him. I don't want to be left here on my own." The worst thing that could happen to you was to be left behind, alive.' When fathers took time off to visit their offspring at Zero School, their children would sometimes run and hide. 'It was considered a mark of pride to have had a father who had been killed at the Front,' recalls a former Red Flower. 'When your father came to visit, you didn't want to be seen in public with him. Why was he there, you wondered, rather than away fighting? Maybe he was a coward.' As each Fighter saw siblings, lovers and friends 'martyred' around him, the invisible scar tissue formed. There is only so much mourning a human being can perform before emotional numbness sets in. 'When people now say "I love someone" or "I care for so-and-so", I just don't know what they mean,' an ex-Fighter once confessed. 'I don't hate people, obviously. I do an awful lot to help those around me, but I feel nothing. I think it started in the 1970s, when so many people I cared for died. I probably need a psychiatrist,' he said with a laugh, 'but here in Eritrea we don't do that.'

What takes the breath away is the extent to which the EPLF determined, in these testing conditions, to carry on regardless. The Derg – which, like Haile Selassie, denied the very existence of an independence struggle – hoped to reduce the *shiftas* to the brutish necessities of survival, modern cave men scrabbling for sustenance. Africa teems with rebel movements with portentous acronyms that amount to little more than armed raiding parties. The EPLF would not let itself go down that route. Maintaining the standards of civilized society in this rugged terrain was not only a means of convincing the mass of the Eritrean population that the Movement was fit to rule, it was a way of showing oneself defiant in the face of overwhelming odds. A la Frank Sinatra, they would do it Their Way.

In the early days, the Fighters exhaustively monitored the airways on their short-wave radios, hoping against hope for some indication their Struggle had been noticed by the BBC, Radio France Internationale or Voice of America. Registering the world's total indifference, the Movement decided it would have to provide its own media. The EPLF printed its own newspapers and political pamphlets, researched and filmed documentaries and ran a mobile radio station that broadcast its version of the campaign. Man cannot live by war alone, so cultural activities were always encouraged, with units expected to compose songs, work on poetry and write plays for staging in Nakfa's underground theatre. Education was another priority, for if one day Eritrea was to become a modern state, the peasantry must be made politically aware. In Nakfa, there was little of the jaw-aching boredom associated with most military campaigns, for the Eritreans became experts at keeping themselves worthily busy. Perched in their mountain eyries, former farmers attended adult literacy classes given by their educated comrades and studied the great political thinkers, looking for lessons that could explain their predicament and reveal the future. If so many Eritreans today show a disconcerting understanding of Henry VIII's clash with Rome and a grasp of the origins of the First World War, it can usually be attributed to those classes. The one luxury the EPLF enjoyed, after all, was time.

No allowances were made for circumstances. One of the Fighters I spoke to had been stationed atop Sulphur Mountain, where putting a foot wrong meant falling to your death. When his unit needed water, the Fighters formed a human chain and, backs pressed against the cliff face, passed the laden jerry cans carefully from one shoulder to another. As there was not enough flat ground for a latrine, Fighters would wrap one arm around a sapling that leant over the void, undo their flies, and relieve themselves into the abyss. 'Not even a snake or a monkey, not even Jesus Christ himself could have survived there,' he remembers. Yet the unit did not skip its three hours of morning study. 'One member of the unit would prepare a subject and we would talk about it: the character of war, Mao's teachings, the Irish question and the issue of Palestine. We had to expand our global outlook.'

The very isolation of the Nakfa experience, the absence of worldly distractions, encouraged a clarity of thought the meditating monks of Shangri-La would have recognized. The OAU had labelled the EPLF a secessionist movement, Washington dismissed it as a bunch of Commies, Moscow wanted it to simply go away. Global rejection created a space and distance in which cool analysis could unfold. Eritrea's successive betrayals by Italy, Britain and America were dissected; the hypocrisy of UN resolutions guaranteeing the right to self-determination examined; the legal justification for Eritrean independence logged with bitter calm. Marx was read with an attention to detail normally reserved for Bible classes. 'We didn't just read it, we savoured it, we digested it, we mulled over the meaning of every phrase. It had the force of a spiritual conversion. Even today, I find myself applying the mental disciplines I learnt then,' recalls a former EPLF ideologue.

Any movement with pretensions to intellectual credibility must be able to hold seminars and debates. The EPLF did just that, inviting foreign politicians to attend congresses where it articulated its ideological differences with the ELF, debated the shortcomings of classical Marxism and hammered out its blueprint for an independent Eritrea. What did it matter if such events were held at night, in the shelter of crags? The fact that they were held at all was a miracle in itself. Nakfa's constraints were never going to stop the Fighters staging a sports convention, however incongruous it might seem, or inviting delegates to attend a symposium on Third World debt. Visiting in 1979, French journalist Olivier Le Brun described the surreal experience of listening to a piano recital given under a thorn tree by a woman Fighter. The performance was interrupted by a MiG bombing raid, but resumed immediately afterwards. No wonder so many of the Western journalists, left-wing politicians and aid workers who visited the trenches returned True Believers, when such quixotic displays of true grit were on offer.

The napalm bomb struck within a few metres of the open-air kitchen, splashing its lethal gel in a phosphorescent starburst. As culinary mistakes go, this was one no cookbook ever thought worth mentioning. While the ground blazed, hospital staff scrambled to shovel dirt on the flaming liquid before it reached the dormitories where wounded patients lay resting. John Berakis had committed a blunder that could have cost his colleagues dear. His preparations for the hospital's evening meal had coincided with a random over-flight by an Ethiopian plane. Spotting the light from the stove fire flickering in the darkness, the pilot dropped his load with impressive precision. 'It lit up just like a Christmas tree,' John remembers, with a guilty laugh. For the rest of the night, staff and patients crouched in the darkness, as one Ethiopian plane after another roared over the narrow gorge, taking turns to try and finish the job.

It is not easy, learning to be a chef when you have been posted to the frontline. The quest for gourmet excellence might, indeed, have seemed a tad counter-intuitive in the bleakness of the Rora mountains. But for John, a man whose brain churned with the relentless energy of a Magimix, pursuing his dream was a way of proving he was still human. 'The Ethiopians' whole aim was to terrorize us so that we couldn't work. So you carried on with your work, regardless.'

Had John been born in the West, I have little doubt that by now he would be running a chain of five-star restaurants, a suave maître d'hôtel gliding to welcome his guests in a sombre silk suit. But John was born Eritrean, and that has made all the difference. Baptized Tilahun, an Ethiopian name, and brought up south of Addis, he spoke Amharic so fluently that all his friends and girlfriends assumed he was Ethiopian. He was careful never to speak Tigrinya in their company and when they talked about Eritrea, that bolshie province up north, he kept silent, pretending to share their views. But something, he knew, was not quite right. Like so many Eritreans of his generation, he felt a creeping unease. There was a sense that he was not like other folk, that he was playing a role that chafed. Once he moved to Addis, the moment of epiphany came.

He had returned home after dark to find a small crowd of worried Eritreans in the family compound. A mentally-retarded boy had disappeared, and concerned relatives had congregated to discuss where the missing youngster might be. The sight of this crowd, gathering after curfew and speaking in a foreign language, was too much for the Ethiopian security services. 'The Derg rounded everyone up, made us put our hands against the wall. Then they beat us with their truncheons and arrested us. All that, just for speaking Tigrinya.' That night, John realized his camouflage was not going to work very much longer. The contradictions had become unbearable. He set off for the north. Doing his best to avoid the main highways, he bluffed his way through a series of police checks until he finally reached the Sahel and joined the Front.

The EPLF system required everyone to be capable of combat, and John did his share. He learnt that it is possible, after your mind has been drenched in the limb-loosening adrenalin of terror, to find a still, calm place where rational judgements are coldly made, the hard core of the soul that constitutes courage. 'In war, you only ever fear the first bullet. Once the shooting starts and you are in the thick of battle, you don't worry any more.' Stationed in one of the valleys east of Nakfa, he was part of an infantry unit which hunted down a force of Ethiopian soldiers helicoptered into EPLF-held territory in an attempt to break the military stalemate. 'For eight hours, we killed them and we ran, we killed them and ran. We killed them until we were out of breath and could run no more.' He had seen friends die, killed by bombs so big they dropped to earth in their own parachutes and, on explosion, left no body parts behind. 'There was nothing to collect, not even a fingertip. They just vaporized.'

John also worked in the laboratory at Tsabra hospital, an underground clinic hidden in a valley on the outskirts of Nakfa, testing blood groups. The Front did not own a refrigerated unit where blood could be safely stored. So the EPLF found its own ingenious way around the problem. By testing Fighters beforehand and cataloguing their blood groups, it created a living, breathing blood bank requiring no storage facilities. When the battle turned fierce and the wounded streamed in, they knew exactly which Fighter to summon for a blood donation. It was while working in the laboratory that John received his own worst wounds. A colleague carelessly forgot to switch off the methane gas before retiring. When John entered to start his night shift the gas exploded, searing his head, upper torso and the arms he threw up to protect his face. His hands and forearms remain a mess of scarring, the skin buckled and twisted by the heat.

But John possessed other useful skills. He had been one of the first candidates to complete a new hotelier course in Addis. He had trained in the Hilton's kitchens – a job he enjoyed so much it barely counted as a chore – and worked as a waiter, donning black coat and white gloves to serve caviar canapés to Haile Selassie, cheekily attempting to catch the Emperor's eye to see if the story that no man could hold his gaze was true (it wasn't). When a consignment of books sent by EPLF sympathizers abroad arrived in Nakfa, the Fighters would sort through it looking for what was useful. Those on hotel management and catering – guides published by training schools in Kenya – ended up on the scrapheap, only to be fished out again by John. He devoured them, page by page, memorizing the jargon. 'When someone was leaving for Sudan and asked me what I wanted I would always say "send me hotel books".' Laying out petri dishes for the laboratory, watching the bacteria spores grow on his preparations, John mulled over the principles of nutrition and hygiene, cause and effect. He was ready to risk his life, endure the bleakness of the Sahel without complaint. But, in the meantime, he would also learn how to prevent a *béchamel* sauce from tasting floury, how to squeeze stock from a scrag of chicken, and what cookery writers meant when they talked about the *purée* and the *gratin*, the *julienne* cut and the *roux*.

When, in 1790, a French officer who had been arrested for duelling found himself confined to quarters for what seemed an endless 42 days, he warded off boredom by launching a mental voyage of discovery around his bedroom, exploring the ideas associated with every humdrum object. When Xavier de Maistre, author of *Voyage around my Room*, was told his confinement was over, he felt nothing but disappointment: the intellectual journey had proved so enriching, he did not want it to end. There was a similar cerebral quality to John's culinary obsession. This was 'virtual' cuisine, a form of mental gymnastics staged almost exclusively in his own imagination, practised on gleaming steel kitchen surfaces he would never possess, using market-fresh foodstuffs he would never receive.

'At the start, when we were moving from place to place, we were quite literally in the Stone Age. To bake bread, you would find a flat rock, prop it up on three stones, light a fire below and wait for the flat stone to heat to a point where you could cook on it. The stones were our pans, our plates and our tables. It was back to the primitive ways.' The position improved slightly when John was assigned to Tsabra. Here, at least, permanent stoves could be constructed, equipment gathered. But he never got a chance to put the lessons of Elizabeth David and Escoffier – John always favoured the classics – into practice. 'Most of the time all we had were lentils, poor man's protein. You boil them, throw in some salt and eat that twice a day. There's not much room for invention.'

If he couldn't deliver on quality, John soon learnt how to provide quantity. From the air, Tsabra was effectively invisible. At first glance, a casual passer-by would have seen only a

V-shaped valley traversed by a clear stream squirming with tadpoles, the odd goat scampering through green sprigs of wild olive. If he lingered, he might have noticed a suspicious number of comings and goings, or registered that the surrounding slopes seemed strangely bare of trees. In fact, this was Nakfa's main referral hospital, a 200-bed facility where doctors received the injured from the trenches and decided which patients needed immediate treatment and which could risk the arduous trip north to Orota. The spot had been chosen because of the river – every hospital needs plenty of water – and because the gorge was narrow enough to present Ethiopian bombers with a challenge. A hidden generator supplied electricity to the maternity ward, operating theatre, lab, dormitories and offices, all built on the same principle. The Fighters had dug deep into the rock, hollowing large rectangles out of the red earth, building up the stone walls until they were thick enough to withstand bombing, then covering the lot with wooden screens heaped with soil and shrubs. It was a big facility and people needed to eat. Juggling his jobs as lab technician and head cook, John regularly turned out meals for up to 500 people.

But his gifts really blossomed when John was reassigned to the laboratory in Orota and was handed the responsibility for feeding the main hospital's 3,000 patients. When he arrived, each department ran its own kitchen, the women walking miles to find increasingly-scarce firewood. Looking at the denuded mountain slopes, John realized the hospital's fuel needs were causing an ecological holocaust. The energy-efficient answer was a central kitchen with a giant, multi-hobbed oven, built according to John's careful specifications. Friends in London were sent a list of the equipment he needed: a dough machine, a pastry mixer, a vegetable slicer, eight 200-litre cooking pans. He had moved on from lentils and salt, introducing poultry to the Sahel, and the arrival of a flour mill meant the Movement could grind its own grain and bake bread. A few more ingredients were becoming available, as Fighters began planting vegetable gardens and raising livestock. Soon the kitchen, staffed by 50 women trained by John, was dispatching containers of hot food and bread to the hospital departments each morning, and fuel consumption had been cut by 80 per cent. In his own way, John had managed to recreate what he had seen and yearned for at the Hilton as a young man: 'I had my modern kitchen.'

Soon, his reputation spread. Tilahun had long since been abandoned in favour of 'The Man'. John was The Man who, if you were getting married, could be relied on to make the open-air banquet a success. He was The Man who could work wonders with a sack of lentils, jerry can of oil and some onions. He was The Man who could be trusted to make sure guests attending a conference on child poverty did not leave grumbling. Life in Nakfa and Orota was not without its frenetic social whirl. War or no war, the workshops and symposiums must go ahead, and it was obvious who could be trusted with the catering: The Man. 'One time I slaughtered 76 goats and 9 cows. The biggest conference I ever catered for was the 25th anniversary of the Revolution, when we had 6,000 delegates,' remembers John. 'People came from all over and slept under the stars.'

He had learnt a few little tricks to tempt appetites. On a really good day, he'd be given a couple of bulls to slaughter and guests would dine on steaks basted in garlic and butter, roasted under the trees on large metal trays. But you can't feed a multitude on steaks, so John became a master of the stew. Even that presented an occasional challenge. People, he had discovered, liked their stews dark in colour. But with wine an unattainable luxury and tomatoes in short supply, this sometimes presented a problem. 'If the sauce was white, people would refuse to eat it. So we would slaughter a goat, drain its blood, mix it with salt to stop it coagulating, and use that as colouring. We'd say it was tomato salsa and then everybody would eat it.'

The limitations of this dour lifestyle must have been exasperating, but John never appears to have experienced second thoughts, not even when concerned relatives arranged a Swedish visa or set up a lucrative job in Hawaii. He had made his decision at the age of 25, and that was that. Not for him the self-interrogation of Kagnew's Vietnam-dodgers, uneasily aware they had ducked their generation's greatest challenge. 'A promise is a promise. You cannot go back on it.' He had quietly calculated the odds and worked out that he was unlikely to see independence. But he had no doubts liberation would come, for others if not for him. 'Once you've done your training and you've been politicized, and you've studied Mao and the struggle of the masses, Lenin and the Russian Revolution, then you know that eventually, you must win. It may not happen in your own lifetime, but eventually, you will win.' Sometimes, he allowed himself the occasional daydream. If – it seemed a very big if – he ever made it through, then he would open his own restaurant and hotel school, training young Fighters to be chefs and waiters, hotel receptionists and chambermaids.

He had expected to die, but his luck held. With the liberation of Asmara – a city he had never set foot in before entering it as part of a conquering force – he was given the job of running the canteens at the old Kagnew base. Then he demobilized and took over responsibility for four UN kitchens and the UN's water bottling plant. A tiny business empire is being created, but John – who rises at 4.00 am each morning to ferry his workers from site to site – gloats not over his profits, but the contribution he is making to Eritrea's new state. 'I have

150 people working for me and each supports a family, each pays his taxes. That thought gives me a lot of satisfaction.' The shelves of his office hold glossy cookery books written by Madhur Jaffrey and Robert Carrier. But pride of place still goes to a pair of well-thumbed hotel management guides, published by a 1970s Nairobi business school, rescued long ago from a garbage dump in the Sahel.

Peopled by such driven citizens, Nakfa represented Eritrea at its best. But as I spoke to ex-Fighters, I began to wonder if it had also contained the seeds of Eritrea at its worst.

What appears to the individual as admirable clarity of thought can seem to the outsider dangerous simplification, a vision stripped of the messy contradictions that mean a situation is rarely as straightforward as it first appears. Isolation allowed the Fighters to hone a steely resolve that enabled them to achieve the seemingly impossible. But if the hermit's life shields you from temptation, it can also stunt your intellectual growth. Rejecting capitalism, with all its vices, came easily to those who had never been exposed to its virtues. 'Had I known about all of this,' exclaimed Eritrea's former vice-president Mahmoud Sherifo, absorbing the bustle of a Western city on his first visit to London, 'I would never have fought so long in the bush.' Insulated from Africa's contemporary reality, it was easy for the Eritreans to make the mistake of assuming they knew all the answers. The awareness of how poorly the colonialists and superpowers had behaved, the bitterness of seeing their natural ally opt for Ethiopia, the knowledge of the continent's casual indifference: it all encouraged the belief that the Movement had nothing to learn from its critics, whether black or white. Like every rejected minority before it, the EPLF convinced itself its very solitude was proof of moral superiority. 'We are certain' was more than just the name of a punishingly steep mountainside, it was the Movement's unstated leitmotif. A leitmotif that hardened like rock during the Nakfa years.

At what point does such purity of purpose cross the line into oppressive authoritarianism? Even those who today pine for a lost golden age acknowledge that individualism was not a quality valued by the EPLF. This was a military organization, after all, and true democracy, with its tolerance of mavericks and loudmouths, is not suited to waging war. At daily meetings, Fighters would publicly pick over each other's revolutionary failings, 'self-criticism' was strongly encouraged. 'There were spies in the Movement who would befriend you, listen to your ideas, pretend to sympathize with your complaints and then, during a meeting, denounce you as "petit-bourgeois" or accuse you of being a "regionalist",' remembers an ex-Fighter. 'People who had taken degrees were made to apologize to the peasantry for their education and privileges.' Such obligatory abnegation fitted in well with the Eritrean national character, the tendency, developed through decades of colonial occupation, to sit in impenetrable silence, accept authority – at least on the surface – and keep one's thoughts to oneself. 'A lot of people thought it was bullshit. The EPLF had been set up by "petit bourgeois" people, after all, most of the leaders had been students at Addis University. But we were taught that the whole world would soon become socialist, so it was up to you to adapt. You learnt to say the right things, keep a low profile and play the game. I went along with it, because I had joined to free my country, and this seemed a price worth paying. But there were some who couldn't stand it, and they deliberately martyred themselves in battle.'

This was the dark side of all the dogged determination, but it was a darkness visiting Westerners were reluctant to recognize. 'At the time one was just swept away by the hard work and efficiency and self-sacrifice of it all. But looking back, you do wonder if there wasn't something rather disturbing about a movement that exercised that level of control,' says Trish Silkin, who visited the front as an anthropologist and aid worker in the 1970s. Dissent, especially dissent that crystallized into direct challenges to Isaias' burgeoning control, was ruthlessly smothered. Even today, ex-Fighters close down whenever the question arises of what happened to the ring-leaders of these internal challenges, made to 'disappear' with typical Eritrean quietness.

Once, on the long trip back from Nakfa, I got to exchanging metaphors with two former Fighters. Our banter was prompted by what is popularly known as the Heart of Tigray, an infamous stretch of road between Keren and Asmara. The nausea-inducing road, as twisted and torturous – so the proverb goes – as the hearts of Eritrea's treacherous neighbours in Ethiopian Tigray, winds its way through bulbous rock formations and the giant candelabra of euphorbia cactus. 'So, if there was a road that symbolized the Eritrean heart, how would it be?' I teased. 'Absolutely straight,' came the cheerful chorus. 'What do *you* think?' 'I think it would be dark, hidden, and very mysterious.'

Yet perhaps Nakfa's most dangerous legacy was not the EPLF's indomitable self-belief, its profound distrust of out-siders or its iron control, but the impossibly high expectations raised in a generation of Eritreans.

During their lessons in the trenches, EPLF ideologues conjured up a vision of Free Eritrea, a prosperous land in which farmers tilled fertile fields, fishermen trawled teeming waters and industrialists tapped long-neglected deposits of gold, potash – even, perhaps, oil. If Eritrea was barren and denuded, they taught their classes, it was only because its forests had been systematically stripped by first Italian developers and then the marauding Ethiopian army. Independent Eritrea would blossom anew. Saplings would be planted, rivers dammed, terraces built. Gazing across what resembled the surface of an asteroid, the Fighters dreamt, like the dying Falstaff, of lush pastures and green bowers, where knobbly trees of uncertain age cast their cool shade. It was a landscape, they came to believe, that had been stolen from them – just like everything else.

That glowing dream of paradise is still captured today in the most everyday of items, all the more poignant for their functional banality. Walk into any Eritrean roadside restaurant, where Christmas tinsel serves as year-round decoration, and you will find yourself sitting at a table decorated with grape clusters and shiny red apples, tumbling alpine torrents and dewy lawns. Perhaps these made-in-Taiwan wax tablecloths are simply the cheapest things on the market. But, like the glossy calendars on the walls, like the murals lovingly painted on the walls of Eritrea's coffee bars – all green glades, quiet pools and rolling meadows – they express Eritrea's vision of Heaven, its Elysian Fields.

I only registered the Utopian quality of that vision one day in a library in Rome. Leafing through some old Italian encyclopedias, I came across photographs of late 19th-century Eritrea, taken before the saw mills and napalm had done their worst. The black-and-white photographs certainly showed thicker tree cover than I was used to seeing. But this was nothing like the green haven lovingly described by my Eritrean friends. Even before the colonial depredations, before the Ethiopian army had got to work, much of the country, it was clear, had already been a dry scrubland of punishing harshness. The realization came as a shock to me. How much more of a shock would it prove for the thousands of Fighters who risked their lives fighting for a land of lost content, a country that had, it seemed, existed largely in their imaginations?

CHAPTER 15

Arms and the Man

'Ethiopia will be destroyed by the very thing that seems her strength and glory: arms.'

Ferdinando Martini

In Eritrea, history always comes tightly compressed, physical evidence of just how many turbulent, world-shaking events have been squeezed into a few narrow centuries. To the west of Asmara, the gates of the 19th-century Italian cemetery which holds the bones of Martini's settlers virtually rub shoulders with the walls of Kagnew Station. And just behind Kagnew stretches a large patch of wasteland which stands as testimony to the last, most lavishly destructive, phase of superpower involvement in the Horn of Africa.

Locals call these abandoned acres 'Tank Graveyard' and at first glance this looks like the scene of some apocalyptic clash, a giant confrontation into which every weapon known to 20th-century military technology was successively hurled. Upturned green jeeps lie on their backs, displaying their axles to the skies as shamelessly as a drunk old woman exposing her knickers. Scores of armoured personnel carriers crouch like brown crabs in the weeds, white butterflies fluttering through their blank view holes. Tanks and cranes, amphibious vehicles

311

and anti-aircraft guns, petrol tankers and mortar launchers – even the odd fighter plane – all lie tumbled in a mess of rusting metal. But, for the most part, the Tank Graveyard holds the remains of hundreds of heavy duty trucks, the basic working tool of any army. Crumpled by explosions, dented by impacts, they have been stacked like airline dinners, five-deep to save space. The highlands wind thrums through the towers built from their twisted chassis, while brown kites shrill their cool, haunting lament from an impossibly blue sky.

The Tank Graveyard is not, as one guidebook to Eritrea claims, a spillover from Kagnew Station. Most of the machinery here is of Soviet make, not American. As the war in Eritrea escalated, and the number of disabled trucks, tanks and personnel carriers littering the province rose, the Derg dragged the damaged hardware donated by its Soviet friends here for dumping. At best, the carcasses could be tinkered with and sent back into battle, although, to be honest, Ethiopian technicians never proved particularly adept at repairs. At the very least, depositing them here would keep them out of the hands of the guerrillas, who were quick to teach themselves the operating principles of captured machinery and then turn it on its former owners. When the Struggle ended, the EPLF completed what its enemy had begun, removing the debris that was cluttering Eritrea's scarce agricultural land.

It is easy to miss when confronted with so many thousands of metal corpses, but the Tank Graveyard has been quietly shrinking as the years go by. Unable to afford the price being asked on the international arms market for new tanks, Eritrea's armed forces have been resurrecting the classics, tapping the pile for old T54s, T55s and T62s. Underneath the brown layer of rust, the thick metal carapaces hold true. The simple Soviet mechanical systems have weathered the passage of time with an ease no state-of-the-art weapons system could match. In three weeks, the 50-year-old Soviet models can be refitted, ready once again to fend off the latest threat from Ethiopia. 'New clothes are obviously better than old clothes. But if you don't buy new, you use the old. This is not just scrap, it's our stock,' says Colonel Woldu Ghebreyesus, the former EPLF Fighter who now heads the army's tank department. He's an unabashed aficionado. 'I've been using these captured tanks for the last 30 years and I've come to really appreciate Soviet technology. I have no criticisms at all.'[1]

The fresh uses to which the contents of Tank Graveyard are still being put underline the bleak message of this stretch of wasteland. Here sits a monument to military oversupply, testimony to excess. Generous to a fault when it came to military hardware, the Soviet Union was to end up sending enough weaponry to the Horn of Africa for not one, not two, but five separate conflicts: Somalia's war on Ethiopia, Ethiopia's war on Somalia, the Derg's battle against the Eritrean rebels, the Eritreans' campaign – using stolen machinery – against the Derg and, finally, most recently, independent Eritrea's two-year border war with Ethiopia. 'Of course, now it is possible to say it was too much,' reluctantly acknowledges Sergei Sinitsyn, who served at the Soviet embassy in Addis in the 1950s and 1970s. 'But in time of war you are swept away by immediate needs and requests. In any case,' he adds, almost as an after-thought, 'if it had not been us supplying, it would have been the Americans.'[2]

Most countries in Africa have been traumatized by the dance of the Cold War, that brutally simplistic era in which what mattered, for any African nation hoping to be picked as part-ner by a superpower, was never good government, financial transparency or enlightened agricultural reform, but whether the leadership concerned chose to mouth the platitudes of Communist orthodoxy or free-market capitalism – each, in its own way, equally out of step with African reality. What made the Horn unique, and left it uniquely damaged, was that midway through the Cold War tango, the two dancing couples – Somalia and the Soviet Union in one corner, Ethiopia and the United States in the other – separated, strode past each other on the ballroom floor, and swapped partners. The fact that such a swap could take place at all exposed the moral vacuity of the pairings. But it also had terrible implications for the nations concerned, where the military stakes were ratcheted to ever giddier heights. During the years in which Washington funded Haile Selassie's military expansion programme, the Soviets dispatched up to $1 billion[3] in weapons and military know-how to Somalia's President Siad Barre, desperate to keep pace with his neighbour. When Moscow suddenly became Ethiopia's new best friend in the late 1970s, its military advisers, ejected by the furious Siad Barre, found themselves facing black Africa's fourth most heavily-armed state – the state they themselves had equipped. There could only be one answer: yet more arms deliveries, aimed at neutralizing the impact of Moscow's previous largesse. The Soviet Union almost fell over itself in its determination to make up for its strategic gaffe. For three months between 1977 and 1978, 225 Soviet transport planes – 15 per cent of Moscow's air force – ferried 60,000 tonnes of hardware to Ethiopia for the war over the Ogaden, one aircraft landing every 20 minutes.[4] The Eritrean front, former EPLF commanders estimate, was destined to receive 800–1,000 tanks, 2–3,000 trucks and an untold number of machine guns, heavy artillery, mortars and the multiple rocket launchers known as 'Stalin's organs'. By the end of its 14-year relationship with the Derg, Moscow had poured nearly $9 billion in military hardware into Ethiopia, working out, at the roughest of estimates, at over $5,400 in weaponry for every Ethiopian man, woman and child.[5] For a developing African nation experiencing one superpower-funded arms race might be regarded as bad enough. Two really verged on the excessive.

Today, many former Soviet policymakers have the grace to feel embarrassed about this deadly double crescendo. 'It was our usual trouble,' ruminates Vladimir Shubin, deputy director of the dilapidated Institute for African Studies in Moscow. 'Whether it is building a hospital or a conference hall, we always tend to do things big, too big.'[6] But like their American counterparts, they still view the Cold War era from what, to the outsider, seems a bizarrely skewed angle. Superpower manipulation? What superpower manipulation? To hear this disingenuous generation of Moscow insiders tell it, the behemoth that saturated the Horn in weaponry was never more than a submissive junior partner in its African relationships, responding to, while never dictating, moves made by unreliable domestic leaders. It was all Siad Barre's fault. He had taken advantage of Soviet naivety, a naivety so deep the Soviets convinced themselves briefly after the Derg veered left that a Marxist Somalia, Marxist Ethiopia and Marxist South Yemen might bury their differences to form a federation of like-minded African states. 'The Somalis fooled us,' says Sinitsyn. 'Siad Barre had promised our ambassador, just a few days before he invaded the Ogaden, that no Somali soldier would ever cross the border. We were interested in both countries, Ethiopia and Somalia. Then the Somalis invaded, and we found ourselves squeezed between two friends. When we decided to come to Ethiopia's support, it was on moral grounds. Ethiopia had been the victim of an act of sheer aggression by a country we supported. We simply had to act.'

It is hard to credit that strategic self-interest counted for quite so little in Moscow's volte-face. With a population 10 times larger than Somalia's, conveniently-positioned Eritrean ports – closer to the Middle East than Somalia's harbours – its role as OAU headquarters and its history as a symbol of the anti-colonial struggle, Ethiopia was always going to be partner of choice for a superpower seeking leverage in the Horn. And the timing for a bold Soviet grab at greater world influence felt right. Defeat in Vietnam had left Washington on the back foot, nervous about intervention abroad, unsure of its place in the world. The sudden collapse of Portugal's African empire had given birth to an independent Angola and Mozambique, led by Marxist governments that looked to the Soviet Union for philosophical guidance and financial help. From Tanzania to Madagascar, Benin to Congo-Brazzaville, Ghana to Algeria, radical African presidents were nationalizing industries and declaring single-party rule. The continent was on the turn, 'progressive' forces appeared to be triumphing. For a super-power pushing against what felt like limp US resistance, in search of berths for its submarines and landing strips for its aircraft, Ethiopia was a prize to be coveted.

In justifying its U-turn to itself, Moscow could also dwell upon the emotional thread that had linked it to Ethiopia through the centuries. This was the one African country with which Russia could boast a long-standing historical connection. As far back as Peter the Great, the Tsars had fantasized about an alliance with Africa's Christian kingdom, land of Prester John. In 1888, a foolhardy Cossack adventurer had actually set up a short-lived colony baptized 'New Moscow' on the Red Sea. During the 19th-century colonial scramble for Africa, Russian emissaries to the Abyssinian royal court had vied with Italian, British and French delegations for Emperor Menelik II's attention. It is even said that Russian military tactics – urged on Menelik by an enterprising Russian captain – played a role in the victory at Adua. Surveying Ethiopia, with its peasant population mired in feudal poverty, its powerful Orthodox church, its decadent aristocracy and restless students, Moscow felt it was looking at a black version of its younger self. This was an African Russia, before Lenin, Trotsky and Stalin had got to work, screaming out for change. It was true that Moscow's new friendship was being forged just as the settling of scores in Addis Ababa reached its bloodiest. 'You would meet an acting governor or a major one day and the next day you would be told "they are no more",' remembers Shubin, who toured Ethiopia in October 1977. 'They were dying one by one.' But the grim reality of the Red Terror merely seemed to confirm the correctness of the parallels being drawn by Soviet analysts. During their revolutions, Russia and France had both experienced such purges, an inevitable part of history's working, it was felt. However unappetizing events in the capital seemed, they had no doubt that Ethiopia was embarking on a fundamental process of structural change. 'These were scuffles at the top. Whatever was going on in Addis, a real revolution was taking place in the countryside, where the land was being nationalized. You could feel it,' insists Shubin.

In the eyes of such men, Moscow was under an active obligation to intervene. Marx had taught his acolytes certain things were destined to pass. Lo and behold, in Ethiopia, his predictions about the rise of the proletariat appeared to be coming true. How could Moscow, keeper of Marx's sacred flame, turn a deaf ear when a socialist new-born asked for help?

'It's difficult to understand now, but at the time many Soviets really believed that the battle between capitalism and socialism was inevitable, unavoidable and, in the final instance, destined to be won by socialism. Going against it would have been like trying to go against the tide of history,' says Anatoly Adamishin, deputy foreign minister under Mikhail Gorbachev. 'Do you know the joke about the husband who decides to murder his wife by screwing her to death? After weeks of constant love-making, she is looking very sprightly, while he is in a wheelchair, utterly exhausted. "Poor thing," he whispers to a friend. "She doesn't realize she has only days to live." We were like that when it came to capitalism. "Poor thing," we kept telling ourselves. "It doesn't realize it's doomed."' He gives a rueful laugh. 'We entered Ethiopia step by step, in stupidity, but we had this conception we were fulfilling an international duty to help people struggling against oppression.'[7]

It was clear that socialism's eventual success, in a country as accustomed to one-man rule as Ethiopia, rested on the shoulders of one individual: the man baptized 'the Red Negus' by the media. Soviet diplomats who met Mengistu when he was still a major jostling for position in the Derg – a period during which he claimed to have survived nine assassination attempts – were favoured with what, in tribute to the Soviet Union's most famous astronaut, was known as a 'Gagarin smile': the beaming welcome of a man who appeared to have a limpid conscience and nothing to hide. They noted the charisma, but were not fooled by the impression of sunny openness. 'Mengistu was a very broad smile, with a very hard man behind,' remembers Sinitsyn, who first met him at a 1974 rally. 'He struck me as a man who had grasped his purpose in life, who knew exactly where he stood. He was very resolute, capable of taking difficult decisions. A man who, as the Americans say, would draw first.'

Mengistu's ruthlessness was built on profound insecurity. In contrast with Haile Selassie, an Amhara aristocrat whose self-belief was reinforced every time he looked in the mirror and registered his own aquiline profile and pale complexion, evidence of that prized Solomonic heritage, Mengistu had grown up nursing a clutch of inferiority complexes, grievances too deep ever to be assuaged. While he towered over the tiny Emperor – not a difficult feat – he was nonetheless acutely aware of his own lack of stature. Soviet advisers chuckled amongst themselves at his built-up shoes, his insistence on being photographed in splendid isolation. ('If you look at the official photos, you'll see Mengistu never allowed himself to be photographed standing next to anyone. He didn't like the comparison,' one former military adviser told me.) They laughed at the way in which an Ethiopian master of ceremonies always arranged for official handshakes to be staged above a table bearing a huge flower bowl, ensuring the resulting photo showed the visitor bowed and at full stretch. 'He was a very small man, small physically and small morally,' sniffs Adamishin.

Mengistu was as conscious of his ethnic origins as he was of his height. He claimed to be half-Oromo and half-Amhara, a perfectly respectable pedigree in race-conscious Addis. But the ruling elite the Derg had removed from power never swallowed the explanation. Curling its collective lip, it noted his dark skin and negro features. Rumours circulated that he came from a backward ethnic group in south-west Ethiopia, that his father had been a slave, that he had had to abduct his beautiful Amhara wife to persuade her to marry him. 'His lips are thick, his bones are African,' ousted royal courtiers would mutter. In public, they referred to him respectfully as 'the Chairman', in private they called him *bariaw* ('the slave'). 'What is the slave up to today?' they would inquire scornfully. Mengistu knew what was said. He had heard it all before, battling his way up through army echelons. But he never stopped minding. 'Why do they hate you so?' the ebony-skinned Zambian president Kenneth Kaunda is said to have asked him, on a visit to Addis. 'Because I look like you,' was Mengistu's sour reply.

A psychologist would have had no difficulty interpreting the bloodletting that accompanied his ascension, the appropriation of land and property and the toppling of a royal dynasty, as a bitter man's revenge on a class that had consistently snubbed him. Humble roots and a massive chip on the shoulder were characteristics Mengistu shared with another socialist leader, a fact Soviet officials registered without following the thought through to its conclusion. 'We often used to comment amongst ourselves on the similarities between Mengistu and Stalin, who as a Georgian also came from a minority and felt an outsider,' says Sinitsyn. With time, the ominous parallel was to prove more accurate than any had foreseen.

Was the early Mengistu really a Marxist? Or was he merely an ardent nationalist who adopted the slogans required to win massive arms deliveries from his new friends? Looking back, one is struck by the extent to which both superpowers made the same mistake in their dealings with Ethiopia: seeing only what they wanted to see. In Washington, policymakers decided that Haile Selassie, a leader who believed himself descended from Solomon, had signed up to Western demo-cratic values. Their Soviet counterparts eyed up Mengistu, a military man of sudden violence – capable of taking a pistol in his hand and personally executing a potential rival – and saw a leader who, however clumsy his grasp of the principles of dialectical materialism, could give shape to Africa's first truly social revolution.

All the initial signs seemed encouraging. It was an article of faith in Moscow that if a country was to undergo permanent revolutionary change, it needed a ruling Marxist-Leninist vanguard party. Mengistu dutifully set the process in motion, announcing the creation of a Commission for Organizing the Party of the Working People of Ethiopia. Addis Ababa became the first African capital to raise a statue to Lenin. Fidel Castro paid a visit. 'I believe in such a thing as African socialism,' Mengistu would tell visitors, a picture of Karl Marx promin-ently displayed behind his desk. He was received by Leonid Brezhnev at the Kremlin – there were to be over a dozen such get-togethers – and the two men hit it off. 'My country is disposed to give you anything except the atomic bomb,' the Soviet leader said. In return, the impressionable younger man offered loyalty and undying gratitude. 'Every time, before I told him anything else, I would say: "Comrade Leonid, I am your son, I owe you everything,"' Mengistu later told an interviewer. 'And I truly felt that Brezhnev was like a father to me.'[8]

While weaponry was always to dominate the relationship, cooperation extended into every sphere, just as it had with America under Haile Selassie. The two countries signed scores of economic agreements. Thousands of Ethiopians won places at Soviet universities, while hundreds of Soviet doctors, teachers, engineers and agronomists were dispatched to Ethiopia, filling the vacuum left by US Peace Corps volunteers. Moscow was granted the Red Sea access it required, mooring its warships off the Dahlak Islands; its military advisers took up positions inside Ethiopia's forces. When sensitive topics came up for international debate – a UN call for Soviet troops to withdraw from Afghanistan, for example, or a decision to boycott the Los Angeles Olympics – Ethiopia voted loyally with the Communist bloc.

But, as the years went by, Moscow began experiencing niggling doubts. It seemed to take an age – 10 long years in fact – for the promised Workers' Party to see the light of day. Officially, power then shifted from the Derg's military committee to the new civilian body. In practice, the party's leadership was made up of prominent Derg members and, just as in imperial days, all real decisions were made by one man: if the names had changed, the song remained the same. By then Mengistu, whose habit of driving past the hovels of Addis in an open-topped red Cadillac was commented on disapprovingly by the Communist Party paper *Izvestiya*, had already cultivated an oppressive personality cult. The traditional foibles of dictatorship – personal riches, easy women – seemed to leave Mengistu unmoved. Power was his aphrodisiac. Before official ceremonies, one of Haile Selassie's gold-brocaded red velvet thrones would be sent ahead to allow Mengistu to survey his subjects in monarchical splendour. An aide would then recite into the microphone: 'There is one man who matters in Ethiopia and that man is Mengistu. Forward with Socialism under Mengistu!', not once, but 10, 20, 30 times. The dutiful applause just kept coming, for no one dared be the first to stop clapping. Aides who in the early days of the Derg had denounced the Queen of Sheba as a whore now floated the notion of Mengistu's Solomonic ancestry: the attending impli-cation of religious predestination needed no spelling out. A rumour spread that he was the illegitimate son of Haile Selassie's brother – royal after all. 'When I met him in the Kremlin for the last time he had changed a lot,' says Sinitsyn, who was then working for the foreign ministry. 'It was clear he had come to regard himself as above everyone else. He was behaving a bit like Haile Selassie, very nonchalant while his entourage bowed and scraped. And already, at that stage, he was beginning to be disillusioned with us.'

Just as Haile Selassie's insatiable demands eventually poisoned his dealings with Washington, Mengistu quickly developed military appetites so vast even the most extravagant superpower could never hope to meet them. Soviet and Cuban support had proved supremely effective against Somalia, forced to withdraw in March 1978. But Mengistu found the problems in the north, where the EPLF and TPLF were now coordinating their attacks on the Ethiopian army, exasperat-ingly difficult to eradicate. He demanded more weapons to finish the job. In 1982 Ethiopia imported $575m in arms, in 1983 $975m and in 1984, $1.2 billion.[9] It was never enough. Aware that the unresolved war in Eritrea risked sabotaging

Ethiopia's revolution, Moscow had started pressing for a political settlement, even going so far as to arrange secret meetings between the Eritrean rebels and the Derg in 1978 and 1980. But the steady stream of Soviet supplies sent Mengistu a very different message. Why bother striking shabby deals when, with hardware such as this, he could simply obliterate the enemy? Resigned to the notion that the Eritrean war might 'continue for generations',[10] he allowed negotiations to peter out. Moscow had fallen into a trap of its own making. Just as its extraordinary generosity once convinced Siad Barre he could seize the Ogaden, it now nurtured the belief in Addis that compromises would never be necessary, the war in Eritrea could be settled by purely military methods. The definition Oscar Wilde once coined for British rule in Ireland – 'stupidity aggravated by good intentions' – aptly described Soviet policy.

Irritated by Mengistu's constant requests, the Soviet Union was also aggrieved by his record on agricultural reform. To Moscow, the case for revolution had always seemed at its most morally unanswerable in Ethiopia's countryside. But land nationalization, which forced peasants to sell at prices fixed by the state, had failed to sate Ethiopia's age-old hunger. In 1984 and 1985, drought hit the country once again, and with it a famine more devastating than the one that preceded Haile Selassie's overthrow. It left 1 million dead, presenting Moscow with a multi-layered embarrassment. Not only did the famine raise a question-mark over the suitability of collective agriculture to the developing world, it threw bleak light on Mengistu's priorities, from the hefty slice of the budget lavished each year on the military to the $55m wasted celebrating the 10th anniversary of Ethiopia's revolution. Humiliatingly, it was Western aid – both governmental and of the kind whipped up by pop star Bob Geldof – that prevented more Ethiopians from dying, not aid from Moscow.

By the time Live Aid was blasting out 'Do they know it's Christmas?' on Western radios, the Soviet Union was no longer the unquestioning, eager partner Mengistu had first dealt with in 1977. In 1982, his beloved Brezhnev had died, taking with him the certainties of yesteryear. Brezhnev's short-lived successor, Yuri Andropov, implicitly acknowledged that Moscow's mission of exporting socialism to the Third World had produced decidedly mixed results. 'It is one thing to proclaim socialism as one's goal, and it is quite another to build it,' he ruminated.[11] Then, in 1985, Mikhail Gorbachev came to power.

Extraordinary as it may seem today, the accession of the man who would eventually preside over the dismantling of the Soviet Union was hailed by Kremlin-watchers at the time as a sign of 'business as usual'. Indeed, Gorbachev initially paid lip-service to tradition, hailing Mengistu as the Soviet Union's most important friend in Africa. But the tone soon changed. At home, the comparatively youthful new Soviet leader spoke of the need for restructuring and openness. When it came to foreign policy, he challenged the very principles on which intervention in the Horn had been based. 'It is immoral to throw hundreds of millions of dollars into the development of homicide when millions starve and are devoid of everyday necessities,' he declared.[12]

Mengistu was cast in a new and uncomfortable role. The high priest was moving on, leaving his protégé as sole defender of a faith now questioned by its founder. Baffled by talk of perestroika and glasnost, Mengistu called the Kremlin to arrange an appointment. 'I needed to know what was going on. I went to Moscow to ask him what those two slogans meant. They were slogans that I didn't understand – and if you ask me, the Soviet people didn't understand them either.' The new general secretary could not have been more reassuring. 'I shall not shift one millimetre from Marxism-Leninism,' Gorbachev promised the Ethiopian leader. 'I am proud of our socialist achievements, and I always will be.' Fine words – but Mengistu sensed, correctly, that his heart was not in them.[13]

What bothered Gorbachev was not that the Soviet Union had consistently backed the wrong horses in Africa, propping up the continent's most disastrous regimes. The brutal truth was that Moscow could no longer afford its self-appointed role as patron of the developing world's Marxist experiments. To revert to Adamishin's analogy, while rosy-cheeked capitalism bounced from strength to strength, the Soviet Union was strapped to its wheelchair, on a glucose drip, grey-faced and drawn. The Soviet economy was stagnating. While Moscow played the part of benefactor, sending Ethiopia hundreds of thousands of tonnes in grain, it was not producing enough to meet its own population's needs. 'Once, when I was on holiday in Havana, I decided to work out how much Cuba cost the Soviet Union,' remembers Adamishin. 'I calculated that for two months of every year, every Cuban could afford to live without spending a single kopeck, thanks to Soviet aid.' He tried raising the issue with Andrei Gromyko, then his boss, but was brushed away. 'I spoke to one of his colleagues. I said, "We have to stop this or we will be desanguinated." He just said, "Don't go there." We were pushing others down the Socialist path while our own country was degenerating with every passing year.' Watching Moscow's deepening predicament, Washington hugged itself in delight. 'There was an attitude of: "It's their Vietnam, let them get bogged down in it,"' a former State Department expert on Africa told me. 'They're stuck on the tar baby, good luck to them.'

The arms race with the US was placing a crippling, unsustainable burden on the Soviet economy. But as long as Moscow continued to fund proxy wars in the Middle East, Africa and south-west Asia, as long as President Ronald

Reagan's label of 'the Evil Empire' held resonance in a frightened West, there could be little hope of mutual disarmament. The veteran Gromyko was replaced by Eduard Shevardnadze, one of Gorbachev's most radical supporters, and around the new foreign minister clustered a generation of iconoclasts who believed détente, rather than confrontation, was the only answer. Breaking with the 'geometer's approach to strategy', men like Adamishin decided that Africa was peripheral to Moscow's concerns. Looking at Ethiopia with fresh, sceptical eyes, they rejected the rigid ideological interpretations of old. 'This was a fight between ethnically-based cliques which had been transformed by the Soviet Union, on the one side, and America, on the other, into a historic struggle between old capitalism and new socialism. It never really was that and it wasn't in Angola, Mozambique or Afghanistan either,' says Adamishin.

Diplomatic feelers were put out to the Somali regime and Mengistu was strongly advised to improve relations with both his neighbours and the Eritrean rebels. In Moscow, anyone interested in Africa would have noticed a telling change in media coverage of the continent. 'The idea that Africa was an intolerable burden on the Soviet Union spread. I was amazed at how far some of our publications went,' says Sinitsyn, who did not share the new, heretical way of thinking. 'There was nothing spontaneous about these articles. This was political manipulation. The public was being prepared for our withdrawal.' One Soviet official, writing in *Pravda* in July 1987, denounced the export of world revolution as an outdated concept. Another floated the idea of a 'Frank Sinatra' policy in Eastern Europe, letting the countries Moscow had once kept on the shortest of leashes do it 'their way'.

The reformists were giving up on Ethiopia. But they were not free to dictate strategy as they pleased. Gorbachev's foreign policy was always a tricky balancing act, in which he struggled to effect radical change while placating uneasy hardliners in the military apparatus and Politburo. Ethiopia, in his view, was hardly worth a showdown with the traditionalists, when there were so many more crucial ideological battles to be staged in Europe, the Middle East and Asia. 'Gorbachev and Shevard-nadze could not permit themselves to be accused of betrayal on every front: Afghanistan, Angola, Cuba and now Ethiopia as well,' says Adamishin. 'They did not want to open another front in the battle against orthodoxy.'

And so Soviet policy came to bear an uncanny resemblance to US policy in the first years of the Derg: its heart was no longer in it, but the military supply machine had acquired its own momentum. Easy to start, it seemed impossible to stop. Even when the time lag involved in closing off previously-agreed contracts is taken into account, what Soviet politicians said in public and what the military delivered often appeared in schizophrenic contradiction. In 1987, a leading member of the Central Committee announced that Moscow was resolutely opposed to 'the transformation of Africa into an arena of confrontation' and pledged his country's commitment to the political settlement of conflicts.[14] That same year, the Derg signed a new $2 billion arms supply agreement with the Soviet Union. Between 1987 and 1991, as Gorbachev issued ever louder warnings of aid cut-offs and Soviet newspapers denounced parasitic African states, Moscow actually sent Ethiopia $2.9 billion in weaponry.[15]

Seemingly oblivious to the tug-of-war over policy taking place above Red Square, Mengistu announced that perestroika and glasnost were irrelevant to Ethiopia and called for more weapons. His warehouses were emptying, his stocks were low, he complained, how could he fight when he was only receiv-ing bullets and caterpillar treads, not the heavy armament he needed? Like a junkie begging his dealer for one last fix, he pleaded with Moscow: if Soviet officials would only meet this one last request, provide this final arms consignment, it would tip the balance in the war. 'He kept telling us that if we helped him he could achieve this military victory,' remembers Adamishin, with real bitterness. 'I remember how he told me, with tears in his eyes: "We may have to sell our last shirt, but we will pay you back. We Ethiopians are a proud people, we settle our debts." Looking back, I almost feel I hate him. Because I believed that what mattered to him was what was best for the country. While really all that mattered to him was his own survival.'

For years, Moscow had thrown hardware at the Eritrean problem. With its help, Mengistu had expanded the regular army from 45,000 to 250,000, militiamen adding another 200,000 to the bloated total. Yet this enormous, well-equipped force of nearly 500,000 could not bring the Eritreans to heel. It baffled the Soviet generals in Moscow, comparing their delivery invoices to the disappointing reports from the field. It made no sense. 'There was this abscess in the north that would not heal,' remembers General Valentin Ivanovich Varennikov, deputy defence minister under Gorbachev. 'We sent weapons and equipment, we sent officers and trained specialists. Yet it never got any better. It seemed intractable.'[16]

What the Kremlin failed to grasp was that there was an element of theatrical posturing to Mengistu's demands. By constantly lamenting a shortage of hardware – in jaw-dropping defiance of the evidence on the ground – the Ethiopian leader was in part addressing his public, providing a convenient scapegoat for the long-running failure to end the Eritrean crisis. But the obsessive harping on a theme also betrayed a dangerously stunted imagination. What Mengistu could not understand, would not understand, was that the reasons for the

Derg's military discomfiture had nothing to do with a lack of weaponry, and everything to do with loss of morale, fifth column activity and his own chronic mismanagement. 'No nation ever benefited from a long war,' the 19th-century Prussian military strategist Karl von Clausewitz once wrote. The same warning applies to an army. Unable to pull off a decisive victory, unable to make peace, Ethiopia's army was rotting from the inside. A battle in north-west Eritrea was about to prove, to anyone who cared to learn the lesson, that the weapons that matter most never come off a factory conveyor belt. They are to be found inside the convoluted whorls of the human brain: a sense of destiny, a momentary forgetfulness of individual existence, the belief a cause is worth dying for.

CHAPTER 16

'Where are our socks?'

'In war, three-quarters turns on questions of personal character and relations; the actual balance of forces counts for only the remaining quarter.'

Napoleon

If you take a map of Eritrea and draw a line from Nakfa to the town of Afabet, the pencil traces the path of the Hidai river, which meanders down a broad valley separating the high plateau from the lowland flats. In peacetime, the dry river bed serves as a highway for turbaned nomads who thwack their moaning troupes of camels along in languid search of water. Vehicles jiggle their teeth-juddering way across the powdery white boulders deposited by rushing torrents, or churn their wheels in hidden sand pits, flailing like waders who have lost their footing. Startled flocks of goats sheer away from spinning front wheels, their white flanks catching the sun. The little boys tending them while away the hours by knocking tiny orange berries from the trees, which they dangle in plastic bags in front of passing cars. Buy these offerings and you will find that the wrinkled fruits do, just, have a discernible flavour of toffee, but not a drop of juice. There is an occasional searing flash of hot colour – the jewel-green flutter of a pair of lovebirds, perhaps, or the yellow chiffon wrap flapping around a Tigre woman who, when she turns to stare at the noise of an engine, reveals a large, intricately-carved gold crescent piercing one nostril. But, for the most part, this is a landscape painted in various shades of dun, a world whose colours have been bleached away by the punishing light.

At the northern end of the valley, the old Italian-built road executes a series of switchbacks and hairpin bends, taking on the challenge of the Roras, cutting through concentric rings of Ethiopian and Eritrean trenches to surface on Nakfa's bomb-blasted plateau. The southern end of the Hidai valley is pinched off by a narrow, V-shaped pass, a natural bottleneck. If you stop your car here to stretch your legs and peel the sweaty T-shirt from your back, you will notice, below you, a brown ribbon threading its intermittent way along the valley floor. Look closer and you will see that the ribbon is made up of individual shapes and that those shapes are, in fact, the rusting skeletons of dozens of trucks, tanks and armoured personnel carriers. Carapaces submerged in a sea of sand, guns askew, they bristle like angry cockroaches.

It was here, at the pass called Ad Shirum, that the outcome of Eritrea's 30-year war of liberation was decided in March 1988. This was not to be the last battle – that would have to wait another three years – nor was it the biggest. But the battle of Afabet was the moment when the war's ending became clear and tangible to both sides, a sudden understanding as unquantifiable as it was undeniable.

The brains behind this particular EPLF operation belonged to Mesfin Hagos, one of Isaias' most trusted military commanders. Mesfin had been pondering the best time for the EPLF to break the stranglehold established by Ethiopia's Second Revolutionary Army. For 10 years, the Nadew Front, strongest of the four Ethiopian commands stationed in Eritrea, had attempted to capture Nakfa. From Kamchiwa, a base at the foot of the Roras, an Ethiopian tank brigade rained its fire up across the trenches and into Nakfa itself, keeping the guerrillas penned inside their mountain stronghold.

Rebel forces at that time numbered only slightly more than half the 15,000–20,000 Ethiopian troops stationed in the area, but the figures, Mesfin knew, did not tell the full story. After a two-year break in major hostilities, a pause the EPLF had used to learn how to operate a haul of captured heavy guns, the guerrilla movement was in prime condition, chafing for action. Now, he was convinced, would be a propitious time to attempt something the EPLF usually carefully avoided: a full-frontal attack against superior forces. Mesfin was convinced the operation, set to begin on March 17, could alter the balance of the war. But he hedged his bets, setting himself a strict 48-hour deadline for the push. 'We should be at Afabet on the morning of the 19th,' Mesfin told a reporter who had been assigned to cover the campaign by the EPLF's magazine.[1] 'If we are not, then we will abandon the whole operation and make a try another time.'

With their divisions positioned in readiness, EPLF commanders gathered their Fighters together and told them of the impending attack. The reaction was one of boisterous jubilation, almost relief. 'It was a great moment,' says Solomon Berhe. Today editor-in-chief of Eritrea's Tigrinya-language newspaper, he was a 24-year-old tank commander at the time. 'We had rested for two years, our firing power was high and we had new artillery. We were missing a war.'[2] If Mesfin's plans proved successful, the claustrophobic existence of trench duty and furtive night-time sorties would be over, the door to the rat's trap sprung. 'People were firing into the sky, women ululated and everyone was feeling very cheerful.' The cheering should have set warning bells ringing. But the EPLF had arranged one of its sports contests in Nakfa the week before. Accustomed to the sight of EPLF units coming and going, and the sound of applause from enemy trenches, the Ethiopians suspected nothing.

The attack began at 5.00 am and was staged simultaneously on three fronts. While its infantry leapt over the parapets, fighting hand-to-hand in the narrow trenches for control of the steep slopes below Nakfa, the EPLF sent its captured tanks and armoured personnel carriers roaring in from the plains to the east in a flanking manoeuvre aimed at putting maximum pressure on the Ethiopian tank brigade at Kamchiwa. For 16 hours, the surprised Ethiopians resisted with extraordinary tenacity. Worried by the number of *tegadelti* the EPLF was losing, a twitchy high command radioed Mesfin to tell him to pull back. Normally the rebels used code to avoid eavesdropping but now, when seconds counted, such stratagems were abandoned. 'I'm in the midst of it, and I can tell you: there's no need to worry,' Mesfin reassured his colleagues. The third time he was told to order a withdrawal, he switched the radio off. Success, he sensed, was within his grasp. As the EPLF infantry reached the valley floor, where the Fighters were almost impossible to pick out amongst the boulders and trees, the position at Kamchiwa became untenable. Ethiopian commanders made a strategic decision to pull back to the garrison at Afabet. Reinforcements and fresh supplies were waiting there and on those open plains, the big guns could be put to effective use. If the brigade could only make it through the pass at Ad Shirum, all would be well, they calculated. They had just taken the worst decision of their military careers.

A convoy of 70 Ethiopian tanks, armoured personnel carriers and trucks loaded with munitions, equipment and men assembled at Kamchiwa. Soon it was trundling south along the sandy river bed, with the EPLF's mechanized division in hot pursuit on one flank and EPLF infantry running alongside on the other. The two armies raced each other to Ad Shirum and there, as the Ethiopian drivers noisily revved their engines for the steep final climb that would take them over the pass, an EPLF tank opened fire, hitting a lorry at the head of the convoy. Another volley, and this time an Ethiopian tank went up in orange flames.[3] And now the contours of the landscape played wonderfully into EPLF hands. For in that natural bottleneck, with the peaks crowding in on three sides, there was no room for manoeuvre, no space for the tanks and trucks behind to circumnavigate the destroyed vehicles. Cursing Ethiopian drivers hit the brakes, twitchy officers shouted warnings down the line to avert a pile-up. The convoy was stuck, a sitting duck. The *tegadelti* licked lips at the quantity of Soviet hardware about to fall into their hands.

It was at this point that what seems, to the outsider at least, the most remarkable event of the Afabet campaign occurred. As news reached headquarters in Asmara that the EPLF was on the verge of claiming a delectable prize – a 70-vehicle convoy – army commanders ordered their jets to take off and head for Ad Shirum. Did the panicking Ethiopians in the convoy look up when they heard the scream of MiGs overhead and think, 'Thank God, we're going to make it after all'? Or did they know enough about their own army, about national honour, the horror of humiliation and their superiors' capacity for hard decisions to sense what was coming? One can't help suspecting that the first the average footsoldier knew of the terrible decision taken at the top was when a well-aimed missile slammed into the column in which they sat.

Ethiopia's bombardment of its own men went on for two solid hours. 'They bombed continuously,' remembers Solomon. 'There were soldiers milling around the tanks and trucks but the pilots didn't care. The flames were rising high and in the trucks the ammunition and missiles kept exploding. Some Ethiopians ran out and were captured by our fighters. But many Ethiopian soldiers burned to death inside. They knew they were surrounded and decided it was better to burn. It was an inferno, a Biblical scene.'

As the smell of roasting human flesh, scorched rubber and melting paint wafted across the valley, a vast ball of grey and black smoke churned and tumbled over the column of frazzled metal, blanking out the sun like an eclipse. In the sudden darkness, Solomon and his fellow Fighters could do little more than gape, aware that the enemy, with this act of defiant self-destruction, had robbed them of a military climax. 'It was a strange feeling. We'd developed a lot of respect for the brigade in the 16 hours it had defended itself, a lot of admiration. We had followed them along the river bed and now this powerful brigade had been destroyed but not through our doing, they were destroyed by their own. For a soldier, it was a very strange end.' Western reporters shown around the site by the EPLF the following day photographed curious patches of black cinders on the ground, some with boots still attached. The Ethiopian soldiers had literally been burnt to a crisp.

It was an ending that highlighted some of the fundamental differences between the two forces. Small, reliant on volunteer enthusiasm, required to ration both its supplies and casualties, the EPLF was incapable of applying such cold-blooded logic. 'The EPLF would have ordered the men in the convoy to do their best to escape,' says Solomon. 'It would never have acted as the Ethiopians did. I met Ethiopian prisoners-of-war afterwards and they told me, "Addis betrayed us."' Ethiopia's top commanders, in contrast, knew they had at their disposal a vast resource of conscripts, whose support for the conflict was considered irrelevant. For these men, the US military motto of 'Leave No Man Behind' would have seemed self-indulgent nonsense, 'Leave No Working Tank Behind' captured their priorities rather better. 'When you lose an area you destroy your equipment – it's a principle of war,' was the matter-of-fact explanation a retired Ethiopian general gave me. 'If you cannot separate your men from their equipment, then you bomb them both together.'

If Ethiopia's grand gesture had taken the edge off the Eritreans' victory, the EPLF had won the battle. Nadew's remaining commanders tried to stop the retreat, shooting mutinous soldiers on the spot, but to no avail. This was a rout, and the EPLF needed to move its mechanized division into Afabet before the Ethiopians could regroup. At Ad Shirum the charred convoy firmly blocked the way. 'Our commanders were saying, "Try to push the burning trucks to one side,"' remembers Solomon. 'We had to tell them, "We wouldn't be able to do it in one month, let alone in one night."' Instead, the tanks went the long way round. Racing back up the valley to Kamchiwa, they traced a loop east along the Red Sea sands, around the mountain range and inland once again. After 12 hours of continuous driving, Solomon and his colleagues rolled into Afabet on the morning of March 19th. The garrison had collapsed. On the plains, Ethiopian soldiers were surrendering in their hundreds, or quietly disappearing into the bushes to blow out their brains. 'We entered easily, with only a little small arms fire. The people came out cheering from their houses, the women were wailing. They gave us water and kissed us.'

Mesfin Hagos had stuck with almost mathematical precision to his deadline for smashing one of Ethiopia's most powerful garrisons. And just as he had predicted, the battle marked a tipping point. Two Ethiopian division commanders had been killed, two-thirds of the army's forces slaughtered and – notwithstanding the convoy's destruction – a vast amount of heavy weaponry seized, equipment that would later enable the EPLF to stage on Massawa's outskirts the biggest tank battle seen in Africa since the Second World War. The road to Eritrea's southern towns had been opened and within a week the Ethiopian army, surrendering one key garrison after another, would be pushed out of north-western Eritrea. 'We moved from defending to continuous attacking. From that moment on, the enemy never recaptured anything of any significance,' says Solomon.

The death knell for the Derg regime had sounded and, for once, the outside world heard it clanging. Basil Davidson, the veteran Africa historian, witnessed the battle as a guest of the EPLF. Carried away, he hailed it as the most significant victory by a liberation movement since the Viet Minh humiliated the French at Dien Bien Phu. For years, EPLF Fighters had scoured the airwaves, waiting in vain to hear the simple fact of their existence acknowledged. 'I can still remember the day we heard the battle of Afabet reported on the BBC,' a former EPLF woman doctor told me. 'That was it. I switched off my radio and never listened again.' Accustomed to blank faces and closed doors on their trips to the West, EPLF delegations found venues being made available for press conferences, journalists returning their calls. From one day to the next, everything had changed. 'After Afabet, the Ethiopians realized the Eritreans were capable of extraordinary things,' says Robel Mockonen, a young Fighter who took part in the Afabet campaign. 'And the Eritreans also realized they were capable of extraordinary things. From then on, it seemed, we could not lose.'

Like an earthquake under the sea, Afabet's shudder reverberated across continents, rumbling all the way to the Kremlin.

The spasm of alarm it triggered in the Soviet Union was prompted by more than mere concern for its Ethiopian ally. While the battle was still raging, word reached Moscow that a dozen Soviet military advisers had been encircled by the EPLF in the mountains. They had hidden in a cave along with a clutch of Ethiopian soldiers, but faced imminent capture.

It was a scenario the Kremlin had long dreaded. During all the years of cooperation, the average Soviet citizen had been kept blissfully unaware of his government's muscular involvement in Ethiopia's war machine. Apart from an occasional mention of 'anti-revolutionary groups' fighting the 'progressive government' in Addis, the Soviet state media, attention focused on Afghanistan, barely covered the Eritrean conflict. While Moscow acknowledged it had experts stationed inside the Ethiopian army, it always defined them as 'advisers', indicating they kept well away from the frontline. If Soviet 'advisers' were taken prisoner by the EPLF, their hands-on role would become impossible to deny. When the battle of Afabet began, Moscow had ordered all its advisers to pull back, in the hope of preventing precisely such an outcome. But the order was issued late and had, in any case, been ignored by at least one foolhardy Soviet officer, itching to play the hero.

Aghast, Moscow scrambled a team of KGB special forces, an elite unit of parachutists, marksmen and scuba-divers trained in hostage-recovery, assassination and the capture of enemy facilities. The unit flew to Addis, then on to Asmara, where the Soviet version of the SAS piled into two Aeroflot civilian helicopters. The choppers set off for the Roras, with two Ethiopian military helicopters flying escort. 'Those disappeared as soon as they were shot at from the ground,' Yevgeny Sokurov, a major in the KGB unit, remembers with a sardonic laugh. Setting down near the cave, the Soviet parachutists began loading up the wounded, many so dehydrated they could no longer walk.

The evacuation turned into a frenzied, fear-fuelled scramble as the EPLF, realizing what was afoot, opened fire. 'The helicopter was overloaded and at that altitude, in the rarefied air, it was having problems taking off. The Ethiopian soldiers were in a panic, running towards us, desperate not to be left behind. We had to shoot at their legs to prevent them boarding.'[4]

It was only when the overweight helicopters flopped into Keren that the Soviet crack team learnt the truth. While they had succeeded in airlifting the errant dozen, elsewhere in the war zone three other Soviet advisers had fallen into EPLF hands. They would spend the next few years as guests of the EPLF, and Sokurov would spend much of that time trying, without success, to track them down.[5] 'It was an embarrassment for Moscow, to say the least,' he acknowledges.

These days, Sokurov, a chain-smoking, ferret-slim, tattooed 52-year-old, works in the Russian film industry as a stunt director. As lean and toned as on the day he flew to Afabet, he drives to film sets in the silver birch woods outside Moscow in a purring black BMW, evidence of how well life in the private sector has treated him. The only member of his unit to break out of the security business, his finely-honed skills are now applied to coordinating battle scenes and staging explosions, talking actors through interrogation scenes and plotting the choreography of hand-to-hand combat. Being his own boss has freed his tongue – 'I answer to no one, so I can say what I like' – and there seems no end to the hair-raising tales he could tell. For, in his prime, Sokurov served in Hungary, Angola and Mozambique and took part in the bloody storming of the presidential palace in Afghanistan, one of Moscow's most infamous foreign interventions. Looking back on his larger-than-life adventures, he sometimes has the surreal feeling they happened to someone else, a simple man of action who trusted his superiors to do the right thing and believed the larger questions were not for the likes of him. 'I'm a product of my time and my country. I come from a military dynasty. I was a military cadet at the age of 10 and I always knew who I would become, I always knew I must serve the state. The important thing was to do my job well.' The pride in professionalism remains, but a yellow miasma of moral doubt now hangs over the question of the service to which his unusual talents were put. 'Helping Mengistu, that arrogant monkey, was pointless – I'm certain of that,' he says, with a quiet passion. 'In Moscow there was a pathological desire to support these thieving, savage, African dictatorships. It was a waste of time.'

For Sokurov, the helicopter escapade marked the beginning of a three-year stint in Ethiopia, a time commemorated by the gold Ethiopian cross that dangles around his neck. Ordered to stay on, he was given the task of training a group of commandos in the hills outside Asmara, a secret unit designed to operate behind EPLF lines. 'They were trained to a very high level. They could cover 70 km in mountainous terrain in 24 hours, stage an operation and return without anyone spotting them,' he boasts. 'They moved just like ghosts.' It was while he was carrying out these duties that he began to grasp why Ethiopia's superior numbers and Moscow's endless supplies had counted for so little at the battle of Afabet and would count for even less in future. 'When I was sent to Afabet, I didn't think the war was a lost cause. But I soon did.'

One of the reasons why Mengistu needed such elite units, it became clear, was because morale in Ethiopia's regular forces was disastrously low, ground down by the conflict's sheer longevity. 'The troops were tired. They didn't see any prospects for peace. They had spent 14 years in the trenches being promised the war would soon be over. Everyone was fed up with the arrogance and stupidity of the Ethiopian commanders, who took on the worst mannerisms of the American sergeants who had gone out to Ethiopia in the 1960s. They walked very slowly, surrounded themselves with bodyguards, treated all army equipment – whether it was a store of petrol or a television set – as their personal property, and didn't have to think at all. From a military standpoint, they were absolutely gormless.'

Another Soviet adviser who registered the morose, mutinous mood in the ranks was Sergei Berets. Now a journalist with the BBC in Moscow, Berets spent two years in the early 1980s working as an army translator near Afabet. 'On the anniversary of the October Revolution, the troops were gathered together and the Russian political officer delivered a long speech about Lenin, Communism, the liberation of the working people and the prospects of world revolution. I translated his words into English and an Ethiopian officer then translated them into Amharic. These speeches could go on for hours. At the end he asked: "Any questions?" and several soldiers shot their hands up. They spoke very passionately for 10 minutes. All I could make out was the word calze – Italian for "socks". They wanted to know when they would be getting a new delivery.'6

Many soldiers had been press-ganged into service, rounded up in market places during the forced relocations at the heart of Mengistu's agricultural programme. Some were no more than teenagers, barely tall enough to carry a rifle. Depressed by open-ended stints in the foxholes of the Sahel, the troops had asked repeatedly for a rotation system to be introduced, but the request had gone ignored. The EPLF used loudspeakers to bombard them with propaganda, and EPLF leaflets circulated freely in the ranks, drumming in a numbing message of looming defeat. While the Ethiopians bombed EPLF prisoner-of-war camps, it was common knowledge that the rebels, who took the view that the Ethiopian footsoldier was as much a victim of a totalitarian regime as any Eritrean, treated their captives with humanity. The effect was insidious. Despite regular executions and jailings, army headquarters could not halt a steady flow of Ethiopian soldiers to the rebel side. Remembering their behaviour in the heat of battle, Sokurov drips sarcasm. 'Ethiopian tank crews had a peculiarity. If their tanks were shot at, even if it was only small arms fire, they would get out of the tanks and run for it, although they would have been safer staying put. The troops' favourite command was "disperse", which meant "run in all directions and meet up in Addis". Then they would be gathered up and sent back to the front. As for the pilots, they were too scared to do any targeting and would just drop their bombs any old how. I was not impressed.'

The conflict in Eritrea had lasted so long, the smallest of structural weaknesses had widened to form crippling handicaps. Under Haile Selassie, the military chain of command had been a clear, top-to-bottom affair. In contrast, the structure the Derg copied off the Soviet Red Army was multi-layered and confused. Each military commander was shadowed by a political cadre and a security officer, who sent their findings, recommendations and unflattering opinions of fellow officers separately back to headquarters. It was a recipe for friction between headquarters and commanders, commanders and cadres, officers and rank and file, with suspicions running so deep that at times effective decision-making seemed impossible.

A retired Ethiopian general I interviewed in Addis, who did not want to be identified, paints a picture of a force on the verge of implosion. 'The higher echelons were fighting the cadres. In principle, we military commanders made the decisions. In practice, the political and security officers had the government's ear. Their reports often did not reflect the reality on the ground, but were given much more importance in Addis than what we said.' If Sokurov has little respect for men like him, the dislike was entirely mutual – in the general's view the presence of Russian advisers only made a disastrous situation even worse. 'The Russian advisers were nothing but troublemakers. They would give instructions without knowing the area, forgetting that perhaps there was no water there, or no road the tanks could use.'

Both the general and Soviet adviser agree, however, on the impact of another morale-sapping phenomenon: the 'honey pot' problem. Any force left in the field long enough will take girlfriends and mistresses, forming complex local bonds. In Eritrea, the beautiful bar girls and waitresses who became intimate with Ethiopian commanders usually worked for the EPLF, passing on careless gossip and boastful pillow talk. 'The Eritreans were very good at using their women,' remembers the general. 'There was a huge amount of infiltration. The infiltrators kept telling the Ethiopian soldiers: "This is not your land, your place, your country", and the soldiers listened.'

Mengistu did not seem to recognize the damage the internecine wrangling and leakage was doing. Instead, in the run-up to Afabet, he made a series of false moves that virtually handed the rebels victory.

The December before the battle, the EPLF launched a canny strike on Nadew's 22nd Division, where relations between the military commander and his political cadre were so bad that soldiers loyal to the one were refusing to obey orders given by the other. The attack left nearly 500 Ethiopians dead and up to $3.3m of military equipment either destroyed or in rebel hands. Appalled by what he saw as an entirely unnecessary reverse, a raging Mengistu flew to the battlefront to hear the explanations of the Nadew Front's commander, General Tariku Ayne, a popular officer with a reputation for speaking his mind. Tariku was clearly a little too frank: he was summoned to Asmara and executed a day later.

If the execution of one of the army's most respected generals was intended to spur his men into action, Mengistu had blundered disastrously, as the delighted EPLF – almost as shocked as Nadew's soldiers – immediately understood. 'The Derg has cut off its right hand with its left,' rejoiced rebel radio. The bitching inside Ethiopian forces continued. Tariku's replacement struggled to establish his authority over the division's commanders. The heads of the 14th and 19th divisions were so busy arguing that when headquarters warned them the rebels had been seen moving weapons around by camel and ordered them to close off a 5-km gap between their forces, they ignored the command. Lost in their hatred for one another, Ethiopia's commanders virtually forgot to hate the enemy.[7]

The government army was sick, and the sickness had turned the steady stream of Soviet supplies into a curse rather than a blessing. Gradually, Sokurov registered what should have been clear to his colleagues back in Moscow years earlier: Soviet deliveries were merely exacerbating the problem, because the more equipment the Ethiopians were sent, the more fell into rebel control. 'We were supplying both sides. The separatists wore our uniforms and used our weapons. The stuff wasn't even being captured – it was being abandoned by Ethiopian commanders after the briefest of skirmishes. I'm not just talking about a few tanks. Entire divisions were being allowed to fall into separatist hands.' Disillusioned members of the Second Revolutionary Army's high command, he concluded, were deliberately sabotaging the war effort. 'No matter how secret the operation, it was becoming known to the separatists. The commanders were supplying the separatists with information about their activities because they wanted to damage Mengistu's regime, even if it meant the slaughter of their subordinates.'

In the wake of the battle of Afabet, Mengistu appeared on national television and publicly admitted, for the first time in a decade, that a war was being waged in the country's northern province. The money spent each year fighting the EPLF could, he acknowledged, have funded 4 universities or 10 large hospitals. Ranked the poorest nation on the globe by the World Bank, Ethiopia was nonetheless spending more than half its government revenue on arms.[8] But Mengistu's message to viewers was the opposite of what his restive generals had hoped: 'From now on,' he declared, 'everything to the battlefront!' Like the doomed Macbeth, halfway across a river of blood, wading back seemed harder than striding forwards. Questioning the foundation on which his power had been built risked bringing the whole structure crashing to the ground.

He signed a peace agreement with Somalia which, by calming the Ogaden front, allowed him to shift reinforcements to Eritrea. In July, he flew to Moscow, begging bowl in hand. The response was one he must have dreaded. If the Kremlin balked at cutting Mengistu off without a penny, it nonetheless slashed his expectations down to size, hinting heavily that it was time to fundamentally rethink his alliances. 'The word in diplomatic circles, which included many of the Communist country emissaries, was that Moscow told Mengistu at that meeting: "We have our own problems and you had better start making other friends," remembers Robert Houdek, posted to Addis as US chargé d'affaires.[9]

Mengistu was running out of time. In early 1989, the Ethiopians suffered an even more humiliating defeat than

Afabet when the TPLF, with EPLF support, captured the northern town of Inda Silase. The rebellion had spread beyond Tigray province to embrace disaffected Amharas and Oromos in central and southern Ethiopia, their movements uniting to form the Ethiopian People's Revolutionary Democratic Front (EPRDF), bent on regime change. The army was being pushed southwards, entire units defecting as it retreated. Sensing the ground shifting beneath their feet, generals in Asmara and Addis staged a concerted coup attempt in May, minutes after Mengistu had left for East Germany, one of the countries he hoped might be persuaded to fill the vacuum left by a withdrawing Soviet Union.

The rebel generals had been in secret communication with the EPLF, preparing a post-Mengistu accommodation. But the putsch failed. Ironically, Sokurov, today's disillusioned stunt man, played a key role in its collapse. Training in the mountains, he heard the news on the radio. He had received no orders from either Moscow, barely aware of what was happening, or Addis and, privately, he had seen enough to sympathize with the coup plotters. Yet, ever the professional, he ordered his elite unit into action in Asmara, occupying the airport and army's staff buildings. 'One of my men shot the commander of the air force in his office. Another rebel general was killed in his car at a checkpoint, trying to escape. They had to use a grenade launcher to get the other one out of his office, where he had locked himself. Afterwards we found warehouses full of military supplies and documents suggesting that after the coup the Eritrean separatists would enter Asmara.'

Twenty-seven commanding officers paid the price for this abortive action. Beaten and tortured to death, their corpses were left for days to bake in the sun as a warning to others. Mengistu, who had flown back to Addis in a hurry, ordered the severed head and mutilated body of General Damesse Bulto, commander of the Second Revolutionary Army, to be paraded through Asmara's streets. 'I felt sorry for them,' Sokurov admits. 'If I'd been a neutral bystander I'd have taken their side. Within a month and a half, another 3,500 suspects were arrested and shot.'

Mengistu was still in control, but he had just executed his best and brightest officers – not an act likely to improve an army's battlefield performance. Where could he turn for the military equipment he persisted in believing could swing the battle his way? The answer was far from clear, given the new warmth developing between the superpowers. In June 1989, Herman Cohen, the US assistant secretary of state for Africa, met Anatoly Adamishin in Rome to discuss US–Soviet collaboration on Africa. During that meeting, Adamishin made it clear that Moscow would be happy to pass the Ethiopian baton to Washington. Worst of all, both the Soviets and the Americans had made diplomatic overtures to the EPLF. For a dictator who had always played the 'my enemy's enemy is my friend' game with such gratifying results, nothing could have been more worrying than this enlightened superpower agreement to pull together.

Boxed in, Mengistu appealed one last time to the ally that had been so generous in the past. How could he sort out Eritrea, he whined to Moscow – blithely ignoring the $725m in weaponry sent in 1988, the $975m he was in the process of receiving for 1989 – when he had no ammunition? His stores were empty, he must have more. Baffled by the discrepancy between the telegrams from Addis and its own figures, Moscow dispatched its top military brass that summer to settle, once and for all, the puzzle of Ethiopia's arms deficit.

'I was sent to really sort it out,' recalls General Valentin Varennikov, then commander-in-chief of the Soviet Union's land forces. 'I gave Mengistu the figures for our shipments –

they looked pretty impressive – he wasn't too pleased by that. I told him, "It can't continue forever like this in Eritrea. Time has shown there can only be a political solution." He just said, "I won't talk to bandits."'[10] Determined to see conditions for himself, Varennikov toured the army fronts, noting the strange intimacy that had developed between the two sides which, increasingly, shared the same covert desire: an end to the war. 'On the Eritrean frontline, I asked why there was no shooting and the Ethiopians said: "We told the rebels the Soviet delegation was visiting today, and asked them to hold fire." I was astonished by that attitude.'

Back in Addis, Varennikov became embroiled in an almost comic cat-and-mouse game of Find the Weapons as he attempted to verify exactly how much Soviet equipment Mengistu had stockpiled around the city, squirrelled away for a rainy day. 'They were trying to be crafty. They would ask us which warehouse we were going to check the following day and then, overnight, they would move all the supplies. But I was an old soldier, I was on to them. In the morning I'd say: "I had a vision in the night. I'd like to visit that other warehouse." I'd find it absolutely bursting with supplies. We would take photos and show them to Mengistu and say, "Look, there's so much stuff you don't even have room to store it."' The general lifts a hand above his head. 'They weren't just supplied, they were oversupplied. Everything was there.'

The Soviet Union's military establishment had always shown more indulgence towards Ethiopia than Gorbachev's reform-minded foreign policy experts had thought wise. When the civilian politicians had scolded and admonished, the generals had come through. Now, finally, the two spoke as one. 'The Soviet Union was like a cow that was being milked by anyone who felt like it, including Ethiopia,' concluded Varennikov. He had a terse exchange with Mengistu before flying out. 'I leave confident that you will end this war,' he told Ethiopia's president. 'If you fail to do so, our leadership may rethink their opinion of you.'

Moscow was effectively finished with Ethiopia. The withdrawal of Soviet experts, started quietly in the wake of the attempted coup, was stepped up a notch and, in the autumn of 1989, Sokurov became the last adviser to quit Eritrea. He rebased in Addis, where he continued training young Ethiopians for an increasingly paranoid regime. But he was no longer happy with his role. 'I felt I was becoming an instrument of manipulation by unscrupulous Soviet politicians and that I could die in the sands of Africa.' Finally, he used a shrapnel wound as an excuse to escape, returning to a Soviet Union itself on the verge of fragmenting into its constituent parts. 'For three years, the Ethiopians wouldn't allow me to leave, so I didn't tell them I was going. Once I'd recuperated I was told the situation in Ethiopia was changing so quickly, there was no point returning. Actually, the situation at home wasn't much better. I felt I had left one country and returned to another.'

Mengistu, the ultimate pragmatist, was not quite finished. He had one last card up his sleeve: a strange, esoteric card, admittedly, but one that might just allow him to snatch triumph from the jaws of defeat.

He began laying the groundwork for a three-way alliance which would provide him with a fresh source of arms, the weaponry he persisted in believing held the key to the war. Its shape emerged when, in August 1989, Cohen became the first US assistant secretary of state to visit Addis in 15 years. The US official found an uncharacteristically friendly and amenable Mengistu, who promised unconditional negotiations with the

EPLF, pledged to open Ethiopia's economy to private investment and pushed for an exchange of ambassadors. Crucially,
the Ethiopian leader pledged forthright action on an issue he
knew was close to the heart of America's influential Jewish
lobby: the Falashas.

Legend has it that the ancestors of the Falashas, who call
themselves the 'House of Israel', were in the retinue that
accompanied the original Menelik on his flight into Africa
bearing the stolen Ark of the Covenant. When Ethiopia
embraced Christianity in the fourth century AD, the Falashas
did not follow suit. They clung instead to the rites of early
Judaism, though their observances diverged so widely from
established orthodoxy that for years Israel's rabbis queried
whether they could be regarded as Jewish at all. Pushed to the
margins of Ethiopian society, accused by Orthodox Christians
of witchcraft, harassed by the state, the Falashas became virtual
outcasts, dreaming of the day they would be granted the Right
of Return.

A long-running scholarly campaign to win them formal
recognition as Jews triumphed in 1975 when Israel's parliament
decreed them descendants of the Tribe of Dan. As conditions in
Ethiopia worsened, securing their evacuation became a rallying
cause in both Israel and the US. During the 1984–5 famine,
Mengistu had allowed Israel to remove 15,000 Falashas, but
tens of thousands still remained. Mengistu now planned to use
them as human hostages in a giant barter that would, with any
luck, remove the need to make any real concessions in peace
talks with the EPLF and its rebel allies.

'For Mengistu, the name of the game was Israeli arms in
exchange for Ethiopian Jews,' recalls Herman Cohen in his
memoirs.[11] Mengistu dusted off the sales patter Haile Selassie
had used to woo foreign supporters in a previous age: in a
hostile Moslem region, a Christian and a Jewish nation must

cling together. Israel would replace the Soviet Union as military supplier and the Americans would give the ingenious deal their blessing. 'Though the [Ethiopian government] knew we could not openly acquiesce in Israeli arms deliveries, they expected a wink and a nod indicating we would not protest if Jerusalem opened the arms faucet,' Cohen writes.

Washington was in a quandary. It had never accepted the notion of Eritrean independence, which it believed undermined the integrity of Africa's borders. But with Mengistu's brutal regime in terminal decline, officials did not want to see a fading conflict suddenly revived by a new injection of weaponry. Yet they were also keen to oblige their closest friend in the Middle East and placate Jewish Congressmen pushing hard for the Falashas. Perhaps, with some skilful manoeuvring, it might be possible to beat Mengistu at his own game, winning both major concessions at the peace table *and* the Falashas' departure without whipping up the war.

Washington decided to turn a blind eye to Israeli military training in Ethiopia and the dispatch of army uniforms, boots and tens of thousands of supposedly 'obsolete' rifles, while warning both Addis and Tel Aviv it would not stand for deliveries of heavy munitions. The warning fell on deaf ears in Tel Aviv, excited by a surge in the number of Falasha arrivals. When, in February 1990, the EPLF finally captured Massawa in a daring speedboat operation, the Ethiopian air force responded with relentless bombing. Survivors reported that the pilots were dropping a 'new kind of bomb' which shredded the port city's crowded streets: Israeli cluster bombs.[12] It took all Washington's powers of persuasion, including a direct intervention by former President Jimmy Carter, to convince the Knesset it meant what it had said.

The unedifying Falasha trade dragged on, but the rebels' momentum could not be slowed. In early 1991, the EPLF careered along the Red Sea coast, coming to rest on the out-skirts of Assab, Keren and Asmara. It sent brigades south to help the EPRDF sever the key artery between Gonder and Addis. With Ethiopia cut off from the sea and a surrounded garrison in Asmara only accessible by air, Mengistu was under almost intolerable pressure to grant the one concession he had always rejected: a referendum on Eritrean independence. He balked, and ordered a halt to Falasha departures, which had been running at a brisk 1,000 a month. 'Give us arms or there will be no further Falasha departures,' he told the Israelis.

As a hardline Marxist regime responsible for at least 500,000 civilian deaths jerked in its death throes, Washington and Tel Aviv made an extraordinary decision. The time had come, they agreed, for an outright bribe. A decorative pretext was required. The payment, Cohen and his colleagues suggested, would be presented as reimbursement for the revenues Ethiopian Airlines, normally used for Falasha emigrations, would lose by ceding its passengers to Israeli-organized charter flights. The US Defence Department agreed to make $20m available. 'It was hidden under the term transportation costs – we didn't want to call it "ransom money",' Cohen told me. The Israelis beat Mengistu down from an initial asking price of $100m to first $58m and finally $35m, which was paid into an Ethiopian government account at the New York branch of the Federal Reserve. On May 24, as the Ethiopian army surrendered in Asmara, the TPLF agreed to delay its entry into Addis long enough to allow nearly 15,000 Falashas to be airlifted to Israel in a 36-hour non-stop shuttle.

The deal had been sealed, but Mengistu was no longer around to reap its rewards. He had boarded a flight to southern Ethiopia, where he was scheduled to inspect a batch of army recruits. Once the plane took off, he entered the cockpit and ordered the pilot to change course for Nairobi. He was bound for Zimbabwe where, under the understanding eye of Robert Mugabe, a former fellow Marxist, he had already purchased a large farm and ensconced his family. The man who had promised he would sell his last shirt to cover his debts left the country owing Moscow $6 billion, most of it for arms.[13] Like Haile Selassie before him, Mengistu defied expectations that he would, in the proud tradition of defeated Ethiopian warlords, commit suicide rather than face capture. Instead, he chose the slow poison of alcoholic exile, complaining bitterly, during rarely-granted interviews, of the way Moscow had turned its back on the true faith and betrayed its favourite African son.

The Falasha operation was the crassest chapter in the story of the world's careless manipulation of Eritrea, final proof of how deep its self-serving cynicism ran. As far as the Israelis were concerned, peace in the Horn came a distant second to its determination to claim a recently-discovered lost tribe. Neither they nor the Americans ever appear to have felt a moment's embarrassment for their readiness to pay off a dictator who, had the rebel advance unfolded at a slower pace, would have used his ransom money to revive a dying conflagration.

The key players would no doubt claim this diplomatic ballet was carefully choreographed to ensure everyone – except Mengistu – got what they wanted. 'I believe we got the better of him,' Cohen wrote. 'Our decision to play along with Mengistu, rather than denounce him openly and thereby end the dialogue, was correct.' But a US Congressional aide who drew up a report on the episode put his finger on the distasteful moral premise at its heart. 'How many Ethiopian lives can be justified for the sake of an Ethiopian Jew having the opportunity to reunify with his family in Israel?' asked Steve Morrison. 'Is this implicitly racist?' And it is interesting that Israeli and American satisfaction with this episode of human trafficking was not shared by the Ethiopian government that replaced the Derg. To general amazement, this destitute African state returned the $35m at the Federal Reserve to Israel, saying it did not believe in accepting 'blood money'.[14]

With Mengistu suddenly out of the picture, peace talks arranged by American officials in a London hotel took on a dramatically new shape. What had originally been envisaged as tough negotiations between the Ethiopian government and the rebels turned into a private agreement over future dispensations between victorious Eritrean and Ethiopian guerrilla movements. In Addis, where Mengistu's former colleagues were begging foreign embassies to grant them asylum, the looting started. 'The doors to Menelik's palace were open, papers were blowing all around and the praetorian guards walked out with mattresses on their heads,' remembers Houdek. 'Addis was full of wandering, leaderless, hungry troops, many of whom didn't even speak Amharic, coming in with their AK-47s. Jesus, it was a volatile situation.' Alarmed, the US ambassador put a call through to Cohen, who now appealed to the waiting TPLF, its forces bolstered by Eritrean tanks, to speed up its entry into the capital and fill the power vacuum.

Today, many Ethiopians sneeringly refer to this as 'Cohen's Coup'. For if they were relieved to see the back of Mengistu, the Amharas, in particular, were far from delighted at the prospect of being governed by the *Woyane*, rebel fighters harking from what, to urbane Addis dwellers, seemed the barely civilized wilds of Tigray. In particular, they cannot forgive TPLF leader Meles Zenawi for the blow he immediately dealt their national pride. Quietly accepting the loss of what every Ethiopian leader had lusted for and fretted over – the Red Sea coastline – Meles agreed to an Eritrean referendum on independence after a two-year pause for reflection, recognition of the pivotal role the EPLF had played in toppling Mengistu. There may have been those in the new Ethiopian administration and diplomatic community who hoped the delay would trigger a rethink amongst Eritreans. But no one in the EPLF was in any doubt which way the vote would go.

The Eritrean contingent stayed long enough in Addis to know their guerrilla allies were securely in position, then turned their tanks around and trundled home to Asmara. They knew they were not welcome. 'When we drove through Eritrean villages, people came out to cheer us. As soon as we crossed the border into Ethiopia the cheers turned to boos,' remembers tank driver Solomon Berhe. 'Once we got to Addis no one dared go out of the compound at night. It was too dangerous.'[15]

Their work was done. Just as Eritrea's longing for independence had played its part in pulling Haile Selassie off his throne, it had eventually destroyed Mengistu. Eritrea had delivered Ethiopia a terrible lesson in how the problems of an unhappy little community, if handled with sufficient clumsiness and brutality, can replicate and swell, turning into something deadly and implacable. A rebel organization that had never added up to more than 110,000 had taken on and beaten a force of 450,000. The Eritreans had triumphed in defiance of troop numbers, export figures and military statistics; in contradiction of logic itself.

The tense days of June offered a curious, terrifying coda to the puzzling story of Mengistu's missing munitions. As the TPLF and its allies struggled to establish their hold on the city, giant ammunition depots scattered across Addis Ababa – the stores Ethiopian officials had gone to such lengths to try and conceal from General Varennikov – detonated one after another, set off by army hardliners in a final gesture of impotent spite. The last, and biggest, explosion sent rockets and shells fizzing into the air for hours, lighting up the pre-dawn sky like a deadly fireworks display. It flattened a nearby shanty town,

reduced at least 100 residents to body parts and sent an apocalyptic mushroom cloud spiralling towards the heavens.
Mengistu had stockpiled so much for the showdown he ultimately fled that when it ignited it felt like a small nuclear
explosion. 'It was like an atomic bomb,' said BBC reporter
Michael Buerk, knocked off his feet by a blast that killed a
television cameraman and mangled another colleague.[16] Here
was ultimate proof that whatever lay behind the Ethiopian
army's defeat, it had never been a shortage of weaponry.

CHAPTER 17

A Village of No Interest

All the presidents of the world have died and gone to Hell. As they sit roasting in the flames, they ask the Devil if they can call home. 'OK,' says the Devil, 'but it'll cost you.' First Bush calls his family in Washington, and they chat away. 'That will be $200,' says the Devil. Then Chirac calls Paris. 'That will be 100 euros,' says the Devil. Finally, it's Isaias' turn. 'That'll be 5 nakfa,' says the Devil. 'How come he gets to pay so little?' wail the others. 'Oh,' says the Devil. 'It's only a local call.'

Eritrean joke doing the rounds in 2004

It was June 1993, and in the Cairo press centre, gathered around the closed-circuit television relaying events from the main chamber, a group of African journalists burst into a surprised cheer.

They had all been assigned to cover one of the most cynicism-inducing of events: the summit of the Organization of African Unity (OAU), that yearly get-together where insincere handshakes were exchanged, 29-year-old coup leaders got their first chance to play the international statesman, and the patriarchs of African politics politely glossed over the rigged elections, financial scandals and bloody atrocities perpetrated by their peers across the table. So entrenched was the OAU's

357

image as a complacent club of sclerotic dictators and psychopathic warlords, few Western newspapers bothered to send reporters to the event. But the speaker who opened this particular conference had been worth turning up for. The man did not appear to know the language of diplomacy. He had just said the unsayable, publicly articulating the private exasperation felt by every journalist covering the OAU's 30th anniversary reunion in Egypt.

Looking around a hall that held the likes of Zaire's Mobutu Sese Seko, Zimbabwe's Robert Mugabe and Kenya's Daniel arap Moi, Isaias Afwerki, president of Africa's newest state, confessed his 'boundless' pleasure at being able to take his seat at the table. But his joy at 'rejoining the family from which we have been left out for so long' was not prompted by respect for an organization which had betrayed its founding principles, he made clear. The OAU, he told a hushed hall, had failed to deliver on its brave pronouncements on human rights and economic development. Africa remained a marginalized continent, scorned by its partners, whose citizens could not walk with their heads held high. 'We have sought membership in the organization not because we have been impressed by its achievements but, as a local proverb goes, in the spirit of familial obligation, because we are keenly aware that what is ours is ours.' Surveying the crush of Parisian designer suits and ceremonial robes, the jaded faces to left and right, the soberly-dressed former Fighter delivered a final rebuke: 'We do not find membership in this organization, under the present circumstances, spiritually gratifying or politically challenging.'

It was as though a bracing wind had swept through the hall. 'He seemed,' recalls journalist Ofeibea Quist-Arcton, who covered the event for the BBC, 'like a leader from a completely different era. Here were all these old dinosaurs, patting him on the back and saying, "Well done, young man" and he was saying, "It wasn't thanks to you." He was the hero of the day. I remember thinking: "Maybe Eritrea has got off to a good start."[1]

It was a speech that effectively set the tone of the new nation's relationship with the outside world. 'I didn't do it for you, nigger', had been the pitiless message of Eritrea's British occupiers. The refrain had been taken up by the Americans who ran Kagnew, the Soviets who funded the Derg and the Israelis who armed Mengistu. Now Eritrea was repaying the compliment. A veteran diplomat might have warned Isaias that telling the OAU to go to hell, on the very day his month-old state took up its longed-for seat, might be a move Eritrea would come to regret. Isaias did not care. His population would never forget the solitude in which it waged its liberation campaign. No one had helped, yet Eritrea had won anyway. So why, now, should it mince its words? 'I didn't fight for 30 years to brown-nose the African leaders who betrayed us,' ran the silent refrain. A few months later, Isaias did it again, castigating the UN in his first speech before the General Assembly. The new kid on the block had delivered his message loud and clear: Eritrea's dealings with the world would be strictly on its own terms.

But it would be wrong to present the Eritrean mood in this period as one of scolding antagonism. A chippiness was certainly there, but it was balanced by bubbling, irrepressible confidence. The long years of solitary confinement, Eritrea's leaders believed, had granted the Movement a unique opportunity to tease out the factors behind Africa's post-colonial slide. Learning from the continent's mistakes, Eritrea could leapfrog the hurdles that had tripped up other developing states and sprint towards the future. 'We have to avoid being influenced by the politics of this region,' Isaias told an interviewer. 'There is nothing positive in the region for us.'[2] If Eritrea refused to resort to mealy-mouthed euphemism, it was not merely because it wanted to remind Africa of past treachery. Buoyed by self-belief, Eritrea was convinced it could serve as an inspirational model: it would show the continent the way ahead. Given that the EPLF had made a fool of every diplomat and historian who ever predicted its obliteration, a little intoxication was perhaps understandable. A thousand Hollywood movies tell the tale of the band of puny no-hopers who take on the mighty bully and confound the sceptics. This David-and-Goliath story was the stuff of dreams.

'You cannot have a society of angels except in heaven,'[3] Isaias warned a journalist who asked about his vision of the future. But in their heart of hearts, every Eritrean in those heady days believed he had been given a chance to do just that: to create the perfect state. The years that elapsed between 1991 and 1998 – although few appreciated it at the time – were to be Eritrea's golden era, when everything seemed possible. In April 1993, the referendum Eritrea had waited so long to see was staged. Only a people raised in knowledge of the UN's shameful record on Eritrea could appreciate the symbolic significance of the fact that when Isaias cast his vote, the UN's special representative was standing at his side as witness and guarantor. The result was a foregone conclusion: 99.8 per cent of Eritreans opted for independence.

As Eritrean academics began drafting a multiparty constitution and the EPLF dissolved itself to form a new political movement – the People's Front for Democracy and Justice (PFDJ) – construction started. It seemed at times that Asmara and Massawa had become little more than massive building sites. It was not just a question of repairing the damage done by Ethiopian bombers. During the struggle, 750,000 citizens had fled abroad. The diaspora was coming home, bringing its savings and its aspirations. Peasants from the dusty refugee camps of Sudan, oil engineers from Texas, dentists from Scandinavia, labourers from Saudi Arabia: they all needed affordable housing, schools for their children, hospitals for their sick. Asmara's creaking Italian factories were crying out for modernization and its demobilized Fighters – men and women who knew only the skills of war – were desperate for work. The task facing the EPLF was daunting. But with the population solidly behind it, the international community falling over itself to make up for past mistakes and a friendly government across the border, what could go wrong?

In May 1998, the halcyon days came to an abrupt end. The national character traits forged during a century of colonial and superpower exploitation were about to blow up in Eritrea's face. Curiously, the spot where Eritrea tragically fumbled its recovery could not have been more nondescript.

Sitting in the Ethiopian government helicopter, I grabbed hold of the hard bench and squeezed with both hands. 'Up,' I muttered to myself, trying to ignore the large yellow tank vibrating a few feet away, sloshing with the fuel that would burn us to cinders if this ageing Russian-made aircraft fell out of the sky. 'Up, up, up. Stay up.' Below us, a grey mountain range curved across the plains like a dinosaur's backbone. Gazing through the porthole at the cement-coloured valley floor and the rills of dried-up streams, curled like bacon rinds under the grill, a lazy thought passed through my mind: 'No one could survive here.' Then I noticed the terracing, standing out on the muscle of the foothills like veins on a bodybuilder, and the green and brown coronas that signalled a tree-circled monastery here, a stockade there, and realized how wrong I was. Tiny metal rectangles, corrugated-iron roofs, flashed the sunlight back up at us as we clattered over the mountains, our toy helicopter's shadow bouncing off the reddish crags. If you lifted your gaze to the southern horizon, the plains of Ethiopia's Tigray province seemed to have been scattered with diamonds, glistening and winking in the sun. People not only survived here, they nurtured every inch of usable land; had done so, in fact, for generations, defying gravity, trapping water, challenging the eroding winds.

The helicopter descended, whipping the red dust into a whirlwind, and then fell quiet, spilling journalists from its belly for a brief walkabout. Back in Addis, rumours had circulated of mineral riches hidden in this land. Only deposits of gold and uranium, surely, could explain what had occurred here. But now that we could see this clump of straw tukuls and thorn-bush compounds for ourselves, none of that made sense. This was the kind of one-hotel, two-bar village in which yellow-eyed goats wandered through front rooms; over-excited urchins, their heads shaved save for one dark lock of hair, ran after new arrivals shouting 'You! You! You!' and the local police-man snored on a trestle bed under a tree. Dazed by the heat, a donkey stood stock-still in the middle of the only street, smiling in that infinitely benevolent way donkeys have. This place was just a pause on the way to somewhere more interesting. Given the choice, any sensible traveller would stop here only long enough to stretch their legs and have a drink, before hitting the road once again.

Only the road no longer led anywhere. Once busy with traders bringing plastic sandals and cheap clothing south and farmers taking grain and livestock north, the track stretched silent and deserted. Events had drawn an invisible, impen-etrable barrier across the scrubby landscape. Shambuko, the Eritrean town at the other end of that road, was as inaccessible as if it had been on another planet. This seemed a village of no interest, yet it was a village of terrible, tragic interest. For this was Badme, claimed by both Eritrea and Ethiopia as their own, a border settlement whose contested destiny had cost more than 80,000 lives and destroyed a dream of Utopia.

Travelling between Asmara and Addis, I often used to think, was like stepping through the surface of a mirror. In Eritrea you would hear one version of events, argued with heartfelt passion and unfailing logic, and come away won over. But fly to Addis, walk through the rippling mercury, and you would hear the same events recounted from an Ethiopian perspective: internally consistent, vigorously argued, also totally convincing. Each version of reality, of course, jarred with its mirror-image on the other side of the looking glass. And so the mind struggled to balance two mutually-exclusive visions of the world until, weary of ping-ponging from one side to the other, tired of trying to be scrupulously even-handed, it barked 'Enough!' And you were left in a sulky grey fug of ambiguity, sure of only one thing: everyone had behaved badly, everyone was to blame.

Nowhere was this truer than with the war that began in Badme in 1998, erupting in an area whose confines had been agreed between Menelik II and the Italians at the turn of the century but had become muddied over time, administrative arrangements on the ground rarely matching what was stipulated in the treaties. From the original trigger incident to the two governments' subsequent handling of the crisis, each country's account of what happened makes perfect internal sense. Unfortunately, each also clashes head on with the other side's version.

According to Ethiopia, a resident was collecting firewood outside Badme on May 6 when he was detained briefly by a group of Eritrean soldiers. When the Ethiopian authorities sent local policemen to discover why Eritrean forces had strayed onto their territory, they were fired on. This incident could and should have been dealt with by a commission the two countries had already set up to deal with precisely such trifling frontier incidents. Instead, Eritrean officials attending talks in Addis slunk back to Asmara without warning and Eritrea launched an unprovoked offensive on Badme on May 12, rolling its tanks into the area in a display of gratuitous aggression.

According to Eritrea, the trouble began when Ethiopian militiamen started setting fire to huts in the Eritrean village of Geza Chi'a, north of Badme. Heading to Shambuko to lodge a complaint, the villagers met an Eritrean army unit, which agreed to send a delegation to discuss the incident with Ethiopian security forces. Instead, the Ethiopians opened fire. When the shooting stopped, an Eritrean major, lieutenant and several others lay dead: the negotiators had been murdered. Far from scurrying away from Addis, Eritrean delegates tried without success to persuade their Ethiopian counterparts the matter needed to be urgently addressed. It was Addis that fatally upped the ante, its parliament declaring war on Eritrea on May 13.

Whatever caused the clash, which escalated with terrifying speed, there was a fundamental difference between this conflict and the one that preceded it. The Armed Struggle, as far as Eritrea was concerned, had been waged against a succession of alien regimes based in distant Addis. At the heart of this new war lay a falling-out between the EPLF and TPLF: former rebel allies, comrades-in-arms from adjacent territories. The two leaderships spoke the same Tigrinya language, shared the same ethnic origins and had survived the same ordeals of famine and military occupation. Their forefathers had worshipped the same Christian God, their women braided their hair in similar ways, they had attended each other's weddings. The war against the *Woyane*, the Eritreans called it, the war against the *Shabia*, Ethiopians said, pronouncing terms which had once been semi-affectionate with startling venom.[4] This was a family quarrel, with all the vindictiveness that implied. And, as with all domestic disputes, the parties to the quarrel showed an almost pathological desire to keep the matter private. The build-up to the May hostilities actually coincided with a visit to the region by Kofi Annan. Yet neither Meles Zenawi, the Ethiopian Prime Minister, nor Isaias saw any need to draw the UN Secretary-General's attention to their little fracas. The quietism of the highlands carried the day.

The open-mouthed astonishment the conflict triggered amongst Western allies was misplaced. The relationship between the EPLF and TPLF, in truth, had never been quite as cosy as foreigners had liked to believe. Italy's colonial occupation of Eritrea had marked a parting of the ways for communities on both sides of the frontier. Inhabitants of Ethiopia's underdeveloped Tigray migrated north in search of work, supplying sophisticated Asmara with its unskilled labour. Eritrean urbanites who had grown up associating Tigray exclusively with maids, street-sweepers and janitors, tended to look down on their neighbours. That Eritrean sense of superiority had been reinforced during the Struggle, in which the longer-established EPLF played the role of mentor and guide to the inexperienced TPLF. In the field, the two movements had squabbled over military practice, relations with the Soviet Union and their vision of the future. They were brothers, certainly, but a touchy younger brother can easily come to hate a patronizing older sibling.

Long before Badme erupted, trouble had been brewing. There had been other clashes, including exchanges of fire, in disputed border areas – Bure and Bada were two examples – during which the Ethiopians had burned Eritrean villages and ousted administrators. Spotting what looked like a pattern of creeping territorial encroachment, Asmara suspected the TPLF leadership of trying to redraw the map. Hated in Addis, the Tigrayans – Eritrean officials argued – were bent on carving out a Greater Tigray in preparation for the day Ethiopia's majority ethnic groups ejected this minority from power. They brandished a 1997 map of Tigray which seemed to confirm their fears: funded by a German charity, it drew a meandering, curving frontier in the west that took a large bite out of Eritrean territory, ignoring the sharp diagonal shown on every other commercial map. If Eritrea surrendered its claim to Badme, everyone would want a piece of it, and the land its Fighters had died to win would be nibbled away, kilometre by kilometre. Ethiopia's leadership met rhetoric with rhetoric. 'Since independence, Eritrea has picked a fight with every one of its neighbours,' Ethiopian officials said, citing a series of disputes with Sudan, Yemen and Djibouti. 'This is a police state, run by a Fascist president, that only knows how to make war.'

Land was not the only issue – economic relations between the two neighbours had also curdled. Chafing over its unfamiliar landlocked status, Addis complained that its goods were being held up at Assab and it was being ripped off by sharp-elbowed Eritrean businessmen selling Ethiopian coffee on international markets. Asmara, for its part, had watched the establishment of giant factories in the Tigrayan capital of Mekelle with a jaundiced eye, suspecting the TPLF of planning to supplant Eritrea as Ethiopia's supplier of manufactured goods. The final blow had been Eritrea's insistence on introducing its own currency, in the teeth of Ethiopian resistance. When the nakfa was launched in 1997, a furious Addis decreed that future trade would have to be conducted in hard currency, requiring Eritrean merchants to take out letters of credit with Ethiopian banks. The fact went virtually unnoticed abroad in all the excitement, but when the fighting broke out in Badme, trade along the once-busy border had already petered away to a mere trickle.

Within weeks, deeds were being committed that could never be forgiven. At the beginning of June, Ethiopia dramatically escalated the conflict by sending the air force to bomb and napalm Asmara airport. When Eritrean gunners shot down one of the jets, they discovered the man at its controls was none other than Bezabeh Petros, an Ethiopian pilot who had been captured by the EPLF during the Struggle and sent home. Returned prisoners, according to the conventions of war, must not go back into battle, yet here was Bezabeh, attacking Eritrea once again.[5] Unable to believe the news, Asmarinos flocked to the airport to see for themselves. 'For 30 years we've been bombed, gassed and murdered by the Ethiopians. They only just stopped blooding us. Now the same pilots, the men who flattened Massawa, are back bombing Asmara. And it's our friends who send them. It's just unthinkable,' a government minister told me, trembling with rage.

Eritrea retaliated by sending its jets screaming over Mekelle, where they struck the military airport but also dropped cluster bombs near a school, killing 12 children. Eritrean officials later said the school strike was a mistake, the work of a pilot with no previous experience of live combat. But Isaias did not do the diplomatically astute thing and issue a hand-wringing apology. Expressions of regret are not an Eritrean forte. ('I didn't spend 30 years at the Front to apologize.') 'It was a terrible accident,' acknowledged this stiff-backed president. 'But this is war.' Mourning a massacre of innocents, appalled Ethiopians were left with the triumphant words of the Eritrean air force chief ringing in their ears: 'They hit us once, I hit them twice.'

It was a revealing comment. The image of a stinging slap kept surfacing in conversations with Eritreans in those tense days. Justifying the Badme operation, a senior Eritrean diplomat likened his country to a man being pinched repeatedly under a table until he smacks his opponent. 'Everyone sees the slap in the face, but no one knows about the pinching that preceded it.'[6] I spoke to an Eritrean merchant with three children waiting to be called up. 'Jesus may have said, "If someone slaps your face, turn the other cheek,"' he told me. 'But here in Eritrea we have our own version. It goes: "Slap my face and I'll hit you back so hard you'll never dream of trying it again."' It was the rationale of a country brutalized into knee-jerk belligerence, the reasoning of a small, threatened society that had concluded, on the basis of hard experience, that violence was the only message outsiders would ever understand. Israel and Rwanda would have understood such thinking perfectly. It was an early hint that Eritrea was destined by its history to bungle the subtle challenges of peace.

A strange fatalism had descended on Eritrea. Gripped by a sensation of déjà vu, residents noted, with resigned impassivity, the fact that Western embassies started evacuating their nationals from Asmara just before the first raid on the airport, a timing that not only suggested a helpful tip-off from the Ethiopians, but spelt out an all-too-familiar message. The going had got tough and Eritrea's newest buddies – the diplomats, aid workers and businessmen who had rushed to promise partner- ships and investment after independence – had scarpered. Just as in the old days, Eritrea was on its own. The realization that the outside world had also labelled Eritrea the aggressor prompted the same shrug of the shoulders: what else was to be expected from the international community that had tolerated the Federation's abrogation, ignored the EPLF's existence and armed the Derg?

There was a feeling of vast sorrow at the thought that, after a brief seven years in which Fighters had exchanged their camou- flage for civvies, they were being asked once again to make the ultimate sacrifice. The Nakfa trenches had been tolerable because Fighters had always believed their offspring would eventually inherit the deep peace that went with unchallenged sovereignty. They had suffered to ensure the next generation would not. Now ageing *tegadelti* dispatched their children to the new front, sick to their stomachs at the sight of the open trucks packed full of tense young faces, trundling west.

But the grief went hand-in-hand with what, to outsiders, seemed a baffling self-confidence. This war would be nasty, but the Eritreans knew they were destined to triumph. The EPLF legend of plucky solitude had them in its seductive grip: how could a people that had bested Haile Selassie and Mengistu, despite the best efforts of the US and Soviet Union, ever lose? Of course, Ethiopia, with its 70 million strong population to draw on for recruits, dwarfed Eritrea, with 4.5 million. But those ratios had held true before and had always proved ultimately irrelevant. It was the EPLF, after all, which had taught the TPLF the secrets of guerrilla warfare and provided the tanks that rolled into Addis. Africa's Sparta knew its own violent origins: no people did warfare better than the Eritreans. 'We will win,' my old friend John Berakis told me with mournful certainty. Grizzled as he was, he stood ready to take up the gun again if his government needed him. 'We have always won. What can the Ethiopians teach us about fighting? We had to teach them how to get rid of their dictator.' A government minister told the same story: 'We could walk through Tigray if we wanted,' he assured me. 'But we're in a dilemma over what to do – do you finish off your friends?'

It was dangerous, delusional folly, not only because the Eritrean army of 1998 was a far flabbier entity than the lean fighting machine constructed during the Struggle, nor because the task of defending a 1,000-km border differed substantially from holding the mountain fortress that was Nakfa. Nations that believe they cannot lose slide into war more easily than states that suspect the contest will be close. When victory seems assured, opportunities for negotiation are neglected; the blurred fudges that allow faces to be saved and compromises struck regarded as beneath contempt. 'I didn't spend 30 years in the bush to compromise.' Below the overweening confidence lay something more ominous in its implications, whose outlines the TPLF had sensed. The EPLF had forced Eritrean sovereignty down the gullet of a wriggling international community by being ready to fight harder, suffer more intensely, hold out longer than its enemy. Modern Eritrea had conjured itself into existence through war; the notion it would have to continue asserting its identity through combat seemed unexceptional. 'We fought the forgotten war, everyone was against us,' an ex-Fighter told me. 'Winning that war meant we came to exist as a nation, we came to be known. So now it's a question of not losing our identity. First we will go to war, then we will negotiate from a position of strength.'

A peace plan put forward by the US and Rwanda in June quickly foundered. In Eritrean eyes, Ethiopia's readiness to launch an air campaign made a mockery of the entire process, which was passed to the OAU, most of whose Addis-based delegates shared the Ethiopian view of Eritrea as hot-tempered regional bully. In retrospect, Isaias' defiant speech at the Cairo summit was beginning to look like a very poor investment.

After a lull in which two of Africa's most famine-prone countries indulged in a multimillion-dollar arms shopping spree, pumped tens of thousands of barely-trained recruits into their armies and signed up Ukrainian mercenaries to fly their planes, fighting resumed in earnest in February 1999. Eritrean predictions that the rest of Ethiopia would leave the Tigrayans to fight alone proved disastrously off-key. Tapping into brooding public resentment over the surrender of Ethiopia's coastline, Meles rallied the nation behind him. His commanders opened a new front on the outskirts of Assab and then sent wave upon wave of soldiers crashing against dug-in Eritrean positions in Badme. With Badme lost, Eritrea's central and eastern fronts under attack and the prospect of recolonization – or at the very least, the instalment of a puppet government in Asmara – looking a distinct possibility, Isaias accepted the OAU's peace terms. Eritrea, the nation that prided itself on needing no one, now looked to the despised UN for its salvation, blue-helmeted peace monitors deploying along the contested border in September 2000.

Engaging over 500,000 troops and displacing 600,000 people, the Badme war won the dubious honour of being not only the worst conflict ever staged between two armies in Africa, but the biggest war in the world at the time, more devastating than the rather better-publicized Kosovo crisis. It created a level of hatred unparalleled even during the Struggle, which found particularly mean-spirited expression. Even under the repressive Derg, Eritreans with no interest in politics and families of mixed origin had been able to earn a living and put down roots in Ethiopia. No longer: Ethiopia loaded more than 70,000 men, women and children – many holding Ethiopian passports – onto buses and dumped them across the border. There was something of a spurned lover's fury behind this mass deportation: 'You want independence? Here, take it, and get out.' In Eritrea, thousands of Ethiopians were made to feel so unwelcome they left of their own accord.[7]

The detail that sticks in my mind, evidence of what miserably petty acts wounded pride can push us to perform, emerged during a UN briefing in a compound outside Assab. The UN unit in the area had been trying to arrange the most basic and humane of services: burying those killed during Ethiopia's abortive push on the port. Nearly a year after the war's end, the bodies still lay baking in the sun. Neither Eritrea nor Ethiopia was ready to lose face by acknowledging the corpses as their own. The UN had been reduced to identifying the dead by their footwear: boots for the Ethiopians, black plastic sandals for the Eritreans.

'Whatever it costs' runs the motto on an Italian bridge erected on the plains outside Massawa. It could have served as the EPLF's maxim, and now the unflinching message had acquired a bitter resonance. It was not so much a question of the numbers of Eritrean dead, although the figure the government eventually announced – 19,000 in just two years – seemed almost unbearably high compared to the 65,000 lost during the Armed Struggle's 30 years. Eritrea had lost the sympathy of its foreign friends, aghast at the shattering of their dream of an African Renaissance. Its economic take-off had belly-flopped. As long as Isaias was in control, Addis made clear it would have no dealings – trade or otherwise – with its neighbour. The Ethiopian market, which had accounted for two-thirds of Eritrean output before the war, vanished. With Addis re-directing its exports to Djibouti and Berbera, Assab became a sun-baked ghost town, the cawing of crows reverberating around the once-bustling port. Eritrea's newly-renovated hotels lay empty: tourism is always the first casualty of war. The ever-loyal diaspora had been squeezed dry during the fighting, digging deep into its pockets to pay for new weaponry. Sighing over a botched opportunity, foreign entrepreneurs put their plans for Eritrea on indefinite hold and looked elsewhere. Even if anyone felt ready to shoulder the risk of investing in Eritrea, they would struggle to find the staff – the nation's brightest and best had been dragooned into a 300,000-strong standing army Isaias was in no hurry to demobilize.

During the fighting, the nation had formed a common front, the Nakfa spirit taking over as a new generation of Fighters suddenly understood what their parents had endured in the Sahel. But now that peace had come, and such a poisonous peace at that, Eritreans looked back over events with critical eyes. Although the government's inner circle would never dream of admitting it in public, where the phrase 'strategic withdrawal' was once more being put to euphemistic use, invincible Eritrea had clearly lost the war. Had the government really explored every diplomatic avenue before sending tanks into Badme? Had that response been proportionate? Why had Eritrea's army been repeatedly caught by surprise during the war, failures of intelligence that played disastrously into Ethiopian hands? The peace deal signed in Algiers in 2000 bore a striking resemblance to the US–Rwanda plan which Eritrea had allowed to slip through its fingers. Had presidential pride cost thousands of lives?

Such questions highlighted another nagging concern: the political status quo. Even before the war's outbreak, Isaias had shown signs of cooling enthusiasm for multiparty democracy. The PFDJ still enjoyed a stranglehold on power and, through its affiliated corporations and banks, what remained of the economy. Those who have risked their lives for their country tend to nurse merciless expectations. When would the promised constitution be adopted? What about elections? Was Eritrea destined, just as the rest of the continent embraced multipartyism, to become a one-party state of the old, discredited variety?

At the centre of this swirling debate stood Isaias. Imposingly tall, fiercely intelligent, naturally austere, he had chosen his path early in life, rebelling against a father who was a committed Unionist. A ringleader amongst the bolshie Eritrean students attending Addis Ababa University, he joined the ELF but lost faith in its capacity to liberate Eritrea. His upper arm bore a scar in the shape of an 'E', carved at a meeting at which three disaffected young ELF members swore to create an effective revolutionary movement. Since the Struggle, in which he manoeuvred his way to the position of secretary-general with Machiavellian skill, he had enjoyed almost saint-like status in the eyes of ordinary Eritreans. When, in the early 1990s, he caught cerebral malaria and was flown to Israel in a coma, an anguished nation held his breath. His return, wasted but alive, became a triumphal procession through Asmara as residents, tears of relief in their eyes, lined Liberation Avenue. 'We allowed ourselves to be misled by the lack of photographs on display, by the president's modest lifestyle,' a woman Fighter who fled into exile later acknowledged. 'Looking back, it's clear that Eritrea did have a personality cult, just like any other African nation.' To a besotted public, Isaias' qualities seemed the quintessence of the Eritrean national character, he was Eritrea Plus. 'The PFDJ is Eritrea and I am the PFDJ,' he once famously pronounced, echoing Martini's vainglorious 'I *am* the colony'. To an outsider, the president epitomized what made Eritrea as maddening as it was magnificent. He was a leader who kept his counsel and nursed his grudges long and hard. Given a chance to ingratiate himself, he could be gruff to the point of rudeness, even at a time when Eritrea needed every friend it could get. An ambassador who delicately reminded Isaias that his country had stood by Eritrea, continuing to supply aid when others faltered, emerged from his tête-à-tête steaming. 'I suppose I should thank you for that,' had been the graceless response. Nothing seemed to dent his belief that he knew what was best for the nation. His critics were dismissed as irritating irrelevancies: 'The dogs bark, but the camel continues to march,' he liked to say.

His was a single-minded, driven personality perfectly fitted to the role of running a guerrilla organization. But Isaias' closest colleagues had started wondering, well before Badme, whether, like Winston Churchill, this was a leader unsuited to the demands of peace. Nominally, a system of executive checks and balances existed. In reality, Isaias had concentrated power in his own hands, shuffling portfolios to prevent ministers forming power bases, duplicating departments until it became unclear where real authority lay. He had squeezed the Front's heavyweights off the ruling party's executive committee in 1994 on the grounds that the PFDJ needed an injection of new blood. He had set up a Special Court, which issued judgements against which there was no appeal. And when Badme erupted, Isaias had neither formed a war council, called a meeting of the party leadership nor summoned the national assembly. The collective approach to decision-making that had characterized the EPLF was abandoned. He kept all the key decisions to himself and, in retrospect, they looked like all the wrong decisions. As for the disarray in army ranks, insiders said, Isaias was to blame. He had undermined his own generals and defence minister by telephoning commanders in the field to receive briefings and issue orders.

Former EPLF cadres will one day have to explain why they failed to rein in Isaias when the first signs of authoritarianism made themselves manifest, why they did so little, during the 1998–2000 war, to voice their unease. One factor – however strange it may sound when talking about battle-scarred warriors – was basic physical fear. Standing well over 6 ft, Isaias, the indomitable Alpha Male, towered over his cabinet colleagues, built in the small and wiry Eritrean mould. He was a hard, dogged drinker and when he lost his temper, he became physical. 'In the EPLF, policy would initially be batted about very informally, over a few drinks. I've seen him head-butt colleagues during those discussions because they wouldn't agree with him,' remembers Paulos Tesfagiorgis, a veteran EPLF activist.[8] The president once brought a whisky bottle crashing down on a cabinet minister's head, and the man sported a plaster for days. As any battered wife can attest, the threat of potential violence works its insidious effect, even when not put into practice. Flinching aides tend to ration their spontaneity. If Gordon Brown knew he risked a black eye when discussing monetary policy with Tony Blair, if Donald Rumsfeld occasionally emerged from George Bush's office nursing a split lip, politics in both countries might take a very different course.

During the Badme war, all had rallied round – Eritrea's survival demanded it. Now the critics put an end to their self-censorship. A group of academics and professionals met in Germany to compile what became known as the Berlin Manifesto, calling for the constitution to be implemented and democratic government restored. In May 2001, 15 high-ranking members of the PFDJ's central council – including such respected figures as Mesfin Hagos, Petros Solomon and Haile Woldensae – wrote to the president, asking him to convene the council and National Assembly to discuss the crisis. The president's response was quietly ominous. 'This morning you sent me a letter with signatures. If it is for my information, I have seen it. In general, I only want to say that you all are making a mistake.' The Group of 15, as they became known, pressed on, publishing their concerns about the slide to one-man rule in an open letter. Asmara's cafés were abuzz; student leaders, dismissed ministers, even the Chief Justice spoke out, and Eritrea's new private press gleefully printed it all. Taciturn Eritrea had never known such openness.

It was to be the briefest of Prague Springs. In an April interview, Isaias assured me he was committed to political pluralism. 'I've been a proponent for the last 15 years. I have lived and fought for these values all my life.' On September 18, he made his move, ordering dawn raids on the homes of the men who had once fought by his side. Only those travelling at the time escaped the police roundup. Eritrea's private newspapers were closed, its most outspoken journalists arrested. The crackdown went virtually unnoticed in the West, its attention fixed on the rubble of the World Trade Towers. But the conclusion the media reached over 9/11 – 'The world will never be the same again' – aptly described what a generation of shocked Eritreans was feeling, for very different reasons.

The Eritrea I visit these days is not the country I knew.

Eleven members of the G-15 languish in detention, their whereabouts unknown. Denied access to lawyers and family, the men and women who formed the EPLF's intellectual core are unlikely ever to face trial. They have already been effectively prosecuted and found guilty of passing military information to the enemy – what else could explain the Eritrean army's defeat? – and of plotting with the CIA to oust Isaias. 'These are not politicians, these are people who betrayed their nation in difficult times,' the president told the BBC. 'A general who betrays his country to the enemy in difficult times is a traitor, not a politician.' It is hard to find an Eritrean who, in his heart of hearts, believes this hackneyed tale of conspiracies and fifth columnists. But external events fuel the view that Eritrea, at this moment in its history, can ill afford the luxury of dissent.

After 12 months of deliberation, in which experts in international law pored over colonial maps, a commission set up in The Hague to settle the border question came up with its decision. Badme, it announced in April 2002, lay in Eritrea, just as Asmara had always insisted. Both governments had agreed the finding would be 'final and binding', but the surrender of this totemic village was too much for Addis to bear. Denouncing the result as 'unacceptable', Meles called on the Security Council to set up an 'alternative mechanism' to decide the border, effectively demanding a second opinion. Ethiopia's forces remain in occupation of Badme and its surrounds and behind a cordon sanitaire erected by the UN, two tense armies bristle at one another in a nervy standoff which carries the constant risk of another flare-up[9].

The UN, which is spending $220m a year keeping its troops deployed in the area, is a guarantor of the peace process and theoretically obliged to enforce the Boundary Commission's findings. By remaining on Eritrean soil, the Ethiopian government is defying international law. Since Addis Ababa relies on injections of foreign aid to feed its population, in theory donor nations enjoy huge leverage over its actions. But just as in the days of Haile Selassie and the Derg, realpolitik carries the day.[10] Given the choice between championing a tiny Red Sea nation with a dodgy human rights record and prickly leader, and maintaining cordial relations with Ethiopia, a regional giant regarded as an ally in Washington's War on Terror, few foreign governments can muster much enthusiasm for the Eritrean cause. For a West applying Bush's simplistic 'You're either for us or against us' criterion, Eritrea is no more than an irritant, just as it was during the Cold War.

Asmara finds itself in a position with horribly familiar echoes. On paper, the legal argument over Badme has been won, just as on paper, Ethiopia's 1962 abrogation of the Federation broke the law. What happens in practice is quite another matter. The limpness of the international community's attempt to persuade Ethiopia to accept the border ruling feels like the last in a long series of betrayals. 'What we cannot understand,' a politician told me in bafflement on my last visit to Asmara, 'is why no one wants us to survive.'

Eritrea languishes in a no-peace, no-war limbo, worst of all possible worlds. Ethiopia seems bent on throttling the economic life out of its neighbour, even if its own Tigray province suffocates in the process. 'The sad thing is that Ethiopia can strangle Eritrea to death without lifting a finger,' a British official told me. While Meles still maintains he has no designs on the coast, his people are less restrained. The curse of the Queen of Sheba has returned with a vengeance in Ethiopia, with opposition parties, civic groups and independent newspapers all arguing that Eritrea's duplicity has proved Ethiopia must secure its own port. 'The arithmetic is irresistible,' argues former attorney-general Teshome Gabre Mariam, a politically-active barrister in Addis. 'A nation of 4m cannot deny 70m people access to the sea. It is a matter of time, but another war is inevitable.' The terrible lessons of the 30-year Struggle, driven by just such a sense of Ethiopian entitlement, appear to have left little trace: a wilful, angry amnesia has set in. Two of the world's poorest nations, whose malnourished people face the constant threat of famine, are quietly restocking their armouries.

The threat from across the border gives Isaias a perfect excuse for his failure to demobilize an oversized army. The requirement to do military service – which keeps 1 in 14 Eritreans in uniform – doubles as a handy instrument of social control. Instead of gossiping in cafés, the nation's restive youth is kept busy terracing mountain slopes or laying new roads in the Danakil. Describing Isaias' hold on the country, foreign officials instinctively clench their hands into the fists of a horse rider pulling in the reins: 'He holds this country tight, very tight.' Occasionally the army stages a raid in Asmara, loading youths who cannot prove they have done their duty onto trucks bound for the bleak training camp in Sawa. Those without military papers are barred from university, have trouble finding jobs, and are denied exit visas. 'Eritrea has become one big land prison,' a former Red Flower told me. The leader who freed Eritrea now holds it captive.

Despite the government's best efforts, people still manage to slip away. Soldiers drive to the border with Sudan and walk, helicopter pilots set course for Saudi Arabia, students at Western colleges extend their foreign residence permits. One group of would-be asylum seekers, being forcibly returned from Libya, became so hysterical at the thought of returning to Eritrea, they actually hijacked the plane taking them home. The shoulder-knocking Eritrean salute – tell-tale sign of a former Fighter – is increasingly to be seen on the streets of London's Camden Town and Washington DC's Adams-Morgan. 'Eritreans have always had two ways of killing something,' an Eritrean scholar told me. 'They never challenge things openly. Either they retire into themselves and say nothing. Or they just leave – go into exile, join the *shifta*, go to the Front. In my family there isn't a single young person who wants to stay in Eritrea, and that's tragic.' This quiet abandonment is the hardest thing for the regime to swallow. For decades, the diaspora dreamt of the day they could return to independent Eritrea. Now those trapped at home jeeringly nickname those with foreign papers '*beles*', or fig cactus, because like that seasonal delicacy 'they come just once a year, and only in summer'. Eritreans are in flight again – not from the Derg, but from their own liberation movement.

The crackdown has taught a nation never prone to chattiness to watch its tongue. I used to come away from Asmara with my notebooks scrawled with names and addresses; asking an Eritrean whether he minded speaking on the record almost felt like an insult. Now acquaintances mutter under their breath, or suggest a drive to Durfo to watch the clouds swirling in over the valleys. There, in the privacy of their cars, they open their hearts. My notebooks are blotched with scribbled-out names and when I write my articles, I resort to the anonymous labels of journalism conducted in a police state: 'an official', 'a former Fighter', 'a minister'. Identities, along with Eritrea's sense of certainty, have swirled away, like ink dropped in flowing water.

Personal opinions have been replaced by jokes, not something I ever associated with Eritrea. Losing most of their impact in translation, they poke fun at Isaias and his remaining cronies with a scornful irreverence unthinkable 10 years ago. One runs as follows: 'An international conference is being staged and every delegation starts boasting about their country. "We have the best engineers," say the Americans, "we can take a man and put him on the moon." "We have the best surgeons," say the Germans, "we can take out a heart and transplant it." "Ah yes," say the Eritreans, "but we have the best doctors. We can replace a man's brain with a coconut and call him president."'

Chillingly, I have begun hearing a refrain I first heard in Bucharest, in the days that followed the dictator Ceauşescu's toppling. 'Whenever more than five of us were gathered together,' a Romanian paterfamilias told me, looking across a crowded lunch table, 'we knew there was a member of the Securitate in our midst.' Eritreans, mindful of the informer network that operated under the Derg, distrust even their nearest and dearest. 'I'm telling you this because you are a foreigner,' an ex-Fighter told me on my last visit, 'but I would not say this in front of an Eritrean, not even my closest friend.' He had only been discussing the mundane problems of adjusting to civilian life, yet even that felt potentially seditious. 'There are many informers inside Eritrea,' says Dr Bereket Selassie, the US-based academic who drafted Eritrea's never-implemented constitution, now in opposition. 'We can tell how effective they are from the way the e-mails from Asmara are drying up.'

Eritrea's leadership is more isolated now than ever it was

during the Struggle, for the True Believers have distanced themselves.[11] Those who once marvelled at plucky little Eritrea's iconoclasm now shrug it off as a 'pariah state'. Bent on proving he cares not a jot for the international community's disapproval, Isaias has grown ever more heavy-handed in his dealings. When Italy's ambassador protested at the jailing of G-15 members on behalf of the European Union, he was expelled: Eritrea was not about to take lessons in democracy from its former colonial master. When Washington, worried about Islamic fundamentalism, was looking for a site for a new military base to police the Red Sea, Eritrea seemed the obvious choice. A US presence would have nipped any Ethiopian designs on Assab in the bud. But Isaias refused on principle to release two American embassy employees arrested during his crackdown and a miffed Washington built its base – a muscular version of Kagnew – in Djibouti instead.

I registered how far the line separating healthy feistiness from self-destructive bloodymindedness had been crossed one torpid day, walking through the alleys of Massawa. Neatly parked by the roadside sat 18 SUVs, covered in a gathering blanket of red dust. Lip-smacking assets in a country with few tarmac roads, the jeeps had been imported as part of a Danish-funded mine-clearing programme and Denmark had always planned to leave them behind as a gift when the job was done. But the government, with its customary lack of tact, had told Copenhagen it had no need of its services. Furious at its brusque treatment, the Danes were making a point of shipping the jeeps out. Denmark didn't need them, but Eritrea would not be getting them either.

With foreign friends gratuitously alienated, its youth in military training and business interest at rock bottom, Asmara seems stuck in one long Sunday afternoon. Since Ethiopian Airlines suspended its services, it has become one of Africa's least accessible capitals, a city whose residents prick up their ears when a flight roars overhead and pinpoint the airline – there are so few – with impressive accuracy. 'This is a good place for old people to retire to,' a frail old Italian, puzzled by my interest in the place, gently told me. 'But there is nothing here for the young.' In a humiliating echo of the past, Eritrea once again plays host to foreign troops: a UN intervention force this time, rather than Kagnew Station. Locals who tut-tutted at the sight of Eritrean girls on the arms of drunk GIs now wince at the louche goings-on in the bars near the UN base. They hate this dependency, but need these visitors both to shield them from the Ethiopians and keep their stagnating economy afloat.

The taxi driver's dirge, that staple of African travels, has finally come to Asmara. The last driver who picked me up at the airport, in that predawn chill when the barks of waking dogs are relayed from one sleepy district to another, drove with the infinite slowness of a man counting every drop of petrol. He would be selling his taxi soon, the old man muttered from a tumbled swathe of turban and shawl. Business had evaporated and the car was costing more to run than he earned. Then he pronounced the words I had heard said about the Belgians in Congo, Portuguese in Angola and British in Zambia, but never dreamed I would hear in Eritrea – all the more heart-rending for being said with such quiet resignation.

'Things were better under the Italians.'

CHAPTER 18

'It's good to be normal'

'When dreams are shattered, they itch like scabies on the buttocks.'

Eritrean proverb

One Saturday morning a small group gathered near London's Gower Street, taking their seats in a room off one of the slightly dilapidated Victorian halls, bequeathed to the nation by high-minded philanthropists, that cluster around Bloomsbury. The meeting's chairman, an exile with an intense, intelligent face, introduced the guest speaker with obvious affection. The white-haired academic talked softly about the efforts opposition movements were making to unite, the need to rally the diaspora and efforts to lobby Western governments. In the audience, human rights campaigners nodded sympathetically, offering information on the latest repressive measures introduced by the authorities. The perennial question of how to grab the attention of a fickle foreign media was raised, and journalists present made a few tentative suggestions.

A sense of terrible poignancy seeped in with the thin winter light, unacknowledged but inescapable. For this meeting on Eritrea, actually called in autumn 2003, could have been staged at almost any time in the 1960s, 1970s or 1980s. The Britons there – many former True Believers – had attended scores of such get-togethers in their day, to discuss identical problems. The speaker, Dr Bereket Selassie, had denounced a government's illegality not once, but a hundred times, in similar gloomy rooms. And the Eritrean chairman, Paulos Tesfagiorgis, had known exile before, when he had been based in Khartoum raising funds for Eritrean refugees in the 1970s and 1980s. First time round, the meeting's participants had been blessed with the energy and optimism of youth, and they had been fighting an alien force, the Ethiopian regime. Now they were past middle age, and the adversary was a system they themselves had helped create: the Eritrean government. Yet here they were, launching themselves once again into the grind of campaign meetings and focus groups, lobbying and leafleting, that constitutes long-distance dissidence. Like a boomerang, history had executed one long ironic arc, returning to knock them off their feet.

Around the world, Eritreans who played a key role in the EPLF, whether as Fighters, activists or merely supportive civilians, are trying to understand what went wrong. Did they, however inadvertently, contribute to the betrayal of the Eritrean revolution? What could, and should, they have done differently?

For Paulos, a small, articulate man with the notched eyebrows of a Christian highlander, the interrogation is particularly severe. In 1974, when still a law student at the University of Wisconsin, he became an actor in what remains a rarely-acknowledged episode in rebel history. The large Eritrean community based in the US, in spasmodic contact with the rebel movement back home, had heard reports of a purge in the factions that would eventually coalesce to form the EPLF.[1] Some Fighters who objected to Isaias' style of leadership had formed a movement dubbed *manqa* ('bat') after its habit of meeting at night. These were the rough and ready days of a movement still finding its feet, and *manqa* complained about poor coordination, supply shortages and the fact that Fighters who dared challenge Isaias' views were often given a good hiding. It wanted greater accountability, increased power-sharing. Many of the suggestions made by *manqa*'s members would later be put into effect, as the Front became better organized. But its ringleaders did not live to see that day. After a year of febrile debate, they were shot by the EPLF.

'People we knew, people who had attended Addis University and had played a key role mustering support for the Move-ment, had been killed. It raised a lot of questions,' remembers Paulos. 'The discussion kept festering, it would not go away, it was creating a lot of disunity.' The diaspora decided to settle the question for good by dispatching Paulos and another young Eritrean to the Sahel. The future goodwill – and sizeable financial contributions – of the North American community would depend on the account the two envoys brought back.

It was Paulos' first visit to the Front and he found it psycho-logically overwhelming. The purge had created an atmosphere of fear and suspicion he could feel but barely understand. In daytime he was kept under strict escort, but at night *manqa* sympathizers sidled up to him to mutter: 'Don't believe everything you hear.' He was awed by the austerity of life at the Front, humbled by the Fighters' sense of purpose. Above all, he was agonizingly aware that while he – a spoilt member of the educated bourgeoisie – was free to return to a cushioned existence in the West, former classmates who possessed no more than the clothes on their backs were staying behind. To question it all would have felt like gross disloyalty.

'In our report we said: "Those executed were guilty of incite-ment, indiscipline and creating division,"' remembers Paulos. 'Our report created a calmness. We were the first people from the North American community who had been there, so no one could challenge us. Our word was the word. Single-handed, we made the Front look fantastic.'

Today, Paulos tortures himself with the thought that he was responsible for what amounted to a whitewash, a ringing endorsement delivered at a time when the young Isaias, facing his most serious challenge to date, might have been either reined in or sidelined. 'We did it completely in good faith. But, looking back, we made a mistake.' Fate has exacted a high personal price for his error. In his fifties, at a time when most men feel they have earned the right to job security, status in the community, a home of their own, Paulos has become an asylum-seeker, doomed to a rootless existence spent sleeping on other people's sofas, negotiating the maze of foreign bureaucracies, dependent on the generosity of friends-of-friends.

It would be heartening to think the foreign powers that meddled in Eritrea with such devastating results might occasionally examine their consciences and records with equal rigour. For while ordinary Eritreans have lessons to learn about how and why their revolution was betrayed, so does the West. If Eritrea today so often comes across as dangerously impervious to criticism and bafflingly quick to anger, she is largely that way because colonial masters and superpowers made her so. An entire society is suffering from a form of post-traumatic stress disorder. Her history of cynical abuse – shared by so many small nations whose gripes prompt irritated yawns in Washington, Moscow and London – should serve as warning as the campaign against Islamic extremism recasts Western foreign policy in brash interventionist mould. If determined enough, guerrillas in plastic sandals can bring down a modern army. A capacity for taking infinite pains can force the most sophisticated occupying power to its knees. When you don't know what you're doing, can't grasp who you are dealing with, best leave well alone.

It would be refreshing to think officials in the State Department, the Kremlin, Whitehall or the UN occasionally remember the parable that is the Eritrean story, but it would be illusory. As with so many former colonies, Eritrea highlights the one-sided nature of memory in an unequal partnership. She is like a girlfriend who remembers every line on the face of the man who abandoned her, nursing each hurtful word of their raging arguments, honing her responses. When, years later, the two meet again, he delivers his most wounding insult yet. While the abuse has scarred her forever, he can barely recall the relationship. Eritrea is now being dealt the final insult: she is being forgotten by the powers that once used her.

The country I once liked to think of as Shangri La has become an unhappy land, but it is also a far more interesting, nuanced place. Once, talking to Eritreans, I had the impression of speaking to a many-headed monster, each of whose mouths chanted the same refrain. Now the Hydra's heads often speak in whispers, but they wear different expressions and none of the opinions they voice are the same. Some believe the government is wrong, but now is not the time to press the point. Some regard Isaias as misunderstood national saviour, some loathe him as the Great Betrayer. Eritreans are becoming rounded individuals, their community a more complex, conflicted society. That is no bad thing.

Writing this book, I used to marvel over the chasm between the stark experiences of the Eritreans I had come to know and the foreigners – often direct contemporaries – who impacted so heavily on their lives. Which life, given the choice, would I pick? A member of the Gross Guys – binge-drinking to obliterate the boredom, uneasily aware that in shirking Vietnam I had balked my generation's ultimate test, certain of tranquil retirement in middle America? Or a Fighter in my too-short shorts, listening to a piano recital under a thorn tree, aware the odds were against me making it through the war? 'We were unique,' Zazz the GI once surprised me by boasting, in a mood of bleary self-congratulation. The adjective seemed rather better applied to the earnest 'shifties' Kagnew's commanders warned their boys against. Give me the Sahel any day, because the choice between blandness and passion seems no choice at all.

Yet Eritrea's story highlights the dangers inherent in that intoxicating, beguiling thing: a sense of purpose. 'For years we felt superior, not just because we won the war but because we had idealism, we had a grand vision,' says Paulos. 'Look at the ex-Fighters. It is only now that they are coming down to the ground and becoming ordinary human beings again.' The last few, chastening years have brought Eritreans earthwards with a vengeance, and even those in government recognize an element of hubris. 'It's good to be normal,' ruefully acknowledges a government minister in Asmara. 'We have gone from thinking we were unique, a people chosen by God like the Israelis, to realizing we too have our faults, we are not so special after all. It's called growing up.' Humility seems unlikely, but Eritreans no longer take it for granted they are a breed apart, no longer assume they know the answers to Africa's problems. As their present becomes murkier, they are losing the black-and-white certainties of the past.

As Isaias accurately predicted, today's Eritrea is no society of angels. The image of a Utopia built up in Fighters' minds has evaporated like the morning mist. Yet I can't write it off as just another numbing Third World disappointment.

If the curse of so many African states has been low expectations, passed from one generation to another like a genetic disease, a generation of Eritreans stands immune. The EPLF spent decades teaching its followers that every man and woman, Moslem and Christian, peasant and urban dweller, was equally valuable. It set up popularly-elected assemblies in the villages, it championed women's organizations, it relentlessly trumpeted the merits of grassroots democracy. That work cannot now be easily undone. Aspirations were created, and the fact that they have been frustrated will not pass unnoticed. The notion of accountability has seeped into a people's psychology, as impossible to uproot as the dream of shady groves and green pastures ex-Fighters regard as the *real* Eritrea. 'We had this idea of equality at the Front, and now it is fixed forever in our minds,' a friend ruminated. Amongst the older generation inside Eritrea, the unarticulated refrain – 'I didn't spend 15/20/30 years at the Front for *this*' – spools through daily life like the subtitles showing under films at Asmara's Cinema Roma. As for the diaspora, its Western-educated, foreign-passport-holding youngsters are coolly appraising, their expectations serving notice to a leadership that has lost its way. 'Do you remember what John Kennedy said? "Ask not what your country can do for you – ask what you can do for your country." Well, I feel precisely the opposite,' a young Eritrean told me on a flight to Asmara. He was returning to teach at secondary school, but his degree from a German business college meant there was no shortage of tempting alternatives if the experience proved frustrating. 'Of course I'm a patriot. Of course I want to do my bit. But not at any price. This has to be a two-way relationship.'

The bumptiousness of such youngsters, like the unforgiving self-examination of older men like Paulos, is a great source of hope. The stroppier they get, the better. Surveying Eritrea's future, I feel nothing like the bleak despair that descends when I try to guess whether Congo will survive as a nation-state, or Sierra Leone's democracy will last the year. Eritreans have already achieved too much, against too many odds, for the country to fail.

Whenever pessimism threatens to set in, I'm always brought up short by the memory of a group of overall-wearing 80- and 90-year-olds, working happily in an abandoned hangar on the edge of town.

I came across them on my first visit to Eritrea, when I drove out to Asmara's grassed-over railway station at the suggestion of the transport minister of the day. With only one winding road linking the capital to Massawa, the government, he said, wanted to rebuild the old Italian railway. It had put the job out to tender, but the estimates offered by Western construction companies were more than an administration allergic to debt could swallow. Eritreans would do it themselves, the government decided. With hard work and application – the qualities on which the EPLF had always depended – the task could be completed on a shoe-string budget. Turn-of-the-century charts showing where the track once ran were dusted off, cannibalized sleepers collected into neat piles. The labourers who had worked as apprentices under the Italians were summoned out of retirement and told to train a new generation of railwaymen. 'It's good to be working again,' one told me. 'When you retire, both mentally and physically, things begin to slip.'

It became something of a showcase project, a picturesque expression of a national fixation with self-reliance. Visiting camera crews adored it. Their lenses lingered over the grizzled labourers as they lovingly oiled down the chuffing black Ansaldo and Breda steam engines, curved rails painstakingly by hand and fired up the disused Italian foundries. Trainspotters around the world, raving over 'the steam story of the 1990s', tracked every development on their websites.

Domestic critics rolled their eyes to heaven. This was self-indulgent folly, they complained, an indication of how the PFDJ was mismanaging the transition from rebel organization to modern government. The Massawa–Asmara line would

probably never be reconstituted and even if it were, the locomotives would carry too little cargo to make a difference. When
the war in Badme blew up, their scepticism seemed confirmed.
With the young apprentices away at the Front, only the
veterans were left, and progress slowed to a snail's pace.

Yet every time I visited, I noticed that the brown ribbon of
track had edged a little further towards the capital, the white
scar that snaked its way around the mountain was that much
shorter. Just as their predecessors had done a century before,
the Eritrean railwaymen paused before tackling the gravity-
defying final climb up to Asmara, gave one last heave, and
scaled the plateau. The railway between Asmara and Massawa,
the engineering feat Martini had regarded as his life's greatest
achievement – now reborn as a symbol of gritty Eritrean
nationhood – is a reality once again.

On one of my last trips to Asmara, I took a ride on the lowland section, watching as a wizened railwayman, who still spoke
functional Italian, expertly spun and loosened the brakes on
the railcar. The diesel locomotive whistled through tunnels and
rattled round bends. As it stopped to let farmers load grain
sacks onto waiting camels and take aboard women carrying
bouquets of upended chickens, the sun sank over the thorn
trees. Lights were a luxury the old network did not run to, and a
dark, starlit peace descended upon us. A confused bat landed in
my lap, wriggled in brief panic, then flew off. Someone turned
on a radio and jangling Eritrean music filled the carriage. By the
time I stumbled off the train, my hair permeated by the smoky
aroma of home-cured leather, the moon was out. Seven hours
to travel 32 km, pause and return: it had undoubtedly been the
slowest train ride of my life. But parts of Eritrea that had been
cut off for decades were trading again, farmers were tending
long-neglected plots in the knowledge they would be able to
get their vegetables to market. The Eritreans had done it their

way – the dogged, counter-intuitive, hardest way – but they had pulled it off.

The following day, my bags packed, notebooks stowed, I was blessed with one of those moments of serendipity which always seem to take place in Eritrea.

We bumped into each other in a snack bar opposite the red-brick cathedral. It was just after lunch, a time of day when most of Asmara, true to its Italian ancestry, pulls down the metal shutters and takes a nap. Outside, the sun slammed down on Liberation Avenue, as relentless as gravity itself. Inside, the blue-veined marble floor was cool. I went to place my order and found my words being echoed by a middle-aged man standing next to the cashier, who had decided – notwith-standing the fact that all three of us were speaking in English – to act as interpreter.

Me to cashier: 'I'll have a *macchiato*.'

Him to cashier: 'She says she'll have a *macchiato*.'

Me to cashier: 'And a doughnut.'

Him to cashier: 'Give her a doughnut too.'

Cashier to me: 'Four nakfa.'

Him to me: 'That will be four nakfa.'

'So,' he said, having established with this friendly, if super-fluous, service, a certain bond: 'Who are you? What are you doing in Asmara?' 'I'm a journalist. From England. I'm doing research.' 'A British journalist? BBC?' He positively beamed. His face was as plump as a cherry and when he smiled, it radiated pure bonhomie. 'Can you help me? You see, I'd like to see some footage the BBC filmed of me when I was a young man. Also, there are two Americans, two old friends of mine, I would very much like to get back in touch with them.'

'A BBC film? Well, they have archives, of course. In London. But I'd need the details to track it down. When was this?'

'It was in 1975. Tom Boudhoud – he stumbled over an

awkward name – and David Strickland. They had been working at Kagnew Station and were taken hostage by the ELF. The BBC did an interview when they were under guard.'

There was something about this story, I sensed, I was failing to grasp. Had this man worked at Kagnew Station? Had he been held hostage alongside the two Americans? What was the connection?

'You say you were their friend?'

'Yes, yes, we became very close. I am curious to know what I was like then, to see myself in my twenties. It's history now, for them and for me.'

Suddenly, the pieces clicked together to form an explanation so magnificently surreal I gave an incredulous laugh. 'You're the one who took them hostage. You were the ELF kidnapper.'

He gently remonstrated with me. 'I was head of the ELF unit. But I think of myself as their friend.' That piercingly-sweet smile again. 'You see, I was simply doing my duty. We were fighting for independence. We were all soldiers together, both the Americans and us, and the Americans were support-ing our enemy.'[2]

'Why on earth do you want to get back in touch with them? Do you want to apologize?'

He ignored the question: it really was a bit crass. 'That was a very special time for me and for them. It was the first time, for both of us. We only kept them for a week, in a little house in the mountains, before the ELF came to take them to Gash Barka, but it felt more like seven years together. At first, they thought we were just villagers who were going to kill them. In fact, I come from one of Asmara's most established families. I've never forgotten Tom and David. We were happy at that time, because we thought we were heroes.'

In a case of reverse Stockholm syndrome, the kidnapper had, it seemed, fallen half in love with his hostages. Here was all the nostalgia of a 50-year-old attending a high school reunion, pining for his lost youth while desiring 'closure', but with the added piquancy lent by the AK-47, the ransom note and the comradeship of camouflage.

We sat together at one of the green banquettes and surveyed the sun-dazed street. He was only passing through. The Communist revolutionary had long since turned respectable businessman and now lived in Sweden, where he worked for a trading company. There were things he urgently wanted to explain, but his Swedish was better than his English and the words were coming with difficulty. He stumbled on, illustrating his argument with rough sketches on a paper napkin.

'In Eritrea, we love the West. But the West has decided to treat us as some kind of enemy. It criticizes our government, its journalists take Ethiopia's side in the war. It is true, we make mistakes, and we will make many more. But what you have to understand is that we are a very young country. We have only been independent for a few years. We are like a child, going for the first time to the . . . the . . . what do you call it?'

'The kindergarten?'

'Yes, a child going to the kindergarten. At the start, his mother has to stay with him. The West must stay with us now. It has to be patient, not beat us like a teacher in a Third World school. Instead of slapping our government and saying: "You did a stupid thing", it should be saying: "He will learn."' He pointed to the cathedral across the way. 'Look at that Italian cathedral. Look at these bars and these cinemas. Look at the way these girls walk around in T-shirts. In Saudi Arabia they would have to cover up, here they are free.'

He searched again for the right words, and found them. 'Our history makes us close. We have an affinity. Do not push us away.'

Chronology

1869 Suez Canal opens. Italian priest Giuseppe Sapeto buys Assab from local sultan.

1870 After decades of conflict, Italy becomes a united nation, with Rome as capital.

1884–5 Europe's colonial powers divide Africa up at Berlin conference. British invite Italians to seize Massawa.

1887 Italian column advancing into Eritrean highlands wiped out by forces of Ras Alula at Dogali.

1889 Abyssinian Emperor Yohannes IV slain in battle against Mahdis. King of Shewa anointed Emperor Menelik II, who signs Treaty of Uccialli with Italians.

1890 Italian King Umberto declares colony of Eritrea, with Massawa as capital.

1891 Ferdinando Martini makes first trip to Eritrea as part of royal inquiry.

1896 Italian attempt to advance into Abyssinia repelled by Emperor Menelik at Adua.

1897 Martini returns to Eritrea as its first civilian governor, moves capital to Asmara.

1911–12 Italy seizes Libya's Tripolitania and Cyrenaica after defeating Turkish forces.

1913 Emperor Menelik dies.

1914 First World War breaks out, Italy enters war on Allied side the following year.

1922 Benito Mussolini becomes Italy's prime minister after March on Rome.

1930 Ras Tafari crowned Emperor Haile Selassie.

1931 Libyan resistance movement brutally crushed by Italians.

1935 Mussolini invades Abyssinia from Eritrea, using chemical weapons.

1936 Haile Selassie flees into exile and Italian troops enter Addis. Mussolini announces creation of Italian East Africa.

1940 Mussolini enters Second World War on Hitler's side. Italian army ejects British from Somaliland.

1941 British forces defeat Italians at Keren and take over administration of Eritrea, Libya and Italian Somaliland. Haile Selassie reinstated.

1945 End of World War Two. Mussolini and his mistress lynched by Italian partisans. United Nations established in New York.

1946 Italy formally renounces all claim to its African colonies.

1948 Four Powers Commission fails to agree Eritrea's future.

1949 UN Commission of Inquiry sent to decide Eritrea's fate.

1950 Korean war breaks out.

1952 Eritrea federated with Ethiopia under UN-brokered deal.

1953 US signs 25-year rights agreement for Kagnew Station.

1960 Haile Selassie survives military coup. Eritrean exiles in Cairo establish Eritrean Liberation Front. Sylvia Pankhurst dies in Addis Ababa.

1962 Eritrean parliament dissolved. Ethiopia formally annexes Eritrea.

1963 First US combat troops arrive in Vietnam. Organization of African Unity sets up permanent headquarters in Addis.

1972–4 Famine sweeps Tigray and Welo.

1974 Emperor Haile Selassie overthrown by the Derg.

1975 Derg announces death of Haile Selassie, aged 84. ELF and EPLF, breakaway Eritrean rebel faction, reach outskirts of Asmara.

1977 Somalia invades eastern Ethiopia. As Red Terror killings start, Derg breaks off relations with US, closes Kagnew Station and joins Communist bloc.

1978 With massive Soviet military backing, Ethiopia pushes back Somali forces and wins upper hand in Eritrea. EPLF forced to retreat into Sahel.

1984–5 Famine sweeps northern Ethiopia, Bob Geldof organizes Band Aid.

1985 Mikhail Gorbachev comes to power in Soviet Union.

1988 EPLF smashes Ethiopia's Nadew Command at Afabet, ending military stalemate.

1989 Berlin wall comes down.

1991 Eritrean and Ethiopian rebels capture Asmara and Addis. Mengistu flees.

1993 Eritreans vote for independence in UN-monitored referendum.

1997 Eritrea launches its own currency, the nakfa. Ethiopian and Eritrean forces clash at Bure and Bada.

1998 Skirmishes at Badme escalate into new war between Eritrea and Ethiopia.

2000 UN peacekeepers sent to patrol temporary security zone between Eritrea and Ethiopia. Eritrean president Isaias Afewerki arrests domestic critics, closes private press.

2002 Boundary Commission announces ruling on border, allotting Badme to Eritrea.

Glossary and acronyms

As there is no precise formula for translating Tigrinya and Amharic into the Roman alphabet, Eritrean and Ethiopian names can be spelt in a variety of ways. When identifying living individuals, I have used the version they preferred. When dealing with historical figures and places, I have chosen the version I guessed readers were most likely to have encountered in the past.

Abyssinia Former name for Ethiopia. It stopped being used after World War II.

Amhara Ethiopia's dominant ethnic group, from which its rulers traditionally came.

Ascari Eritreans recruited to serve in Italy's colonial army. Between 1890 and 1941, around 130,000 Eritreans served as *ascaris*, fighting at the battle of Adua, in Somalia and in Libya. They also played a crucial role in Italy's 1935 invasion of Ethiopia.

ASA Army Security Agency. Branch of the US' highly secretive National Security Agency in Fort Meade, where government communications are enciphered and foreign communications monitored. It ran Kagnew Station until 1972.

Berbere Rich mix of spices used in Eritrean and Ethiopian food.

Derg Popular term for the Provisional Military Administrative Council, a shadowy group of dissident army officers that emerged in Ethiopia in early 1974. The Derg overthrew Haile Selassie and shocked the public by then killing its own chairman, executing 59 former government members and cutting ties with the US. By 1977, having eliminated his rivals, Mengistu Haile Mariam emerged as its undisputed leader.

Eritrean rebel movements Eritrea's armed struggle was launched by the Eritrean Liberation Front (ELF), founded in Cairo in 1960 by Moslem Eritreans. Although it had strong links with Arab states, the ELF initially recruited both Christians and Moslems. But in the early 1970s a group of Christian highlanders broke away, going on to form the Eritrean People's Liberation Front (EPLF). Civil wars between the two movements played disastrously into Ethiopia's hands. But by 1980 the EPLF had eclipsed the ELF, whose remnants were pushed into Sudan. Both movements embraced Marxism, although the EPLF watered down its Communist message as victory approached.

EPRDF In 1989 Ethiopia's ethnically-based opposition movements merged to form the Ethiopian People's Revolutionary Democratic Front, dominated by the TPLF. The coalition captured Addis Ababa in 1991 with Eritrean support and set up Ethiopia's current government.

Isaias Afewerki President of Eritrea and former EPLF secretary-general. Born in a district of Asmara in 1946, he interrupted his engineering studies at Addis Ababa University to join the ELF. He was sent to China for training but grew disillusioned, setting up the EPLF with a group of close associates.

Kebessa Eritrea's central highlands, inhabited by Christian, Tigrinya-speaking people. Include the historic provinces of Hamasien, Akele Guzai and Seraye.

Meles Zenawi Prime Minister of Ethiopia. Born in Adua in 1956, he interrupted his medical studies in Addis Ababa to join the TPLF, rising to the post of chairman.

Mengistu Haile Mariam Of modest origins, Major Mengistu emerged as leader of the Derg in the mid-1970s and went on to establish one-man rule. He brutally suppressed opposition during the Red Terror, embraced Marxism and won massive military backing from Moscow for his wars on Somalia and Eritrean separatism. Sometimes dubbed 'the Red Negus', he fled to exile in Zimbabwe in 1991.

Oromo A pastoral people, the Oromo began migrating in the mid-16th century into southern and southeastern Ethiopia, where they gradually integrated with the Amhara nobility. They are now the country's largest ethnic group.

PFDJ In February 1994, with independence achieved, the EPLF dissolved itself and launched a political movement, the People's Front for Democracy and Justice.

Ras Ethiopian title for 'prince' or 'duke'.

Ras Tafari Son of the governor of Harar province, Ras Tafari became regent to Empress Zawditu, acceding to the imperial throne following her death. He took the name 'Haile Selassie', 'Power of the Trinity' in Amharic.

Shabia 'People's' or 'popular' – used to denote the EPLF.

Shewa Central Ethiopian province, historically the heart of Abyssinia's Christian empire. It recovered that role in the late 19th century under Menelik ll, and now holds the modern capital of Addis Ababa.

Shifta Traditionally applied to those who rebelled against their feudal lords. It was later used to denote armed bandits operating in the countryside.

Tegadlai Freedom fighter. Plural is *tegadelti*.

Tigray North-eastern Ethiopian province, bordering Eritrea. Political centre of the ancient Axumite empire. Its people are known as Tigrayans. They share their Tigrinya language, highland customs and Orthodox Christian faith with inhabitants of the Eritrean kebessa.

TPLF Tigrayan People's Liberation Front. Marxist-Leninist movement that sprang up in Ethiopia's northern province in the mid-1970s. Relations with the longer-established Eritrean rebel groups across the border were strained by ideological and tactical differences, but in 1988 the TPLF and EPLF agreed to cooperate on military strategy, a move that led to the Derg's defeat.

Woyane used to denote the TPLF.

Notes

Chapter 1 The City Above the Clouds
1 Two photographic books capture the wonders of urban Asmara: Edward Denison, Guang Yu Ren and Naigzy Gebremedhin, *Asmara – Africa's secret modernist city*, Merrell, 2003; Sami Sallinen, *Asmara Beloved*, Kimaathi, 2004
2 Presidential interview with Scott Stearns of Voice of America and Richard Dowden of *The Economist*, April 9, 1999

Chapter 2 The Last Italian
1 Cesare Correnti, addressing the Italian Geographical Society, April 18, 1875. Maria Carazzi, 'La Societa Geografica Italiana e l'esplorazione coloniale in Africa: (1867–1900)', *La Nuova Italia*, 1972, pp 144–57
2 Pellegrino Matteucci took part in an 1879 expedition, subsidized by northern Italian industrialists, to assess Abyssinia's commercial potential. His conclusions were damning. 'Allowing Italy to nurse any illusions about a country's wealth, if it doesn't exist, strikes me as anti-patriotic,' he recorded in his memoirs: Angelo del Boca, *Gli Italiani in Africa Orientale, Vol 1*, Mondadori, p 93
3 Ferdinando Martini, *Nell'Affrica Italiana*, Fratelli Treves Editori, Milan, 1891, p 332
4 Foreign Minister Pasquale Mancini, January 27, 1885: *Gli Italiani in Africa Orientale, Vol 1*, p 182
5 *Dizionario Biografico De Guberatis*, Firenze, 1879
6 Former army officer Cesare Pini, who served in Eritrea, recalled a 'flood' of death sentences in Asmara following the Massawa scandal.

Haunted by those executions – he said he bore personal witness to around 40 – he was struck by the fortitude shown by the condemned: 'I never saw a single one of those blacks, even those who were very young, almost boys, humiliate himself, rage or weep: not once!' *Gli Italiani in Africa Orientale, Vol 1*, p 447

7 Achille Bizzoni, *L'Eritrea nel passato e nel presente*, Milano, Sonzogno, 1897

8 For more on this episode, *see* Massimo Romandini, 'Il "dopo Adua" di Ferdinando Martini, governatore civile in Eritrea', *Studi Piacentini*, 20, 1996, pp 177–204 ; Massimo Romandini, 'Da Massaua ad Asmara: Ferdinando Martini in Eritrea nel 1891', *La Conoscenza dell'Asia e dell'Africa nel XIX Secolo*, vol III, 1989, pp 911–33; Robert Battaglia, *La Prima Guerra d'Africa*, Einaudi, 1958; *Gli Italiani in Africa Orientale, Vol 1*, pp 435–61

9 For more on colonial theories of biological determinism, *see* Sven Lindqvist, *Exterminate all the Brutes*, Granta Books, 1992

Chapter 3 The Steel Snake

1 The idiosyncrasies of Eritrea's system have won it a keen following amongst railway experts around the world. For details, *see* 'Railways Administration in Eritrea', Imperial Ethiopian Government, 1965. 'Eritrea – Rebirth of a Railway', a video by Nick Lera, Locomotion Pictures, tracks the rehabilitation project; www.trainweb.org/eritrean has a site dedicated to the subject and a book by Jennie Street is due to be published by Rail Romances in 2006

2 Richard Pankhurst, *The Ethiopians: a History*, Blackwell, 2001

3 *Gli Italiani in Africa Orientale, Vol 1*, p 740

4 Ferdinando Martini, *Il Diario Eritreo, Vol 1*, Vallecchi Editore, 1946, p 1

5 *Il Diario Eritreo, Vol 1*, Foreword

6 ibid

7 ibid

8 ibid, p 165

9 ibid, p 29

10 ibid, p 60

11 ibid, p 23

12 ibid, p 165

13 ibid, p 159

14 ibid, p 89

15 *Il Diario Eritreo, Vol 2*, p 121

16 *Gli Italiani in Africa Orientale, Vol 1*, p 758

17 *Il Diario Eritreo, Vol 3*, p 328

18 *Il Diario Eritreo, Vol 1*, p 3

19 ibid, p 94

20 *Il Diario Eritreo, Vol 3*, p 3

21 ibid, p 424

22 *Il Diario Eritreo, Vol 4*, pp 48–9

23 *Il Diario Eritreo, Vol 3*, p 248

24 *Il Diario Eritreo, Vol 2*, p 472; *see also*, Massimo Romandini, 'Il Problema Scolastico Nella Colonia Eritrea: Gli Anni 1898–1907', *Africa*, September 1984

25 Massimo Romandini, 'Ferdinando Martini ad Addis Ababa', *Miscellanea di storia delle esplorazioni geografiche*, IX, Genoa, 1984, pp 201–43

26 *Il Diario Eritreo, Vol 4*, p 606

27 Author's interview

28 Emilio de Bono, *Anno XIII: The Conquest of an Empire*, Cresset Press, 1937; Angelo del Boca, *The Ethiopian War 1935–1941*, University of Chicago Press, 1969, p 4

29 *The Ethiopian War 1935–1941*, p 21

30 *The Ethiopian War 1935–1941*, p 210

31 Giorgio Maria Sangiorgio, see *Gli Italiani in Africa Orientale, Vol 3*, p 219

32 Richard Pankhurst, 'The Legal Question of Racism in Eritrea during the British Military Administration', *Northeast African Studies*, vol 2, part 2, 1995; Richard Pankhurst, 'Fascist Racial Policies in Ethiopia 1922–1941', *Ethiopia Observer* 12, pp 92–127

33 Martino Moreno, head of political affairs at the Ministry; see *Gli Italiani in Africa Orientale, Vol 3*, p 239

34 Araia Tseggai, 'Historical Analysis of Infrastructural Development in Italian Eritrea 1885–1941', Part 2, *Journal of Eritrean Studies*, vol 1, no 2, 1987; see also *Gli Italiani in Africa Orientale, Vol 3*, pp 236–23

35 *Gli Italiani in Africa Orientale, Vol 3*, p 239

36 *Gazetta del Popolo*, May 21, 1936. Once again, top-level attempts to prevent interbreeding proved stunningly unsuccessful. In 1950, the Associazione Meticci dell'Eritrea estimated the number of half-castes

at 25,000, although that number, confusingly, included Eritrean mothers

37 Author's interview

Chapter 4   This Horrible Escarpment

1 'Retrospect' – Lecture by Lieutenant-General Sir William Platt, Khartoum, 1941
2 AJ Barker, *Eritrea 1941*, Faber and Faber, 1966
3 The tale of the Italian officer who staged the cavalry charge is told in Sebastian O'Kelly, *Amedeo: A true story of love and war in Abyssinia*, HarperCollins, 2002
4 Imperial War Museum, London; Sound Archives 7373/3 A
5 Archibald Harrington, Imperial War Museum; Sound Archives 8332/7 A
6 'Retrospect'
7 Author's interview
8 *Eritrea 1941*, p 101
9 Peter Cochrane, *Charlie Company*, Chatto & Windus, 1977, p 64
10 Imperial War Museum, 7373/3 A
11 *Eritrea 1941*, p 136
12 'Retrospect'
13 Author's interview
14 *Charlie Company*, p 72
15 Imperial War Museum, 7373/3 A
16 General Nicola Carnimeo, *Cheren*, Casella Editore, 1950, p 212
17 'Retrospect'
18 Alberto Rovighi, *Le Operazione in Africa Orientale, Vol 1*, Officio Storico SME, Rome, 1995, p 256; *Gli Italiani in Africa Orientale, Vol 3*, p 433
19 'The Abyssinian Campaigns', The Official Account, London, 1942, p 46
20 MAJ Trimmer, West Yorkshire Regiment, letter of September 1942
21 Imperial War Museum 7373/3 A

Chapter 5   The Curse of the Queen of Sheba

1 GKN Trevaskis, *Eritrea, a Colony in Transition*, Oxford University Press, 1960, p 21
2 Commander Edward Ellsberg, a former US naval officer, was sent

to Eritrea in 1941 to clear the scuttled ships from Massawa port. He must have been in charge of the shipyard labourers from whom Cicoria learned his skills. Ellsberg was nonplussed by the scene that met him on Asmara's streets. 'Apparently every Italian officer captured in the East African campaign the year before was out, magnificently caparisoned, strutting along the Viale Mussolini . . . Every one of these prisoners of war was armed – clinging from his waist was an automatic pistol protruding from its holster! There were enough armed Italian officers in sight easily to take over the country.' Edward Ellsberg, *Under the Red Sea Sun*, Dodd, Mead and Sons, 1946

3   The Jewish prisoners were highly inventive, digging tunnels, donning mocked-up British army uniforms, requisitioning buses and disguising themselves as Arab women in their attempts to escape across the border. David Cracknell, deputy police commissioner in Asmara at the time, told the author his men once fished Yitzhak Shamir from the tank of a water container in which he had been hiding, hoping to pass through police checkpoints unnoticed. Cracknell was proud of the fact that 106 of the 107 men who escaped from Sembel internment camp were recaptured. Eliyahu Lankin was the only man to slip through his clutches, reaching Djibouti via Addis. 'I'll never forget that name,' Cracknell said

4   Author's interview

5   Richard Pankhurst, 'The Legal Question of Racism in Eritrea during the British Military Administration', *Northeast African Studies*, vol 2, part 2, 1995

6   The political scene was always more complicated than this summary suggests. Woldeab Woldemariam, regarded as the founding father of the Eritrean independence movement, was a Christian highlander who started his political career with the Unionist Party. Moslems were not the only citizens suspicious of Ethiopia, and not all those who believed in Union were Christians

7   Wilfred Thesiger, *Arabian Sands*, Penguin, 1985

8   For an exhaustive account of the theories surrounding the Ark's location, see Graham Hancock, *The Sign and the Seal*, William Heinemann, 1993. Hancock concludes, somewhat surprisingly, that the Ark may well be in Axum

9   Translation used by Miguel Brooks, *Kebra Negast, The Glory of the Kings*, The Red Sea Press, 1995

10 Author's interview
11 SKB Asante, *Pan-African Protest: West Africa and the Italo-Ethiopian Crisis 1934–1941*, Longmans, 1977, p 60
12 Report of the United Nations Commission for Eritrea, General Assembly, Fifth Session, Supplement No 8 (A/1285), New York, 1950, p 46
13 'The Ethiopian Revolution and the Problem in Eritrea', cited in David Pool, 'The Eritrean Case – Ethiopia and Eritrea: the precolonial period', Research and Information Centre on Eritrea, Rome, 1984
14 ibid
15 Massimo Romandini, 'Ferdinando Martini ad Addis Ababa', *Miscellanea di Storia delle esplorazioni geografiche IX*, Genoa, 1984

Chapter 6   The Feminist Fuzzy-Wuzzy
 1 FO 371/108261, Public Record Office, Kew
 2 FO 371/80957
 3 Colonel Maurice Petherick MP, February 6, 1945; FO 371/46070 J 563
 4 First issue of *New Times and Ethiopia News*, May 5, 1936
 5 Letter to Emmeline Pethick-Lawrence, February 1936. Richard Pankhurst, *Sylvia Pankhurst: Counsel for Ethiopia*, Tsehai Publishers, 2003, p 38
 6 Author's interview
 7 Author's interview
 8 *Sylvia Pankhurst: Counsel for Ethiopia*, p 20
 9 Kwame Nkrumah, *Ghana: The Autobiography of Kwame Nkrumah*, Thomas Nelson and Sons, Edinburgh, 1957
10 In Jamaica, followers of Rastafarianism went even further, hailing Ras Tafari as the living God. While Haile Selassie was always reluctant to publicly offend the Rastafarians by denouncing their faith, as head of the Ethiopian Orthodox Church he, like most Ethiopian Christians, would have regarded it as sacrilegious
11 Author's interview
12 *Sylvia Pankhurst: Counsel for Ethiopia*, p 126; FO 371/24639
13 *Sylvia Pankhurst: Counsel for Ethiopia*, p 113
14 FO 371/35841
15 *New Times and Ethiopia News*, December 9, 1944
16 FO 371/46070
17 *Sylvia Pankhurst: Counsel for Ethiopia*, p 236; FO 371/63217

18 In Patricia Romero, *E Sylvia Pankhurst: Portrait of a Radical*, Yale University Press, 1987, the author interviewed Musa Galaal, a young Somali student who met Sylvia in the 1950s. He found Sylvia so 'pro-Ethiopia she considered all of Africa to be Ethiopia'. Unaware she was championing a rigidly Amhara version of history, she insisted all Somalis were, in fact, Ethiopians. 'What could you be that is better than being an Ethiopian?' she asked him, causing great offence

19 FO 371/108261

20 In an interview with the author, Spencer recalled several disputes over infrastructure with the British authorities. 'What made me quite angry was that they took the dry dock installation and shipped it off to Pakistan. I said "Surely this belongs to Ethiopia?" but they shipped it off anyway.' He said most of his information on the British dismantling in Eritrea came from Sylvia, but he found her radicalism alarming. 'She was really an aggressive person, far too aggressive.'

21 Author's interview

22 FO 371/96811

23 In *Eritrea: A Colony in Transition: 1941–1952*, Oxford University Press, 1960, author Trevaskis, who served as a British colonial officer in Eritrea, lets slip the fact that after 1941 Eritrean gold production plummeted from 17,000 to 3,000 ounces a year following British and American removal of mining equipment whose replacements were 'difficult and often impossible to obtain' and that the Eritrean fishing industry was similarly crippled by the requisitioning of irreplaceable equipment

WO 230/131 5343528 – This file contains a fascinating extended correspondence over the future of the ropeway. It shows British officials in Asmara campaigning energetically for the pylons and cable to be sold to companies in Burma, Sudan and Palestine, only to be reined in repeatedly by civil affairs officers in London who warned that as a 'caretaker' administration, Britain would stand accused of looting if a sale went ahead. The tireless persistence with which the British courted foreign buyers says legions about attitudes in Asmara. In the event, the ropeway remained in place despite their efforts, only to be sold by Haile Selassie's son-in-law after Federation

See also, FO 371/102667 for Foreign Office correspondence on Sylvia's claims, ADM1/19588 for the original 1946 British military debate about disposing of Massawa's assets and FO 371/96804 for evidence of Haile Selassie's unhappiness at the destruction

24 'Four Power Commission of Investigation for the Former Italian Colonies, Volume 1: Report on Eritrea', 1948, FO 371/69360

25 FO 371/27558, FO 371/27541, WO 230/57. These files' contents are cogently summarized in Richard Pankhurst, 'Post World War II Ethiopia: British military policy and action for the Dismantling and Acquisition of Italian Factories and other Assets, 1941–2', *Journal of Ethiopian Studies*, vol 29, no 1, 1996

26 E Talbot Smith in early 1943; *see* Harold Marcus, *The Politics of Empire. Ethiopia, Great Britain and the United States. 1941–1974*, The Red Sea Press, 1995, p 15

27 KC Gandar Dower, 'The First to be Freed: British Military Administration in Eritrea and Somalia', Ministry of Information, 1944

28 ibid

29 Dr Catherine Hamlin, *The Hospital by the River*, Macmillan, 2001

30 Author's interview

Chapter 7 'What do the baboons want?'
1 John Spencer, *Ethiopia at Bay: A Personal Account of the Haile Selassie Years*, Reference Publications, 1984, p 354

2 *Ethiopia at Bay*, p 23

3 *Ethiopia at Bay*, p 134

4 *Ethiopia at Bay*, p 135

5 Report of the UN Commission for Eritrea, Fifth Session, Supplement No 8 (A/1285), Lake Success, New York, 1950, p 3

6 First Confidential Report, February 17, 1950, Eritrea I, Series 2, Box 1, File 1, Acc A/269, UN Archives, New York

7 March 2, 1950, Eritrea I, Series 2, Box 1, File 1, Acc A/269

8 FO 371/80984

9 FO 371/80985

10 In *Eritrea and Ethiopia. The Federal Experience*, Transaction Publishers, 1997, historian Tekeste Negash challenges the notion that *shifta* activity was funded from Ethiopia, presenting it as a purely Eritrean phenomenon. His thesis, however, clashes with the views of British officials of the day such as Stafford and Trevaskis and Italian diplomats. It is also rejected by older Eritreans and Ethiopians

I interviewed who lived through those years and were in no doubt Addis Ababa lay behind a deliberate campaign of destabilization

11 FO 371/80984

12 FO 371/80985, Letters from Brigadier FG Drew and Frank Stafford

13 Fourth Confidential Report, March 7, 1950, Eritrea I, Series 2, Box 1, File 1, Acc A/269

14 ibid

15 FO 371/80985, Letter from Brigadier Drew, March 7, 1990

16 FO 371/80985, March 16, 1950

17 Fourth Confidential Report

18 Sixth Confidential Report, April 1, 1950

19 Yohannes Okbazghi, *Eritrea, A Pawn in World Politics*, University of Florida Press, 1991, p 147

20 Report of the UN Commission for Eritrea

21 June 7, 1990, Eritrea II, Series 2, Box 1, File 2, Acc A/269, UN Archives

22 In *Ethiopia at Bay* (pp 232–9), John Spencer gives an exhaustive account of the back-room deals that preceded the drafting of the key UN Resolution

23 Final Report of the UN Commissioner in Eritrea, General Assembly, Official Records: Seventh Session, Supplement No 15 (A/2188), Vol 4, p 4

24 FO 371/81043

25 Final Report of the UN Commissioner in Eritrea, p 11

26 *Ethiopia at Bay*, p 243

27 December 30, 1950 Imperial address

28 Author's interview

29 Author's interview

30 Final Report of the UN Commissioner in Eritrea, p 21

31 Final Report of the United Nations Commissioner in Eritrea, General Assembly, Official Records: Seventh Session, Supplement No 15 (A/2188), Vol 4, p 50

32 On page 236 of his autobiography, Spencer claims that during consultations in New York, he and Aklilou were privately assured the UN 'would be divested of all further jurisdiction' once the Federation was introduced. But in his interview with the author, Spencer told a different story. He acknowledged Matienzo stipulated that any change to the Federation would have to be approved by the UN and that Ethiopia's eventual abrogation was 'illegal'

33 In a February 26, 1952 public address, for example, Matienzo said the panel of legal experts found 'that as the Federal Act is an international instrument, the regime established under that Act cannot be altered without the concurrence of the General Assembly'. There are many such mentions in his draft reports and speeches of the period, stored in the UN Archives. *See also* S-0721-0009-01, S-0721-0003-09, S-0721-0003-10, S-0721-0003-11

34 *Ethiopia at Bay*, p 244

35 Author's interview

36 Author's interview

37 Author's interview

38 Author's interview

39 Author's interview

Chapter 8   The Day of Mourning

1 WO 97/2817

2 Tekeste Negash, *Eritrea and Ethiopia. The Federal Experience*, Nordiska Afrikainstitutet, Uppsala, 1997, p 83

3 FO 371/102655; *Eritrea and Ethiopia. The Federal Experience*, p 82

4 British officials, who described Tedla Bairu as a 'megalomaniac', also regarded him as corrupt. A February 19, 1954 report to London states that the chief executive had accepted 'property and a coffee plantation in Ethiopia from the Emperor and acquired a large concession near Cheren which had been purchased . . . with funds loaned by the State Bank of Ethiopia under the recommendation of the Ethiopian authorities'. Bairu was also drawing a $10,000 salary, a huge amount in Eritrea at the time. Eritrea VII, Jan–July 54, Series 2, Box 2, File 4, Acc A/277, UN Archives

5 Andargachew Messai, March 28, 1955 opening address at the First Regular Session of the Eritrean Assembly

6 *Journal of Eritrean Studies*, vol IV, nos 1 and 2, 1990

7 Ethiopian and Western histories routinely refer to a parliamentary 'vote' in favour of dissolution. This infuriates Eritreans, who insist the motion was passed by 'acclamation', a method obviously susceptible to manipulation. Having interviewed two men who were inside the chamber that day, I am satisfied no vote was ever staged. The same conclusion is reached by Bocresion Haile, author of *The Collusion on Eritrea*, 2001, who interviewed many of those present that day

8 Richard Johnson to State Department; James Firebrace and Stuart Holland, *Never Kneel Down: Drought, Development and Liberation in Eritrea*, The Red Sea Press, 1985, p 21. Looking back on the event nearly four decades later, Herman Cohen, US assistant secretary of state for Africa, had no hesitation in describing it as a 'unilateral coercive takeover by the Ethiopian regime'. 'It had all the trappings of a Stalinist operation, with troops surrounding the building while parliament was in session ... In retrospect it was disgraceful that the United States did not protest,' he wrote in *Intervening in Africa: Peacemaking in a Troubled Continent*, Macmillan, 2000

9 WO 97/2817; *Eritrea and Ethiopia. The Federal Experience*, p 80

10 Cordier letter April 5, 1954. Eritrea VII, Jan–July 54, Series 2, Box 2, File 4, Acc A/277, UN Archives. The words 'not sent' are scrawled on this letter. Whether it was ultimately sent or not, the draft sheds devastating light on Cordier's thinking

11 *Eritrea and Ethiopia. The Federal Experience*, p 131

12 SO265 EOSG/OSG Jan 1, 1961–Dec 31, 1973, UN Archives

13 Between 1985 and 1991, this role was performed by Dr Bereket Selassie, who labelled it 'Mission Impossible'. The future author of Eritrea's constitution would occasionally attach himself to the Somali delegation to gain access to the UN General Assembly and generally did all he could to irritate the Ethiopian delegation. He was arrested at least three times for his pains

14 Author's interview

15 'Stavropoulos', Box 67, S-0466-0128, 1963, UN Archives

Chapter 9 The Gold Cadillac Site

1 John Rasmuson, *A History of Kagnew Station and American Forces in Eritrea*, Public Affairs Office, 1973, p 40

2 Author's interview

3 Anthony Cave Brown, *Bodyguard of Lies*, WH Allen and Co, 1977, p 357

4 Harold G Marcus, *The Politics of Empire: Ethiopia, Great Britain and the United States 1941–1974*, The Red Sea Press, 1995, p 84

5 US Senate Hearing before the Subcommittee on African Affairs of the Committee on Foreign Relations, August 4, 5 and 6, 1976, p 36

6 Evelyn Waugh, *Waugh in Abyssinia*, Penguin, 1986

7 John Spencer, *Ethiopia at Bay*, Reference Publications, 1984, p 161

8 US Senate Hearing, p 26; John Spencer told the subcommittee: 'The

United States had indicated to the Ethiopian Government in advance of the December, 1950 Resolution adopted by the General Assembly, that once Ethiopia re-assumed sovereignty over Eritrea, it (the US) would want to conclude an agreement by which the United States would take over the large communications center there, just outside Asmara.'

9 Bereket Habte Selassie, *Eritrea and the United Nations*, The Red Sea Press, 1989, p 37

10 Author's interview

11 History has exposed the absurdity of this thesis. The current Eritrean government enjoys excellent relations with the Israelis, with whom it feels it has a great deal in common

12 Author's interview

13 Tom Farer, author of *War Clouds on the Horn of Africa*, Carnegie Endowment for International Peace, 1979; 'I don't think it makes any difference at all to Western strategic interests, which include Israeli interests, whether or not there is an independent Eritrea which is oriented toward the Arab world,' he testified before the US Senate Hearing in 1976

14 *The Politics of Empire*, p 104

15 Author's interview

16 Haile Selassie did not forgive the Crown Prince for meekly falling in with the coup plotters' plans. Asked why he had not designated Asfa Wossen as heir to the throne, the Emperor is reported to have replied: 'Why should We? He has already been on the Throne!' *Ethiopia at Bay*, p 317

17 *The Politics of Empire*, p 135

18 ibid, p 153

19 ibid, p 178

20 Terrence Lyons, 'Great Powers and Conflict Reduction in the Horn of Africa', *Cooperative Security: Reducing Third World Wars*, (ed, I William Zartman and Victor A Kremenyuk), Syracuse University Press, 1995 p 245; *see also* Terrence Lyons, 'The United States and Ethiopia: The Politics of a Patron–Client Relationship', *Northeast African Studies*, vol 8, nos 2–3, 1986

21 Bahru Zewde, *A History of Modern Ethiopia 1855–1991*, James Currey Ltd, 1991, p 186

22 Author's interview

Chapter 10 Blow Jobs, Bugging and Beer

1   Unless otherwise indicated, all quotes in this chapter come from author's interviews

2   Stroppy American behaviour appears to have been a long-running feature of life in Kagnew. In January 1957, the base was the scene of perhaps the NSA's only strike, triggered by servicemen's unhappiness over new regulations. James Bamford, *Body of Secrets: How America's NSA and Britain's GCHQ eavesdrop on the world*, Arrow Books, 2002, p 161

3   Victor Marchetti and John Marks, *The CIA and the Cult of Intelligence*, Jonathan Cape, 1974, p 226

4   US Defence Department news release, January 1964

5   John Hallahan on www.topsecretsi.com

6   Bizarrely, Kagnew radio's pop hits – the Doors, the Beach Boys, Procol Harum – were to end up as musical accompaniment to life in the EPLF trenches. They were taped off the station by appreciative Eritrean students, who took the cassettes with them into the Sahel

7   Website address: www.kagnewstation.com

8   Alex de Waal, *Evil Days: Thirty Years of War and Famine in Ethiopia*, Human Rights Watch, 1991, p 44

Chapter 11 Death of the Lion

1   Amina Habte Negassi, 'The Massacre of Besik-Dira and Ona', paper presented at the First International Conference on Eritrean Studies, July 2001, University of Asmara

2   *Ethiopia at Bay*, p 335

3   *New York Times*, 1974

4   *Ethiopia at Bay*, p 335

5   Author's interview

Chapter 12 Of Bicycles and Thieves

1   Not his real name

2   This was the assassination that Tzadu Bahtu, who played the part of lookout, served time for. Tortured for 55 days, he was sentenced to 10 years in prison as an accomplice but escaped during the Sembel operation

3 Dan Connell recounts this story in detail in *Against All Odds –
A Chronicle of the Eritrean Revolution*, The Red Sea Press, 1997,
pp 10–11

Chapter 13 The End of the Affair

1 Marina Ottaway, *Soviet and American Influence in the Horn of Africa*,
Praeger Publishers, 1982, p 100
2 National Security Decision Memorandum 231, August 14, 1973
3 Paul Henze, 'Arming the Horn 1960–1980', Wilson Center Working
Paper No 43, Washington DC, Smithsonian Institution, December
1982

In *Ethiopia, the United States and the Soviet Union* (Croom Helm,
1986), David Korn says Washington approved a $100m military
programme for 1974–5, which compared with the annual $10m
normally granted Haile Selassie. 'Altogether from 1974 to 1977 the
United States supplied Ethiopia with approximately $180m in arms,
in dollar value approximately one and a half times more than everything it had furnished up to 1974.'
4 *Ethiopia, the United States and the Soviet Union*, p 8
5 Author's interview
6 *Evil Days*, p 50
7 Basil Burwood-Taylor was held by the EPLF in a mountain ravine
for four months. Like so many kidnap victims in Eritrea, he retains
surprisingly fond memories of his ordeal. 'In retrospect, being kidnapped was a wonderful experience,' he told me. He returned to
Asmara in 1998 and shared a drink with one of his former abductors.
'There are no hard feelings. They know they are always welcome in
my house.'
8 Yohannes Okbazghi, *Eritrea, a Pawn in World Politics*, University of
Florida Press, 1991, p 229; 'United States policy toward Ethiopia',
American Foreign Policy Basis Documents 1977–1980, doc 662,
p 1233
9 *Against All Odds*, p 23
10 Author's interview
11 Robert G Patman, *The Soviet Union in the Horn of Africa: the
Diplomacy of Intervention and Disengagement*, Cambridge University
Press, 1990, p 274; Y. Bochkarev, 'What a Hope!', *New Times*, 20,
1986, pp 16–17

12 For more on the disastrously mixed messages Washington sent Somalia during this key period in which it was gearing up to invade the Ogaden, see *One Hundred Years of American–Ethiopian Relations* by David H Shinn, former ambassador to Addis; see also, *Ethiopia, the United States and the Soviet Union*

13 Author's interview

14 Author's interview

Chapter 14 The Green, Green Grass of Home

1 All quotes come from author's interviews. Ex-Fighters find talking about themselves awkward. A few of those I spoke to were happy to be identified; most, while having nothing to hide, felt deeply uncomfortable with the notion. Rather than use some names and omit others, I therefore decided to keep all contributions to this section anonymous

Chapter 15 Arms and the Man

1 Author's interview

2 Author's interview

3 Marina Ottaway, *Soviet and American Influence in the Horn of Africa*, Praeger Publishers, 1982, p 67

4 Oleg Gordievsky and Christopher Andrew, *KGB. The inside story of its foreign operations from Lenin to Gorbachev 1941–1990*, Hodder and Stoughton, 1990

5 My calculation is a very conservative one. It is based on annual figures from 'World Military Expenditures and Arms Transfers', compiled by the US Arms Control and Disarmament Agency (ACDA). Since these exclude the amount the Soviet Union spent on military advisers and training, and arms exports from Warsaw Pact nations made with Moscow's blessing, the real sums for Soviet military support are far larger than those given

6 Author's interview

7 Author's interview

8 Riccardo Orizio, *Talk of the Devil: Encounters with Seven Dictators*, Secker and Warburg, 2003

9 'World Military Expenditures and Arms Transfers'

10 Robert G Patman, 'Soviet–Ethiopian Relations: The Horn of Dilemma', chapter 5 of 'Troubled Friendships: Moscow's Third

World Ventures', edited by Margot Light, Royal Institute of International Affairs, 1993, p 115

11 Robert G Patman, *The Soviet Union in the Horn of Africa: the Diplomacy of Intervention and Disengagement*, p 277

12 M Volkov, 'Militarisation versus Development', *Asia and Africa Today*, no 5, 1987, p 9

13 *Talk of the Devil*

14 Lev Zaikov, Secretary of the CPSU Central Committee, September 10, 1987

15 'World Military Expenditures and Arms Transfers'

16 Author's interview

Chapter 16 'Where are our socks?'

1 The reporter, Alemseged Tesfai, later became a respected Eritrean historian and playwright. His account of the battle of Afabet is to be found in *Two Weeks in the Trenches*, The Red Sea Press, 2002

2 Author's interview

3 The EPLF tank man whose accurate shot had brought the convoy grinding to a halt at Ad Shirum was fêted as a hero, but he did not enjoy his status long. During the Massawa campaign, he bent too quickly to reload the tank's gun and his head received the full impact of the recoil, killing him

4 Author's interview

5 The three Soviet advisers were released in March 1991, after Moscow acceded to EPLF demands that it remove its ships from the Dahlak Islands and stop airlifting supplies to Asmara. Interpreter Alexander Kuvaldin, who now lives in Minsk, told me he spent his time in captivity giving EPLF fighters English lessons and teaching them how to drink shots 'Russian-style'. Being held prisoner had left him with some psychological problems, he said, 'but when I look back I don't remember any bad things from that time'.

6 Author's interview

7 For a detailed account of the Ethiopian army's failings in this period, *see* Gebru Tareke, 'From Af Abet to Shire: the Defeat and Demise of Ethiopia's Red Army 1988–89', *Journal of Modern African Studies*, vol 42, 2, June 2004

8 Harold G Marcus, *A History of Ethiopia*, University of California Press, 1994, p 213

9 Author's interview
10 Author's interview. Varennikov later took part in the abortive August 1991 attempt to unseat Gorbachev. Exonerated for his role in the attempted coup, he has now entered Russian politics
11 The Falasha episode, and the ultimately unsuccessful US attempt to broker a peace agreement between Mengistu and the EPLF and TPLF, is recounted in Herman J Cohen's *Intervening in Africa: Peacemaking in a Troubled Continent*, Palgrave, New York, 2000
12 Alex de Waal, *Evil Days*, Human Rights Watch, September 1991. In late 1989, Washington Jewish Week published a leaked US Congressional staff memo confirming that 100 cluster bombs had been supplied to the Ethiopians in late 1989
13 In May 2001, Russia President Vladimir Putin agreed to cancel 80 per cent of Ethiopia's outstanding debts and said the balance would be rescheduled through the Paris Club
14 The Promised Land proved a mixed experience for the Falashas. Shunted into isolated development towns in Israel, many ended up doing menial jobs. The community complains that its members are discriminated against because of its skin colour and are regarded as second-class Jews. In January 1996, the discovery that their blood donations were being routinely thrown away because of AIDS contamination fears triggered a wave of Falasha rioting. Nevertheless, the Israeli and Ethiopian governments agreed in January 2004 to complete the repatriation of Ethiopia's remaining Falashas
15 Author's interview
16 Interview on 'The World at One', BBC Radio 4, June 4, 1991

Chapter 17 A Village of No Interest

1 Author's interview
2 Paul Henze, *Eritrea's War*, Shama Books, 2001, p 24
3 Dan Connell, *Against All Odds – A Chronicle of the Eritrean Revolution*, The Red Sea Press, 1997, p 272
4 *Shabia* means 'the masses' in Tigrinya, while *Woyane* translates as 'popularly-based'. Depending on context and tone, the terms can be either matey or insulting, rather like the 'Yank' Britons use to refer to Americans or the 'Pom' Australians apply to Britons
5 Eritrea has since admitted that Bezabeh Petros died in custody
6 Patrick Gilkes and Martin Plaut, 'War in the Horn. The Conflict

between Eritrea and Ethiopia', Royal Institute of International Affairs, 1999, p 22

7 Amnesty International: 'Ethiopia/Eritrea. Amnesty International witness cruelty of mass deportations', January 29, 1999; 'Ethiopia and Eritrea: Human rights issues in a year of conflict', May 21, 1999; Human Rights Watch: 'The Horn of Africa War: Mass Expulsions and the Nationality Issue' June 1998–April 2002

8 Author's interview

9 On November 25, 2004, as this book was going to press, Ethiopia appeared to soften its position, saying it accepted the boundary commission finding, however 'illegal and unjust'. But Prime Minister Meles Zenawi stressed the acceptance was 'in principle' and Ethiopia expected 'give and take' in implementation, suggesting his country has yet to recognise the binding nature of the border ruling.

10 Britain, for example, said in February 2004 that it would triple its bilateral aid to Ethiopia to £53m. Announcing the rise, Hillary Benn, international development secretary, specifically ruled out using the aid as a way of pressurizing Addis into accepting the boundary ruling, saying London hoped the stalemate could be resolved through dialogue

11 One notable example is Dan Connell, a journalist who covered the EPLF liberation campaign in impressive detail. 'Enough! A Critique of Eritrea's Post-Liberation Politics', presented at African Studies Association in Boston in November 2003, summarized his disquiet over events

Chapter 18 'It's good to be normal'

1 For more details of the *manqa* episode, *see* David Pool, *From Guerrillas to Government – The Eritrean People's Liberation Front*, James Currey, 2001. Two years later, Isaias faced another challenge by a group of Christian highlanders pushing for greater democratic accountability. Over a dozen members of the 'rightist movement', as it was known, were executed. Like the *manqa* affair, the episode has been air-brushed out of the leadership's memories of this period

2 David Strickland and Thomas Bowidowicz, two Americans working at Kagnew, were kidnapped in September 1975 by the ELF, which threatened to kill them unless the US halted arms supplies to Ethiopia and closed Kagnew. They were subsequently released

Other sources

Bamford, James – *The Puzzle Palace*, Penguin, 1983

Bereket, Habte Selassie – *Eritrea and the United Nations*, The Red Sea Press, 1989

Duncan, W Raymond and Ekedahl, Carolym McGiffert – *Moscow and the Third World under Gorbachev*, Westview Press, 1990

Erlich, Haggai – *The Struggle Over Eritrea 1962–1978*, Hoover Institution Press, 1983

ESFA – *The Federal Case of Eritrea with Ethiopia*, Mogadishu, 1979

Gandar Dower, Kenneth – *Abyssinian Patchwork, An Anthology*, Frederick Muller Ltd, 1949

Goodman, Melvin – *Gorbachev's Retreat*, Praeger Publishers, 1991

Harding, Jeremy – *Small Wars, Small Mercies: Journeys in Africa's Disputed Nations*, Viking, 1993

Henze, Paul B – *Horn of Africa: From War to Peace*, Macmillan, 1991

Henze, Paul B – 'Arming the Horn 1960–1980', Wilson Centre Working Paper No 43, Washington DC, Smithsonian Institution, December 1982

Killion, Tom – *Historical Dictionary of Eritrea*, Scarecrow Press, 1998

Lefebvre, Jeffrey – *Arms for the Horn: US Security Policy in Ethiopia and Somalia 1953–1991*, University of Pittsburgh Press, 1991

Levine, Donald – *Greater Ethiopia: The Evolution of a Multi-Ethnic Society*, University of Chicago Press, 1974

Lewis, Jon (editor) – *The Mammoth Book of Battles: The Art and Science of Modern Warfare*, Robinson Publishing, 1995

Longrigg, Stephen – *A Short History of Eritrea*, Clarendon Press, 1945

Luard, Evan – *A History of the United Nations. Volume 2 The Age of Decolonisation 1955–1965*, The Macmillan Press Ltd, 1989

Lyons, Terrence – 'The United States and Ethiopia: The Politics of a Patron–Client Relationship', *Northeast African Studies* Volume 8, Numbers 2–3, 1986

Mack Smith, Denis – *Italy, A Modern History*, University of Michigan Press, 1969

Mack Smith, Denis – *Mussolini*, Granada, 1981

Mack Smith, Denis – *Mussolini's Roman Empire*, Longman, 1976

Martini, Ferdinando – *L'Eritrea Economica*, Istituto Geografico de Agostini, 1913

McGiffert Ekedahl, Carolym and Goodman, Melvin – *The Wars of Eduard Shevardnadze*, C. Hurst and Co, 1997

Mockler, Anthony – *Haile Selassie's War: The Italian–Ethiopian Campaign 1935–1941*, Random House, 1984

Negash, Tekeste and Tronvoll, Kjetil – *Brothers at War: Making Sense of the Eritrean–Ethiopian War*, James Currey Ltd, 2000

Negassi, Amina Habte – 'The Massacre of Besik-Dira and Ona', University of Asmara, Department of History, July 2001

Ottaway, Marina – *Soviet and American Influence in the Horn of Africa*, Praeger Publishers, 1982

Pankhurst, E. Sylvia and Pankhurst, Richard – *Ethiopia and Eritrea: the last phase of the Reunion Struggle 1941–1952*, The Walthamstow Press Ltd, 1953

Prouty, Chris and Rosenfeld, Eugene – *Historical Dictionary of Ethiopia and Eritrea*, Scarecrow Press, 1993

Pugh, Martin – *The Pankhursts*, Penguin, 2001

Romandini, Massimo – *Visita a Dogali*, L'Universo LXI 3, 1981

Romandini, Massimo – *Le Comunicazioni Stradali, Ferroviarie e Marittime dell'Eritrea durante il Governatorato Martini*, Africa (Rivista Trimestrale di Studi e Documentazion dell'Istituto Italo-Africano), Anno XXXVIII – No 1

Sbacchi, Alberto – *Legacy of Bitterness, Ethiopia and Fascist Italy, 1935–1941*, The Red Sea Press, 1997

Segre, Vittorio – *La guerra privata di tenente Guillet*, TEA, 1997

Tesfagiorgis, Abeba – *A Painful Season and a Stubbon Hope: The Odyssey of an Eritrean Mother*, The Red Sea Press, 1992

Ullendorff, Edward – *The Ethiopians*, Oxford University Press, 1965

Acknowledgements

Discretion is a quality dear to Eritreans. Few, I know, would want to be publicly thanked in these pages. But those who welcomed me into their homes, accompanied me on my trips, pointed me in interesting directions and alerted me to my errors know who they are. I am enormously grateful.

Apart from my editor, Mitzi Angel, and literary agents Pat Kavanagh and Joy Harris, I owe special thanks to Clive Priddle, who originally commissioned the book and maintained his interest in the project long after moving on. My mother helped by reading through all of Martini's Eritrean diaries, my father meticulously edited the text, my sister Jessica provided weekend respite, my brother-in-law Julian kept my computer running.

I have tapped the intellectual riches of many experts. In Addis Ababa, Richard and Rita Pankhurst were generous with their insights and Teshome Bokan Gabre Mariam was kind enough to offer detailed comments on an early draft. Professor Massimo Romandini, probably the world's leading expert on Martini, was a wonderful correspondent, Martin Plaut kept me up-to-date with current events on the Horn, Nick Lera and Jennie Street told me what was what on the railway.

During my research trips I relied on the hospitality of many: Caroline Lees and Alan Macdonald in Asmara, Judith Matloff in New York, Peter Whaley in Washington and the Ostrovskys and Frys in Moscow deserve particular mention. Thanks are also due to the Kagnew veterans who welcomed me to one of their reunions in Florida and to Zazz, Zio Bob and Tom Indelicato for their running commentary on early drafts and life in general.

ACKNOWLEDGEMENTS

John Caveney, researcher at the *Financial Times*, was a wonderful asset and Olga Shevtsova, the *Times*' fixer in Moscow, saved me from confusion. The Society of Authors helped by paying for one of my many trips to Eritrea.

Having already dedicated one book to Michael Holman, I won't repeat myself, but he was the reason why, Against All Odds and Never Kneeling Down, I finished the book.

Index

425

BC 3/06